SO-AJA-916

SAFETY SYMBOLS	HAZARD	PRECAUTION	REMEDY
Disposal	Special disposal required	Dispose of wastes as directed by your teacher.	Ask your teacher how to dispose of laboratory materials.
Biological	Organisms that can harm humans	Avoid breathing in or skin contact with organisms. Wear dust mask or gloves. Wash hands thoroughly.	Notify your teacher if you suspect contact.
Extreme Temperature	Objects that can burn skin by being too cold or too hot	Use proper protection when handling.	Go to your teacher for first aid.
Sharp Object	Use of tools or glassware that can easily puncture or slice skin	Practice common sense behavior and follow guidelines for use of the tool.	Go to your teacher for first aid.
Fumes	Potential danger from smelling fumes	Must have good ventilation and never smell fumes directly.	Leave foul area and notify your teacher immediately.
Electrical	Possible danger from electrical shock or burn	Double-check setup with instructor. Check condition of wires and apparatus.	Do not attempt to fix electrical problems. Notify your teacher immediately.
Irritant	Substances that can irritate your skin or mucous membranes	Wear dust mask or gloves. Practice extra care when handling these materials.	Go to your teacher for first aid.
Chemical	Substances (acids and bases) that can react with and destroy tissue and other materials	Wear goggles and an apron.	Immediately flush with water and notify your teacher.
Toxic	Poisonous substance	Follow your teacher's instructions. Always wash hands thoroughly after use.	Go to your teacher for first aid.
Fire	Flammable and combustible materials may burn if exposed to an open flame or spark	Avoid flames and heat sources. Be aware of locations of fire safety equipment.	Notify your teacher immediately. Use fire safety equipment if necessary.

Eye Safety
This symbol appears when a danger to eyes exists.

Clothing Protection
This symbol appears when substances could stain or burn clothing.

Animal Safety
This symbol appears whenever live animals are studied and the safety of the animals and students must be ensured.

Glencoe

SCIENCE

An Introduction to the Life, Earth, and Physical Sciences

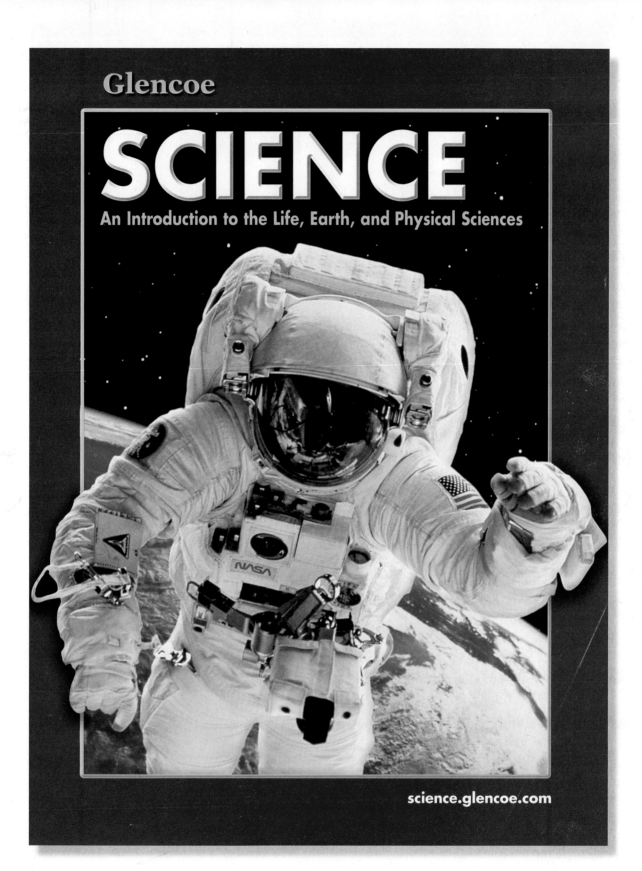

science.glencoe.com

 **Glencoe McGraw-Hill**

New York, New York   Columbus, Ohio   Chicago, Illinois   Peoria, Illinois   Woodland Hills, California

A GLENCOE PROGRAM

Glencoe Science

Student Edition
Teacher Wraparound Edition
Study Guide for Content Mastery SE and TE
Reinforcement SE and TE
Enrichment & Critical Thinking/Problem
 Solving SE and TE
Activity Masters
Chapter Review
Chapter Review Software
Lab Manual SE and TE
Science Integration Activities
Cross-Curricular Integration
Science and Society/Technology Integration
Multicultural Connections

Performance Assessment
Assessment: Chapter and Unit Tests
Lesson Plans
Spanish Resources
Teaching Transparencies Package
Section Focus Transparencies Package
Science Integration Transparencies Package
MindJogger Videoquizzes and Teacher Guide
English/Spanish Audiocassettes
ExamView Pro Test Bank Software
Vocabulary PuzzleMaker
Interactive Lesson Planner
Virtual Labs and Quizzes—
 Interactive CD-ROM

The Glencoe Science Professional Development Series:
 Performance Assessment in the Science Classroom
 Laboratory Management and Safety in the Science Classroom
 Cooperative Learning in the Science Classroom
 Guide to Using the Internet in the Science Classroom
 Home and Community Involvement in the Science Classroom
 ELL Strategies for Science
 Dinah Zike's Teaching Science with Foldables

Cover: Science Source/Photo Researchers; inset NASA/The Stock Market

Glencoe/McGraw-Hill
A Division of The McGraw-Hill Companies

Copyright © 2003 by the McGraw-Hill Companies, Inc. All rights reserved. Except as permitted under the United States Copyright Act, no part of this publication may be reproduced or distributed in any form or by any means, or stored in a database or retrieval system, without prior written permission of the publisher.

Send all inquiries to:
Glencoe/McGraw-Hill
8787 Orion Place
Columbus, OH 43240

ISBN 0-07-830616-7
Printed in the United States of America.

2 3 4 5 6 7 8 9 10 071/055 08 07 06 05 04 03

Authors

Dan Blaustein
Science Teacher and Author
Evanston Intermediate School
Evanston, Illinois

Louise Butler
Principal, Normandy
School District, retired
St. Louis, Missouri

Wanda Matthias
6th Grade Science
Teacher, retired
Plano, Texas

Bryce Hixson
President, Loose in the Lab, Inc.
Sandy, Utah

Contributing Authors

Carolyn Randolph, Ph.D.
Vice President of Outreach
and Research
South Carolina Governor's
School for Science and
Mathematics
Hartsville, South Carolina

Leonard Rodriguez
Assistant Principal
First Avenue School
Arcadia, California

Contributing Writers

Mary Dylewski
Science Writer
Kassel, Germany

Ralph M. Feather, Jr., Ph.D.
Science Department Chair
Derry Area School District
Derry, Pennsylvania

Rebecca Johnson
Science Writer and Author
Sioux Falls, South Dakota

Nancy Ross-Flanigan
Science Writer
Detroit, Michigan

Consultants

Chemistry

Anne Barefoot, A.G.C.
Physics and Chemistry
   Teacher, Emeritus
Whiteville High School
Whiteville, North Carolina

Cheryl Wistrom, Ph.D.
Assistant Professor
Chemistry Department
St. Joseph's College
Rensselaer, Indiana

Physics

Albert E. Acierno
Supervisor Science Education
   K-12, retired
Columbus Public Schools
Columbus, Ohio

Safety

Jay A. Young, Ph.D.
Safety and Chemical Safety
   Consultant
Silver Spring, Maryland

Earth and Planetary Sciences

Professor Larry Lebofsky, Ph.D.
Professor
Department of Planetary
   Sciences
University of Arizona
Tucson, Arizona

James B. Phipps, Ph.D.
Instructor
Grays Harbor College
Aberdeen, Washington

Assessment

Audrey B. Champagne, Ph.D.
Professor of Chemistry
Department of Chemistry
Professor of Education
Department of Educational
   Theory and Practice
University at Albany
State University of New York
Albany, New York

Multicultural

Lorraine Cruz-Lugo
Middle School Teacher
Chicago, Illinois

Karen Muir, Ph.D.
Lead Instructor
Department of Social and
   Behavioral Sciences
Columbus State Community
   College
Columbus, Ohio

Life Science

William R. Ausich, Ph.D.
Chair and Professor
   Geological Sciences
Department of Geological
   Sciences
The Ohio State University
Columbus, Ohio

Maryanna Quon Warner
Science Teacher
Del Dios Middle School
Escondido, California

Reading

Nancy Farnan, Ph.D.
Graduate Programs
   Coordinator
School of Teacher Education
San Diego State University
San Diego, California

Elizabeth Gray, Ph.D.
Adjunct Professor
   Department of Education
Otterbein College
Westerville, Ohio

Reviewers

Linda Bodie
Pace Middle School
Milton, Florida

Teckla Dando
Troy City Schools
Troy, Ohio

Celeste Dellinger
Atlanta Middle School
Atlanta, Texas

Connie Denk
Starling Middle School
Columbus, Ohio

Lenore Gallagher
Northeast Middle School
Bristol, Connecticut

Fred George
Dr. T.F. Reszel
    Middle School
North Tonawanda,
    New York

Penny Hamisch
Greenfield School
Greenfield, California

Barbara Hartgrove
Everett Middle School
Columbus, Ohio

Michael Henson
Beery Middle School
Columbus, Ohio

Patrick J. Herak
Westerville South High
    School
Westerville, Ohio

Phyllis Herzog
Dominion Middle School
Columbus, Ohio

Mario Inchaustegui
Academy of the
    Americas
Detroit, Michigan

Kevin Kerr
Fort Lincoln School
Mandan, North Dakota

Mike Mansour
John Page Middle School
Madison Heights, Michigan

Stephanie Molesky
Davidson International
Baccalaureate
    Middle School
Davidson, North Carolina

Rebecca Morris
H. L. Harshman
    Middle School
Indianapolis, Indiana

Debbie Panebianco
South Charleston
    Middle School
Charlotte, North Carolina

Lashawn Porter
Louie Welch
    Middle School
Houston, Texas

Ramonita Torres
Thomas Giordano
    Middle School 45
Bronx, New York

Mike Walsh
Carmel Junior High
    School
Carmel, Indiana

Teacher Activity Testers

Life Science

Diana Such
Science Teacher
St. Michael's School
Worthington, Ohio

Physical Science

Tom Speece
Science Teacher
Big Walnut
    Middle School
Sunbury, Ohio

Earth Science

Debbie Huffine
Science Teacher
Noblesville
    Intermediate School
Noblesville, Indiana

Contents

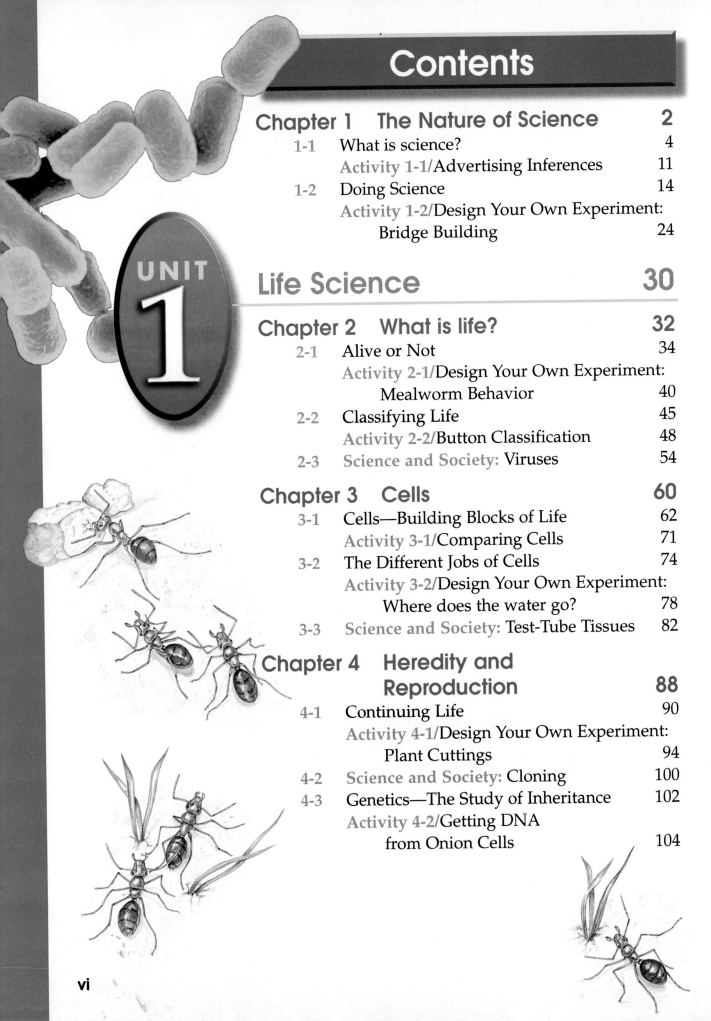

UNIT 1

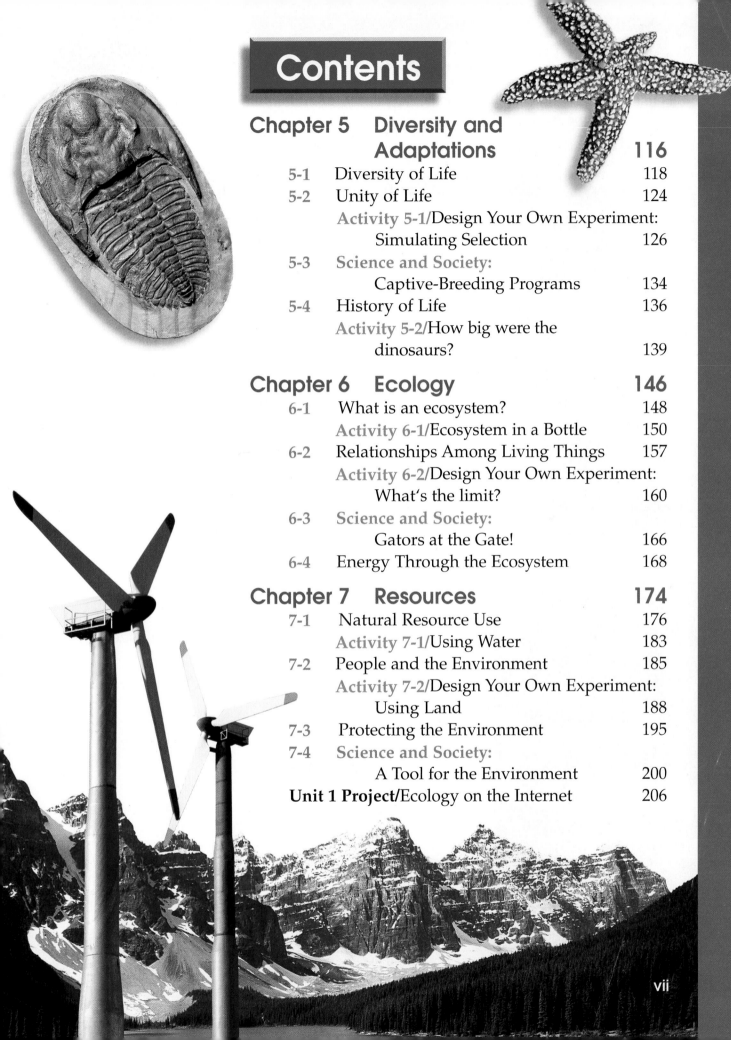

Contents

Contents

UNIT 2

Physical Science 208

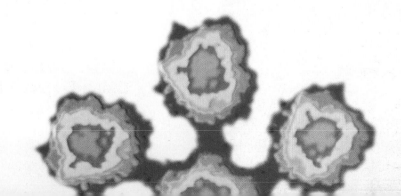

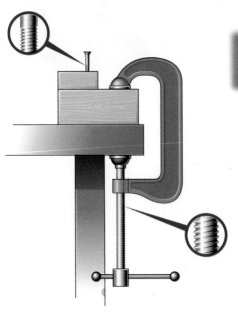

Contents

Contents

Contents

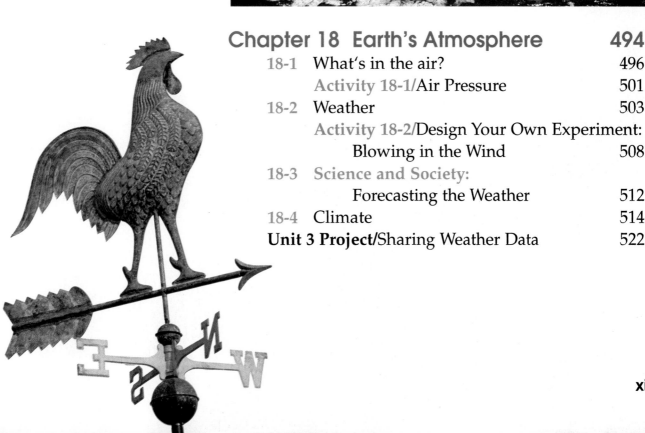

Contents

Appendices

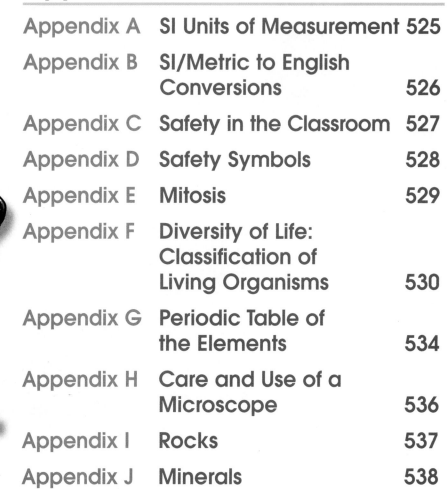

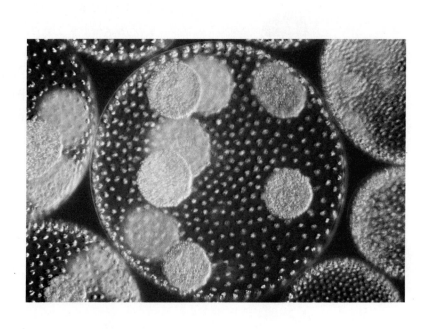

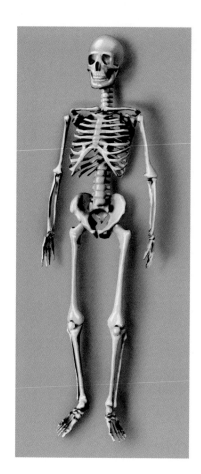

Contents

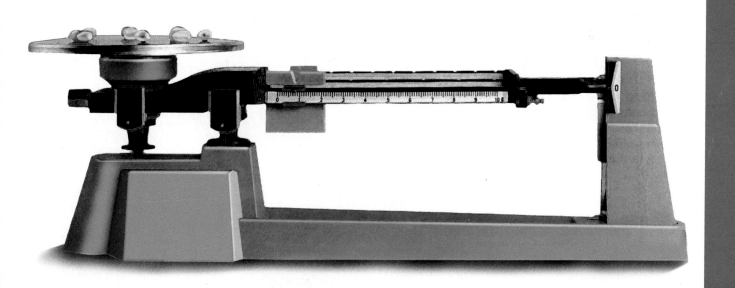

Activities

Activities

MiniLABs

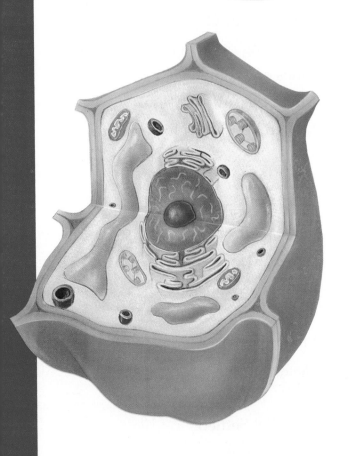

MiniLABs

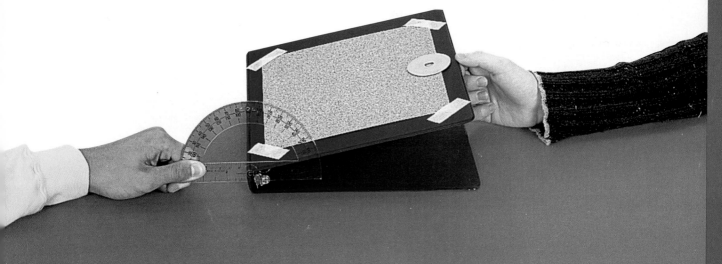

Explore Activities

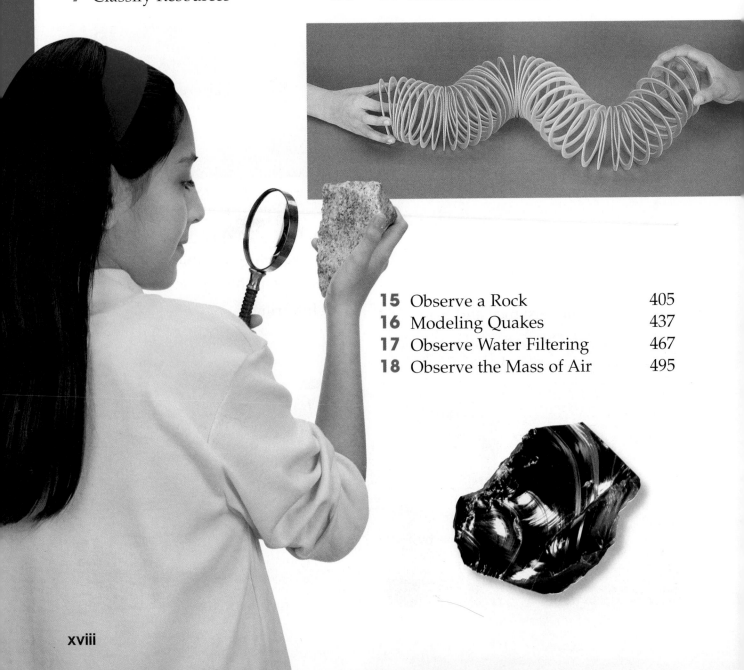

Problem Solving

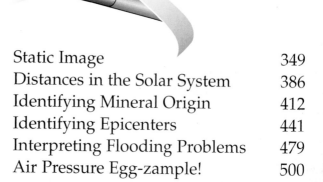

Using Technology

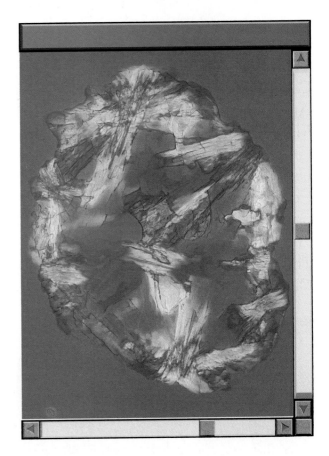

Skill Builders

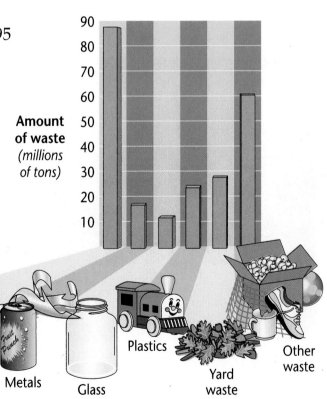

Amount of waste (millions of tons)

Paper products

Metals

Glass

Plastics

Yard waste

Other waste

People and Science

Science Connections

Science and the Arts

Science and History

Science and Language Arts

Chapter Preview

Skills Preview

▶ **Skill Builders**
- communicate
- design an experiment

▶ **MiniLABs**
- observe
- infer

▶ **Activities**
- observe
- infer
- hypothesize
- collect data
- communicate
- compare results
- make a table

Chapter 1

The Nature of Science

Y ou hold your breath as you speed over the edge of the gorge. The only thing separating you from the river 200 meters below is a spidery network of steel and cables, and a whole lot of air. How can the bridge hold so much weight? Bridges are amazing achievements. They connect worlds by making travel across rivers, canyons, and bays possible. Bridges cross roads, train tracks, and even small gullies and ditches. You'll see how they are also great examples of how science affects everyday life.

EXPLORE ACTIVITY

Draw a Bridge

1. Find a bridge close to your home or school. You may find a pedestrian bridge, a highway overpass, a railroad bridge, or a bridge across a small ditch.
2. In your Science Journal, draw a detailed picture of the bridge that shows how the bridge is made. Record any other observations you make such as what the bridge is used for and what you think it is made out of.

Science Journal

In your Science Journal, write a paragraph that describes how you think the people who built the bridge knew that it would be strong enough.

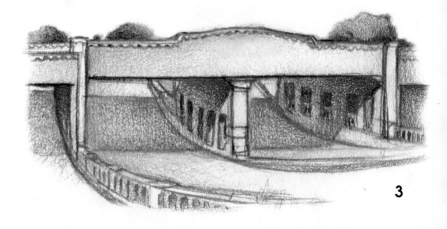

1•1 What is science?

What YOU'LL LEARN

- How science can help you solve problems
- The skills used in science

Science Words:
science
technology

Why IT'S IMPORTANT

Learning about science helps you find out more about the world around you.

Science and Scientists

What do you think of when you think of a scientist? Do you think of someone alone in a laboratory, mixing chemicals and doing strange experiments? Maybe you imagine some person with wild hair and a lab coat. That's the image many people have of scientists. However, nothing could be further from the truth. Anyone who tries to learn something about the world is a scientist. That means anyone, including the people who designed the bridges shown in **Figures 1-1** and **1-2,** can be a scientist. Even you can be a scientist. The best way to understand science and what scientists do is to jump in and do it. Let's follow a science class as they explore the nature of science.

It was the first day of school and Gabby was unsure about what to expect in her new science class. As the rest of the students sat down and Ms. Quon took attendance, Gabby wondered what they were going to study.

"All right everyone, settle down. We need to get started," said Ms. Quon. "We've been given a special project and I want to start right away. Have any of you heard that the LEP Corporation donated some property for our school to use as an outdoor education center?"

FIGURE 1-1

Some bridges are designed for people to walk across. Others are designed to carry the weight of cars and trucks.

FIGURE 1-2

Some bridges help ease the flow of traffic. *How does the highway overpass shown here differ from the "walking bridge" shown in Figure 1-1?*

"Yes, it's great," Pablo responded, "but we'll be in high school before it's ready to be used."

"Right now, it's just an empty lot next to some old warehouses," added Sydney. "It only has a few trees and a polluted stream. Some gift that was!" He paused. "What does the lot have to do with us?"

Ms. Quon laughed. "It has a lot more to do with you than you realize. In fact, it's going to be a big part of your grade!"

Science as a Tool

Gabby raised her hand. "Ms. Quon, I thought that this was science class. What does a vacant lot have to with science?"

"Maybe we'll come up with a way to blow it up so we can start over!" joked Jim.

"No, we're not going to blow up anything. There's a lot more to science than explosions," Ms. Quon said. "We've been given a special assignment. We're going to use science to build a bridge over the stream."

"Wouldn't a hammer and nails work better?" Sydney asked, grinning.

"Science can be as good a tool as a hammer, Sydney," she explained.

Gabby wondered aloud, "What's so hard about building a bridge? All we need is some wood and nails, and we put them together to make a bridge. Why do we need science?"

"Well, do you know everything you need to know before you build your bridge? For example, how can we build a bridge that's strong enough to hold an entire class?" Ms. Quon pulled some books on bridges out of her backpack. "Are you sure wood is the best material to make your bridge out of? How high should your bridge be to keep it from being washed away in a flood? Where along the stream should we build the bridge?"

"I guess there's a lot we need to know before we start," replied Gabby. "But I still don't see how science can help. We haven't even had science yet. We don't know enough about engineering to answer those kinds of questions."

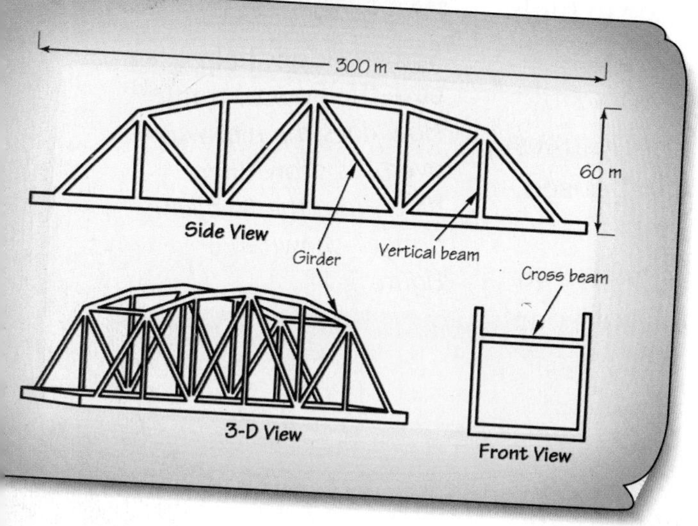

Using Science Every Day

"Science can help us answer those questions. Science is not just about knowing a bunch of facts and information. Science is a process. It's a way of finding out about the world around us." Ms. Quon began unrolling a poster. "It can help you solve problems by giving you a way to find out more about the problem. In fact, I'm sure you already use science every day to help you solve problems."

Using Prior Knowledge

"I don't think so, Ms. Quon," replied Tinho. "This is my first science class."

Ms. Quon thought for a moment. "Tinho, how did you decide which clothes to wear today?"

FIGURE 1-3

Planning is the first step in building a bridge. Engineers test different designs before construction begins.

"Well, I knew it was going to be warm, so I wore a short-sleeved shirt."

"How did you know it was going to be warm today? Did you listen to a weather report?"

"No, I know that it's always warm here in early September."

"I knew it! You did use science! You used what you've experienced for the last few years to guess what the weather would be like today. You made a prediction," said Ms. Quon. "But you didn't stop there. You made another prediction. You guessed that the best shirt to wear on a warm day is a short-sleeved shirt. How did you make that prediction?"

"I know that shirts with short sleeves are more comfortable to wear when it gets hot than shirts with long sleeves. We don't have air-conditioning in our school, so I knew that I would need to be as comfortable as possible."

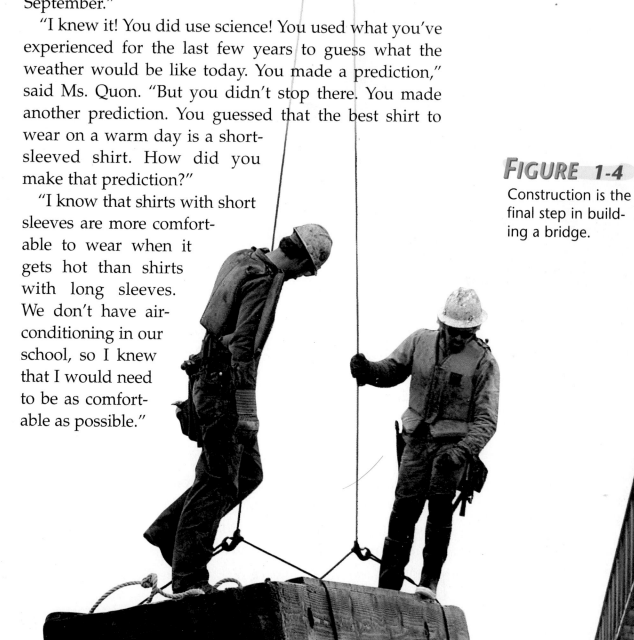

FIGURE 1-4

Construction is the final step in building a bridge.

FIGURE 1-5

Learning to play a musical instrument requires a lot of practice. *How is this similar to learning the skills used in science?*

"Great! You used what you already knew about clothes to predict that the best shirt to wear today should have short sleeves. You then tested your prediction by wearing a short-sleeved shirt." Ms. Quon paused for a second to tape the poster to the bulletin board. "What will you wear tomorrow if your prediction is correct?"

"I'll wear a short-sleeved shirt, of course," Tinho said.

"What if the weather was not what you predicted and it was cold today?"

"I'd probably wear a warmer shirt tomorrow."

"That's right, Tinho. Depending on the results of your test, you either keep doing the same thing, or you try something different. That's using science!"

Practicing Science Skills

"If we already use science, why do we have to take this class?" asked Jim.

"You're in the band, aren't you, Jim?" Ms. Quon asked.

"Yes, I play the trumpet," he answered.

"Even though you can play the trumpet now, you keep practicing and taking lessons so you can play even better, don't you?"

"Yes," Jim said. "I still have a lot to learn about playing the trumpet."

"Science is the same way," said Ms. Quon. "As I said, science is a tool, and you need to practice using that tool so you can get better. That'll help you learn more about the world we live in and help you to solve harder problems—such as building a bridge."

Science Skills

As Ms. Quon's class will learn, science is not a collection of facts. **Science** is the process of trying to understand the world around you. But what do we mean when we say science is a process? That means that science is a set of skills that you can use to help you find out more about something or to solve a problem. But wait! Before you go out and start "doing science," it's a good idea to practice the skills you need to do it well. Just as athletes and musicians practice the skills necessary to do their craft, science has skills that you can practice every day.

When many people think of science, they think of complicated machines, computers, and robots. As you are learning, however, science is a way to find out about the world around you. The gadgets and gizmos some think of as science are really technology. **Technology** is the use of knowledge learned through science. **Figure 1-6** shows an example of technology.

FIGURE 1-6
This virtual reality computer helps scientists test machines before the machines are constructed.

USING TECHNOLOGY
Science and Technology

In the late 1800s, before cars became common and traffic jams were a daily event, a New York engineer named John Roebling was given the task of building a bridge between Brooklyn and Manhattan.

To make sure that the bridge would be safe and strong, Roebling developed a new technology using a cable-weaving design. This is the design he came up with: four cables, each 38 centimeters thick, would hold up the bridge. Each cable would be made of 19 strands of wire. Each strand, in turn, would be made up of 278 long wires. Roebling designed the wires to be long, continuous strands so that the builders did not have to link wire ends together. This way, the wires would not break apart.

In 1883, the Brooklyn Bridge was completed. More than 100 years later, it is still considered a technological wonder— and a good example of how a new technology can impact millions of lives.

SCIENCE *Online*

Visit Glencoe Science Online at *science.glencoe.com* to learn more about the technology of bridges.

FIGURE 1-7

The water in this pot is actually cold—dry ice is making it bubble.

The Skill Handbook

The important skills used in science are described in the Skill Handbook on pages 541-561. The Skill Handbook can be a helpful reference as you learn and practice the skills of science.

Some of the skills may seem easy but still require practice. For example, observing is a skill that may seem simple. After all, everyone knows how to watch something, right? However, observing is more than just watching. You have to make sure you observe carefully and write down exactly what you see.

You also have to know the difference between an observation and an inference (IHN fuh runtz). An inference is an explanation of why something happened. This isn't as easy as it seems. See the pot in **Figure 1-7?** It looks like it is full of boiling water. Would you believe that the water is actually cold? Dry ice is making the water bubble as if it's boiling hot, so you infer that the water is hot. You are trying to explain what you observed—the bubbling water. However, all you can *observe* is that the water is bubbling. You *infer* that the water is hot because the water is bubbling. An observation is what you see. An inference is how you explain what you observe.

In science, it's important to make sure you know exactly what you observed. Once you know what you observed, you can begin to explain, or infer, what happened.

Mini LAB

Observing and Inferring

Practice the skill of using observation to make an inference with this activity.

1. Look at the illustration on this page. It is a part of a larger illustration.
2. Record in your Science Journal everything you can observe about the illustration.
3. Use your list of observations to make inferences about what is happening in the illustration.

Analysis

1. What do you think is happening in the illustration?
2. Compare your inference with the entire illustration on page 27. How close was your inference to the illustration?

Advertising Inferences

Imagine you're flipping through your favorite magazine and you see an ad showing a skateboard with wings. Would you infer that the skateboard could fly? In this activity, you'll use advertisements to practice the science skills of observing and inferring. Do the products really do what the ads lead you to infer?

What You'll Investigate
What observations and inferences can you make from advertisements?

Procedure
1. **Select** three ads from those supplied by your teacher.
2. In your Science Journal, **make a table** like the one shown below.
3. For each ad, **list** your observations. For example, you may **observe** that there are athletic people pictured in a soda ad.
4. What inferences does the advertiser want you to make from the ad? **Make inferences** that relate your observations to the product that the ad is selling. The soda ad, for example, may lead you to infer that if you drink that soda, you will be athletic.
5. **Share** your inferences and advertisements with others in your class.

Conclude and Apply
1. **Analyze** the inferences you made. Are there other explanations for the observations?
2. **Create** your own ad to sell a product. Think about what people will observe in the ad and what you want them to infer from it.

Goals
- Make inferences based on observations.
- Recognize the limits of observations.

Materials
- magazine advertisements

	Observation	Inference
Ad 1		
Ad 2		
Ad 3		

Communicating

Another important science skill is communication. Look at **Figure 1-8.** Scientists depend on one another to share what they have learned. One way to practice doing this is by using a Science Journal. A Science Journal is much more than a place to write down your observations. It's also a place to express your ideas about what you're investigating, make a sketch, or explore different opinions on a subject. Your Science Journal allows you to practice communicating your thoughts and ideas.

Throughout this book, you'll get many opportunities to practice observing, inferring, and many other science skills. By practicing these skills, you'll become better at solving problems and you'll learn more about the world around you. In the next section, you'll get a chance to start practicing these skills as you learn some ways to use science to solve problems.

FIGURE 1-8

These scientists are sharing the results of an experiment with one another.

Section Wrap-up

1. What is science?

2. How can science help you solve a problem?

3. **Think Critically:** Why do you think communication is an important skill in science?

4. **Skill Builder**
   Communicating Choose a science skill from the Skill Handbook, such as observing or classifying. In your Science Journal, write a paragraph describing the skill. If you need help, refer to Communicating on page 542 in the **Skill Handbook.**

Science Journal

Research the Internet for information on computer probes, or if your school has computer probes, have your teacher show you how to use them. In your Science Journal, write about ways you can use this technology to help you collect data.

People & Science

Amanda Shaw, International Science Fair Contestant

Q Ms. Shaw, tell us about competing in an international science fair.

A My project was about plants and global warming. Global warming is a rise in global temperatures that may be due to increases in certain gases in the atmosphere. I went to a regional competition where I won first place. Then I competed against hundreds of other kids in the International Science and Engineering Fair (ISEF).

Q What did you test?

A Many scientists think that increased amounts of carbon dioxide in the air may cause global warming. But plants take carbon dioxide out of the air. I predicted that plants would have a reversing effect on global warming—that is, not cause temperatures in the atmosphere to rise.

Q How did you test your prediction?

A I used glass jars to model the atmosphere. I had one jar with just air inside, one with carbon dioxide, one with a plant, and one with carbon dioxide and a plant. In sunlight, the jars containing carbon dioxide got warmer than the jar with air. But

the jar with the plant and carbon dioxide didn't get as warm. That showed that the plant reversed the carbon dioxide effect.

Q How else are you pursuing your interest in science?

A After the ISEF, I worked as a summer intern at Johns Hopkins University. I worked with scientists who study childhood diseases.

Q Do you enjoy working with scientists?

A I sure do! It seems like we learn something new every day. When we make a new discovery, the scientists in the lab get so excited! And I get excited when I think that with every discovery, we might be closer to helping somebody.

Career Connection

Scientists from many fields study global warming.

• meteorologist • geochemist • ecologist

Choose one of these careers and research the education and training required for it. Share your findings with your classmates.

1•2 Doing Science

What YOU'LL LEARN

- How to use science to solve problems
- How to design your own experiments

Science Words:
hypothesis
variable

Why IT'S IMPORTANT

Using scientific methods will help you solve many types of problems.

Solving Problems

Now, it's time to start practicing your science skills as you learn some ways to use science to solve problems. **Figure 1-9** shows one method to solve problems. Let's see how Ms. Quon's class uses this method to solve the bridge problem.

"All right," said Ms. Quon, "we've been given the job of designing a bridge to go across the stream at the site for the new outdoor education center. Where do we start?"

"I think we should go look at the site," said Tinho.

"Why would you start there? What would you look for?" questioned Ms. Quon.

"I don't know. Maybe we could get some ideas by seeing the place," he answered.

"You might be right, but what should we do to make the best use of our time there?" asked Ms. Quon.

"We should think about it some more and make a plan about what we want to find out by going there," offered Gabby.

Getting Organized

"That's right," said Ms. Quon. "Whenever you try to solve a problem, you need to start by making sure you know exactly what the problem is. Then you develop a plan for how to solve it. Let's start by defining the problem. What do you think it is?"

Sydney raised his hand and said, "I think the problem is: How should a bridge be built to cross the stream at the outdoor education center?"

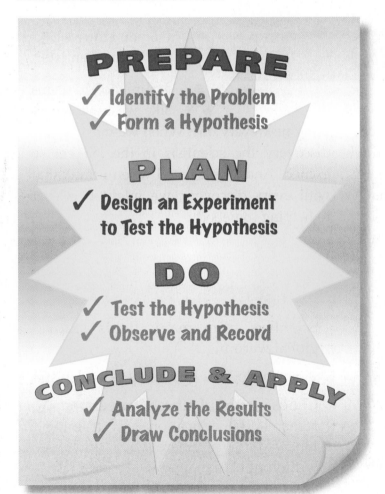

PREPARE
- ✓ Identify the Problem
- ✓ Form a Hypothesis

PLAN
- ✓ Design an Experiment to Test the Hypothesis

DO
- ✓ Test the Hypothesis
- ✓ Observe and Record

CONCLUDE & APPLY
- ✓ Analyze the Results
- ✓ Draw Conclusions

FIGURE 1-9

This poster shows one way to solve problems using scie..ills.

"That's a good start," said Ms. Quon. "What do we already know about the problem?"

"Well, we know that the bridge needs to be strong enough to hold people safely," suggested Tinho.

"And the bridge must be able to survive high water in the spring," said Sydney.

"Wonderful!" said Ms. Quon. "We also know that the site is safe, thanks to studies done by the city to make sure there are no harmful chemicals or wastes dumped there. Now we can ask a question based on what we already know."

"I think we have a lot of questions, like how strong it should be, or what we should make the bridge out of," said Tinho.

"That's right, sometimes we need to break a problem down into smaller problems that are easier to answer. But we have to start somewhere. What do we want to find out first?" asked Ms. Quon.

"First, let's decide where we're going to put the bridge," said Gabby. "That should be easy to figure out."

The class came up with several possible places to put the bridge. Ms. Quon smiled. "Hmm. It looks like this question isn't as easy as we thought. How can we decide which of the suggested places is best?"

Sydney suggested, "Well, we need to look at how the rest of the site is going to be used—you know, like where the trails may be. That may help us decide."

Ms. Quon agreed. "Other classes are working on different parts of the outdoor education center. Maybe we need a way of communicating what each class is doing so we can see how our project fits into the bigger picture."

SCIENCE
Online

Go to Glencoe Science Online at *science.glencoe.com* to learn more about bridge design.

FIGURE *1-10*
Gathering information at the library or on the Internet can make your problem-solving tasks easier.

FIGURE 1-11
Problem solving involves making careful observations and gathering information.

"We also need to look at the stream. Maybe there are trees, or the shape of the stream's banks may be an issue," said Tinho.

"What if we end up with two equally good choices?" asked Ms. Quon. "That is, what if both meet all of the requirements that we come up with?"

"We could make some sketches of different locations and ask people which one they prefer," suggested Gabby.

"Yeah, we could do a survey," agreed Tinho.

Carrying Out the Plan

The class made a list of requirements that the bridge location should meet. Then they went to the site and, like the student in **Figure 1-11,** took photos of the stream and made careful observations of each of the possible bridge locations. They chose the two best sites, based on the requirements that they had decided on. Then, Gabby and Tinho drew a map of the site that showed the two possible bridge locations. Using these drawings, the class surveyed people in the school and the community to find out which location they thought would be the best. The class counted the results and found a clear winner.

"Well," said Ms. Quon, "it looks like we have a location. What's the next step?"

"We need to find out how to build a bridge that's strong enough," said Sydney.

"How should we solve this new problem?" asked Ms. Quon.

"We can look at other bridges to get ideas about how we could build ours," said Tinho, pointing to the pile of books that Ms. Quon brought in.

Using Models

Ms. Quon flipped through one of the books. "There are a lot of different bridge designs. I think many of them might work for our bridge. How will you decide which design to use?"

"We could test them to see which ones are strong enough," said Gabby.

"Won't it be hard to test different bridges? How are we going to build all of these bridges?" asked Ms. Quon.

"We can build smaller ones and test them that way," suggested Sydney.

"That's a good idea," Ms. Quon said. "That's called using models. If you need to experiment with something that takes too long, is too fast, too big, or too small, you can test it using a model. We can't build full-size bridges to test, but we can build models of them and see which designs are strong enough. After we experiment and find out which design we will use, we can then experiment with materials in the same way.

"Before we experiment, though, we need to know how to plan an experiment to make sure we find out the information we need."

Problem Solving

Flex Your Brain

Solving problems requires a plan. This plan may be a simple thing that you do in your head, or it may be something more complicated that you actually write down. Below is a process called *Flex Your Brain*, which is one way to help you organize a plan for solving a problem.

Solve the Problem:
Use the *Flex Your Brain* chart to explore different styles of bridges.

Think Critically:
Why does *Flex Your Brain* ask you to share what you learned?

Flex Your Brain

1. **Topic:** _____
2. **?** *What do I already know?* 1. ___ 2. ___ 3. ___ 4. ___
3. **Q:** Ask a question
4. **A:** Guess an answer
5. **How sure am I ? (circle one)** Not sure 1 2 3 4 Very sure 5
6. **?** *How can I find out?* 1. ___ 2. ___ 3. ___ 4. ___
7. **Explore**
8. Do I think differently? → yes no
9. **?** *What do I know now?* 1. ___ 2. ___ 3. ___ 4. ___
10. **SHARE** 1. ___ 2. ___ 3. ___

Planning an Experiment

Experiments are often part of a plan for solving problems. They give you a chance to test your ideas as you go through the process of solving the problem. **Figure 1-12** shows one way to test experiments. As Ms. Quon's class is finding out, experiments will help them learn more about bridges as they try to solve the problem of building the bridge across the stream.

Throughout this book, you'll have opportunities to design your own experiments. As you can see in Activity 1-2 on pages 24-25, the two-page activities in this book are divided into four major parts. First, you'll *Prepare* for the experiment. Then, you'll *Plan* the experiment. Next, you'll *Do* the experiment. And finally, you'll *Conclude and Apply.*

FIGURE 1-12

Car designers test the safety of automobiles using crash-test dummies.

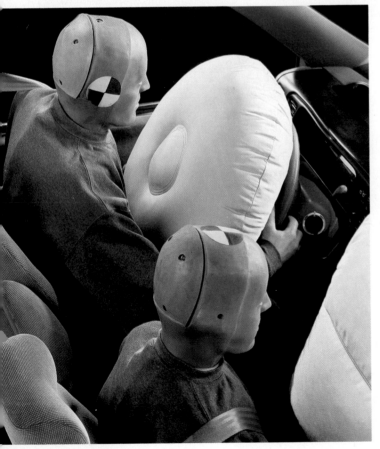

PREPARE

Preparing for an experiment involves getting a clear idea of what you want to find out. You start by thinking about the problem that you are trying to solve. What do you already know that may be helpful in solving the problem? Organize this information and use it to form a hypothesis (hi POTH uh sus).

A **hypothesis** is a statement that can be tested about a problem. A hypothesis about the strongest material to make the bridge may be "Wood is the strongest of the materials we can use to build the bridge." Notice that this hypothesis can be tested. You can do an experiment that tests the strength of different materials. After doing the experiment, you may find that your hypothesis was supported. Or, you may find that there was a stronger material and your hypothesis was not supported. When this happens, you'll have to develop a new hypothesis.

FIGURE 1-13

The amount of water added to the plants is the variable in this experiment. *Based on the photographs, what would you conclude about the effects of water on plants?*

A At the beginning of the experiment, similar plants received the same amount of sunlight and were planted in the same type of soil. The plant at top left received the most water. The bottom plant received none at all.

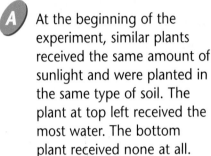

B Three weeks later, by controlling other factors and changing only one variable—water—the results of the experiment clearly show the effect of water on plants.

PLAN

Once you've developed a hypothesis, it's time to plan a way to test the hypothesis. Planning an experiment may take longer than the experiment itself. Careful planning helps ensure that your experiment is safe and will help you get useful results.

Variables

One of the most important considerations in an experiment is controlling variables. **Variables** are factors in an experiment that can change. **Figure 1-13** gives an example of testing variables. For another example, think of bridges. If you are testing different bridge-building materials to see which is the strongest, the material you use is one variable. You may test wood, steel, or plastic. You would test them all the same way. But you test only one at a time. Each time you run the experiment, you change which material you test. Other variables in such an experiment might be the shape of the material, how you measure the material's strength, and how you test the material.

You have to make sure, though, that you change *only one* variable at a time. This is called "controlling the variables." To make sure you are testing which material is the strongest, each material you test should have the same shape, and you should test each one the same way. If you change more than one variable, you can't be sure which variable caused the results.

Number of Trials

Another thing to decide when planning an experiment is how many trials to have, or how many times you need to repeat the experiment to make sure the results are useful. Look at **Figure 1-14.** If you're testing the strength of wood, do you test just one piece of wood? What would happen if the wood you tested had a crack in it to begin with? Can you trust that the results from the cracked-wood test will show the strength of another piece of wood? What if you made a mistake measuring the strength of the wood? These kinds of errors are always going to occur. That's why it's a good idea to run more than one trial of an experiment. If there's an error, it will be balanced by the other trials.

Sometimes you may need to run ten trials; other times you may need to run 100. However, some experiments are either too costly or take too much time to run more than two or three times. When this happens, you'll have to think about the possible errors when you analyze the results.

FIGURE 1-14

To balance the possibility of errors in an experiment, it's a good idea to run several trials of the experiment. *Is this always possible? Explain why or why not.*

Strength of Different Bridge Materials

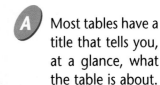

A Most tables have a title that tells you, at a glance, what the table is about.

	Trials			
	1	2	3	4
Wood	broke	didn't break	didn't break	didn't break
Plastic	broke	broke	broke	didn't break

(Materials)

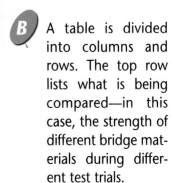

B A table is divided into columns and rows. The top row lists what is being compared—in this case, the strength of different bridge materials during different test trials.

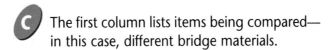

C The first column lists items being compared— in this case, different bridge materials.

FIGURE 1-15

Data tables help you to organize your observations and test results.

Materials

After you have planned the experiment, you must make a complete list of the materials you will need to do the experiment. Try to think of alternate materials in case something you need is not available. Be resourceful. You don't have to have fancy scientific equipment to do experiments. You can learn a lot using things you have around your house and classroom.

Data Tables

A data table organizes observations in columns and rows. **Figure 1-15** shows how you might set up a data table for one particular experiment. Because you will need to record your observations during the experiment, it's a good idea to make any data tables you'll need ahead of time. This will help you make sure you make all of the observations necessary to make a conclusion about your hypothesis.

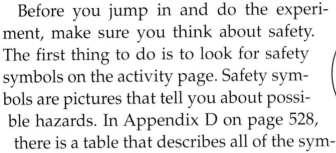

DO

Before you jump in and do the experiment, make sure you think about safety. The first thing to do is to look for safety symbols on the activity page. Safety symbols are pictures that tell you about possible hazards. In Appendix D on page 528, there is a table that describes all of the symbols used in this book. While you may not think you need goggles and aprons all the time, accidents happen—and when they do, you'll be glad you were protected. Follow safety rules whenever you are doing activities. **Figure 1-16** gives more safety tips.

Doing experiments can be a lot of fun, but it's important to be careful when doing them. Not paying attention to what you're doing could seriously injure you or one of your classmates. Remember, safety is no accident! Here are a few important safety rules that you should follow:

1. Before beginning any lab, understand the safety symbols shown in Appendix D.
2. Follow all safety symbols.
3. Always slant test tubes away from yourself and others when heating them.
4. Never eat or drink in the lab, and never use lab glassware as food or drink containers.
5. Never inhale chemicals, and don't taste anything.
6. Report *any* accident or injury to your teacher.
7. When cleaning up, get rid of chemicals and other materials as directed by your teacher, and always wash your hands after working in the lab.

Another thing you must do before you begin the experiment is to have your teacher approve your plan. He or she will be able to find any problems with your plan and suggest ways to improve it. There's nothing worse than spending time carefully doing an experiment only to have to start over because you forgot to include an important step in your plan.

FIGURE 1-16

Lab equipment, like the test tube shown above, must be handled properly. Goggles like the ones shown below should be worn during all lab activities. *Look at the safety symbol on this page. What does it tell you?*

Recording Observations

As you do the experiment, record all of your observations in your Science Journal. **Figure 1-17** shows an example of a Science Journal. Observations can include measurements and descriptions of what happened and sketches or drawings. For example, if you were testing the strength of bridge materials, you might observe that a piece of plastic held a certain amount of weight without breaking. However, you may observe that it bent a lot. You should also note any mistakes you made during the experiment.

CONCLUDE AND APPLY

After you've carried out the experiment and recorded all of your observations, it's time to look at the results and figure out what they mean. This is when you start asking *why* things happened. You'll need to look at all of the observations you wrote down and even compare your results with those of other groups that did the same experiment. Making tables and graphs can also help you see what happened in the experiment.

If you tested the strength of bridge materials, you might decide that wood is the strongest material you can use. That's called drawing a conclusion. A conclusion is a decision based on the results of the experiment. It may either support or not support your hypothesis. If you hypothesized that wood was the strongest material you could use to build the bridge, then you could say that your hypothesis was supported.

Next, you'll get a chance to design your own experiment.

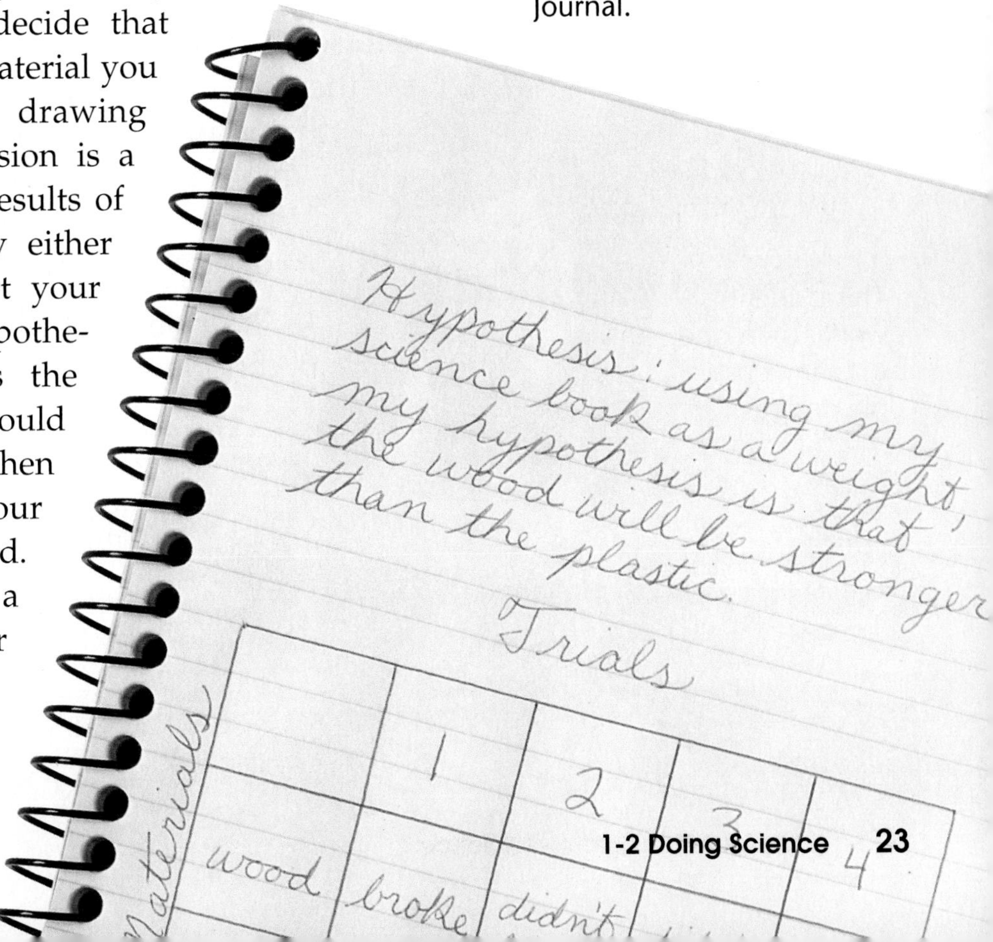

FIGURE 1-17

Record observations and conclusions in your Science Journal.

Hypothesis: using my science book as a weight, my hypothesis is that the wood will be stronger than the plastic.

Trials

Materials	1	2	3	4
wood	broke	didn't		

Design Your Own Experiment
Bridge Building

One of the problems Ms. Quon's class needs to solve is what design they should use to build their bridge. In this activity, you will plan and do an experiment to design and test a model bridge. You only get to use certain materials, your bridge must cross a gap 50 cm wide, and it must be strong enough to hold your science book. Good luck!

Possible Materials
- drinking straws
- craft sticks
- string
- glue
- tape
- scissors
- meterstick
- books

PREPARE

What You'll Investigate
How can you build a model bridge that will cross a 50-cm gap and hold your science book?

Form a Hypothesis

As a group, look at the materials your teacher has supplied. Think about ways to use them to build a model bridge that meets the requirements stated above. How can the materials be used to make bridge shapes you have seen? Try putting together a few pieces to see how they join together and how strong they are. Sketch some designs, and discuss the good and bad points of each. You might need to combine some of the best parts of different designs. Decide on a design that will meet the requirements.

This student is working on one of many possible designs.

State your hypothesis. For instance, "The bridge design that our group decided on will be strong enough to hold a science book across a 50-cm gap." Write your hypothesis in your Science Journal.

Goals

Design a model bridge.

Construct and **test** a model bridge.

Safety Precautions

PLAN

1. **Decide** on a way to test your group's hypothesis. **List** the materials you will use in each step of your experiment. **Describe** how you will connect the materials.
2. Sketch your design. Look for any spots in the design that may be weak. How would you strengthen those spots?
3. How will you **test** the strength of your model? How many times will you repeat the test?
4. **Make a data table** to record your observations.

DO

1. Make sure your teacher approves your plan and your data table before you begin.
2. Write down the observations you make in your Science Journal.
3. **Build** your model using the materials provided by your teacher.
4. **Test** your bridge to see if it meets or exceeds the requirements.

CONCLUDE AND APPLY

1. Based on your results, did your **model** bridge meet the requirements?
2. **Compare** your model with the models of all the other groups. Which designs performed best in the tests?
3. What was different about the strongest bridge models? How did these differences make the models stronger?
4. **APPLY** Based on your observations of all the models, how would you change your design? **Draw** the new design.

Communicating the Results

Ms. Quon's class performed many experiments, testing many designs and materials for their bridge. When they were finished, they had decided on a bridge design that would be perfect for the new outdoor education center. As a class, they wrote a report describing the design and how they came to that conclusion. They included graphs and tables that summarized their observations from the various experiments. When it was all finished, they presented the report and a model of the bridge to the school board. The board was impressed with how well the class had communicated their work. The design was approved.

"You've done a fabulous job planning this bridge," said Ms. Quon. "Unfortunately, we don't have much of a budget to actually build it."

"It looks like we have another problem to solve," sighed Gabby.

"Yes, but I'm sure you can do it!" smiled Ms. Quon.

"You mean, you *hypothesize* that we can do it!" laughed Gabby.

FIGURE 1-18
Sharing the results of your experiment with others is part of doing science.

Section Wrap-up

1. How can science help you solve problems?

2. Why do scientists sometimes use models to test the design of objects?

3. **Think Critically:** Valerie is going to do an experiment testing how much flour is needed to make the cake with the best texture. Which variables should she control?

4. *Skill Builder*
   Designing an Experiment to Test a Hypothesis Design an experiment that tests the hypothesis "Wood is a stronger material than plastic or aluminum." Describe the steps you would follow and list the variables in your experiment. If you need help, refer to Designing an Experiment to Test a Hypothesis on page 553 in the **Skill Handbook.**

Using Computers

Graphics Software
Use a graphics program to design a poster that shows the importance of a particular science skill. Use pictures to help show the skill. Present your poster to your class.

Read the statements below that review the most important ideas in the chapter. Using what you have learned, answer each question in your Science Journal.

1. Science is the process of learning and studying things in the world around you. *How is science like a tool?*

2. There are many ways to solve problems. Science allows you to break the problem down into steps that you can follow to solve the problem. *Describe the first step in solving any problem.*

Side View

300 m

60 m

Girder   Vertical beam

Cross beam

3-D View

Front View

3. When designing an experiment, following a set of steps such as *Prepare, Plan, Do,* and *Conclude and Apply* will help you get the answers you need to solve the problem. *Describe what a hypothesis has to do with designing an experiment.*

BAKER

PRINTER

Using Key Science Words

hypothesis
science
technology
variable

Answer the following questions in complete sentences. Use the terms in the list above in your answers.

1. What should you do before you begin an experiment?
2. What is the difference between science and technology?
3. What is a variable?
4. What does a scientist do?
5. Define *hypothesis*.

Checking Concepts

Choose the word or phrase that completes the sentence.

6. Science is _____.
   a. a way to help solve a problem
   b. a tool
   c. the process of trying to under-stand the world around you
   d. all of the above

7. The use of scientific discoveries is called _____.
   a. the scientific method
   b. a hypothesis
   c. technology
   d. a variable

8. Comparing and contrasting is an example of a _____.
   a. science skill
   b. hypothesis
   c. conclusion
   d. control

9. You should never _____ in a science lab.
   a. wear safety goggles
   b. eat or drink
   c. wear an apron
   d. report any accidents to your teacher

10. A _____ is a statement about a problem that can be tested.
    a. conclusion   c. variable
    b. hypothesis   d. data table

Thinking Critically

Answer the following questions in your Science Journal using complete sentences.

11. Give one example of how Ms. Quon's class controlled variables.

12. You observe a flock of geese flying south. What inference can you make?

13. Give an example of a hypothesis.

14. Describe how you have used technology today.

15. Is every problem solved using the same steps? Explain.

Developing Skills

If you need help, refer to the description of each skill in the Skill Handbook.

16. Forming a Hypothesis: Enrique has an aquarium. He bought a new fish and placed it in his aquarium with his other fish. Two days later, it died. Form a hypothesis about why the fish died. Design an experiment to test your hypothesis.

17. Concept Mapping: Scientists use different methods to help solve problems. Complete the events chain on this page that shows the basic steps in solving a problem.

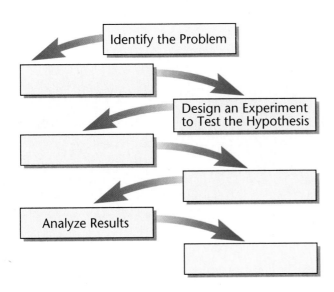

Identify the Problem

Design an Experiment to Test the Hypothesis

Analyze Results

18. Forming a Hypothesis: Your volleyball team has lost three games in a row. Form a hypothesis about what can be done to help your team play better.

19. Separating and Controlling Variables: Suppose you hypothesized that your volleyball team would play better if it practiced more. You decide to test your hypothesis. After adding one more afternoon practice to your schedule, your team wins a game! What were the variables in your test? Which variables did you control?

20. Sequencing: Maggie wants to see how many kinds of birds live in a nearby park. Sequence the following steps in the order she should follow to solve her problem.
   a. Observe birds.
   b. Analyze the results.
   c. Research the kinds of birds that typically live in her area.
   d. Hypothesize how many birds live in the park.
   e. Record her observations.

Performance Assessment

1. Bulletin Board: Think about the steps involved in solving a problem. Make a Solving Problems with Science bulletin board. Include the major steps of solving problems using science. Write a summary of what to do at each step, and include drawings or photographs that help explain the process.

2. Oral Presentation: Design an experiment that investigates the effects of sunlight on small plants. Prepare an oral presentation that describes your experimental plan and give it to your class.

UNIT 1

Life Science

What's happening here?

Seeds can often be found some distance away from their parent plants. Most plants don't move, so how do they spread their seeds? Some seedpods explode, shooting their seeds far away. Other seeds hitch rides on air currents or passing animals. Milkweed pods burst open, releasing seeds with plumes of silky hairs. These hairs allow the seeds to ride the wind. Cocklebur seeds, as shown in the smaller photo, may travel miles on animal fur. How does traveling help seeds? By being spread far from their parents, the seeds do not have to compete for water and sunlight as they grow into new plants.

SCIENCE Online

Visit Glencoe Science Online at *science.glencoe.com* for links to information about milkweed seeds, cockleburs, and other seeds. Find out how other plants spread their seeds. In your Science Journal, write about how plants use wind, water, and other organisms for seed dispersal.

Chapter Preview

Skills Preview

▶ **Skill Builders**
- make tables
- develop multi-media presentations

▶ **MiniLABs**
- infer
- classify

▶ **Activities**
- hypothesize
- design an experiment
- classify
- compare and contrast

What is life?

First, there's just a tiny crack in the eggshell. Then, the opening becomes larger. Soon, you see a beak, then bulging eyes, and you hear a weak squeak or two. Finally, a wobbly, wet chick comes out! An egg is a protective place for the developing chick. It also stores things such as food and water that are needed by the chick for development. Just weeks after being laid, the "lifeless" egg hatches.

The baby chicken is alive. You are, too. So is the grass you walk on. But is a sidewalk alive? What about a bicycle? You probably don't think about the fact that you are surrounded by living and nonliving things. Have you ever thought about the difference between a living thing and a nonliving thing?

EXPLORE ACTIVITY

Compare Living and Nonliving Things

1. Look around your classroom. In 15 seconds, how many living things can you find? How many things can you find that are not alive?
2. Look carefully at each thing you found. Think about each one.

Science Journal

In your Science Journal, list the items you saw. Next to each object, write down why you think it's a living thing or a nonliving thing.

2•1 Alive or Not

What YOU'LL LEARN

- Five ways to decide what is alive
- The main needs of living things
  Science Words:
  trait
  organism
  environment
  cell
  development

Why IT'S IMPORTANT

You will understand why you are thought to be alive.

Identifying Life

Have you ever seen a living stone? Sounds like a joke, doesn't it. But there really are such things as living stones. *Lithops* (LIH thops) is the plant pictured in **Figure 2-1.** It is sometimes called a living stone or a flowering stone because, even though it is a plant, it looks like a stone. Although *Lithops* appear to be nonliving, they are just as alive as flowers, trees, and other plants with which you are familiar. Most plants and animals are easy to recognize as living things, but sometimes it's not so simple. Does something have to move to be alive? Does it have to breathe as you do?

Must certain activities take place inside plants and animals if they are to be considered living? Or must they share common physical features, or traits? A **trait** is a specific feature of something. For example, human traits include brown eyes, red hair, and the ability to walk upright. No single trait can tell you whether something is alive. Over time, however, scientists have observed common traits that are used to identify something as a living organism. An **organism** is a living thing that has all of the traits of life. **Figure 2-2** shows different organisms that have the traits of life. Look at the list of some traits of life shown below.

FIGURE 2-1

There are more than 50 different kinds of *Lithops* plants. Each type looks a lot like a real stone. They're small, hard, and rounded. Their colors are just like rocks. *How are living stones really alive?*

Some Traits of Life
- organisms respond
- organisms move
- organisms show organization
- organisms reproduce
- organisms grow and develop

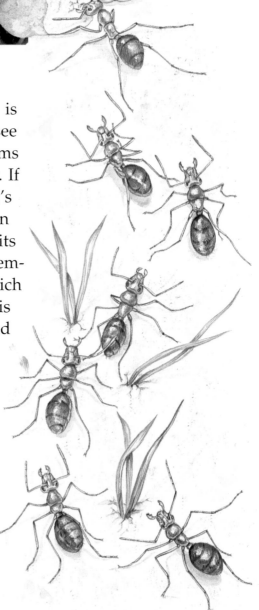

Organisms Respond

What might you do to find out whether *Lithops* is alive? Maybe you would gently touch it and see whether anything happens. The first trait of organisms is that they respond to things in their environment. If you touch a living stone, you are checking the plant's ability to react to its environment (en VI ur munt). An organism's **environment** includes everything in its surroundings—other organisms, water, weather, temperature, soil, sound, and light—anything with which it comes in contact. Reacting to the environment is what squirrels do when they move away from loud sounds, and when plants grow toward light.

An organism can also respond to changes inside itself. For example, body temperature or disease can cause changes inside an organism. Sometimes, your body does not work the way it should. It must then make changes to return to "normal." Body temperature increases a little when it's hot or when you're working hard. In response to this internal change, you begin to sweat. Your face gets red as tiny blood vessels in your skin swell with blood. Both of these responses help cool your body so it can maintain its normal temperature.

Mini LAB

Inferring Body Responses

How does your body react to internal changes?

1. Find your pulse by placing the index finger of one hand on the thumb side of the wrist of your other hand, just below the base of the thumb.
2. While at rest, count how many pulse beats you feel for 10 seconds, and then multiply this number by 6. This is your pulse rate before exercise.
3. Jog quickly in place at your desk for one minute.
4. Determine your pulse rate immediately after exercise, then at 5 minutes and 10 minutes after.

Analysis

1. Did your pulse rate change while you exercised? If so, how?
2. How did your body change 5 and 10 minutes after exercising?

Organisms Move

Movement is the second trait of organisms. Birds glide through the air. Snakes slither on the ground. Dogs, cats, and people walk and run. These are all ways that animals move from place to place. As shown in **Figure 2-3,** all organisms move in some way, but they don't all move to different places. Many plants move merely by bending toward the sunlight.

Living Things Show Organization

All living things are made up of parts that are organized—the third trait of organisms. If you could look at a small piece of a *Lithops* plant under a microscope, you would see hundreds of tiny compartments set in a pattern that looks like a brick wall. These little compartments are the plant's *cells.* The **cell** is the smallest unit of life in a living thing.

All organisms are made up of cells. An example of cells is pictured in **Figure 2-3.** You, like elephants and whales, are made up of trillions of cells. But as you'll learn in the next section, many other organisms are made up of only one cell, such as bacteria. In all organisms, cells do the same basic job. They take in materials and release energy and waste products.

Living things are organized in other ways, as well. For example, plants and animals have parts that perform certain jobs. Some plants have leaves, stems, flowers, and roots for carrying out many different plant activities. You, as an animal, have a mouth for eating, legs and feet for moving around, and ears for hearing.

Organisms Reproduce

A fourth question you might ask when you're trying to decide whether or not something is alive is "Where did it come from?" All living things reproduce. Reproduction means that organisms make more of their own kind. Animals such as cows and deer give birth to live offspring, while some animals such as turtles and birds lay eggs. But no matter how organisms reproduce, reproduction is necessary for life to continue.

Living organisms reproduce in many different ways. For example, *Lithops* and some other plants produce flowers. Flowers usually contain male and female parts needed for reproduction. In Chapter 4, you'll learn more about reproduction and discover how it enables all groups of organisms to survive.

FIGURE 2-3

How do these different organisms demonstrate the traits of life discussed—movement, showing organization, and reproduction?

Organisms Grow and Develop

FIGURE 2-4

Frogs show major changes during their development. Fertilized eggs hatch into legless tadpoles that live in water. These tadpoles develop legs and lungs and move onto land as young adult frogs.

If you could watch a *Lithops* plant over a long time, you would see that it doesn't always look like a stone. *Lithops* plants grow and produce flowers for part of the year. All of the changes that take place during the life of an organism are known as its **development.** Most organisms show growth and development—the fifth trait of organisms. The activities that take place inside a chicken egg are an example of growth and development. **Figure 2-4** illustrates growth and development in frogs. You, too, are a product of growth and development. You began life as one cell. As you developed, the number of cells making up your body increased and you became larger. Development took place. Some of your cells developed into skin cells, while others changed into muscle cells. Still other cells became nerve cells and bone cells.

Development isn't easy to see in all organisms. Many single-celled organisms, such as bacteria, don't seem to develop. The single cell grows a little bit in size and then reproduces itself by splitting in half. Even though bacteria do not seem to develop, many changes are going on inside the cell as it carries out its different life activities. **Figure 2-5** illustrates the five traits of life using different organisms.

A An adult African elephant grazes on grass in an African plain. The elephant *responds* to its sense of smell to lead it to food. *How does the elephant show organization?*

FIGURE 2-5

These organisms are showing different traits of life.

B A female leopard *responds* to her cub's "play attack" by pulling out of the way. The cub shows a lot of *movement* when it leaps up, using its tail for balance. The patterns on the leopards' coats provide camouflage. *What other traits of life are shown in this picture?*

C A honeybee flies toward a clover flower and collects food from it. The shape of the flower is *organized* to ensure that pollen gets on the bee at the same time. At the next flower, the bee transfers the pollen, which fertilizes the flower. The flower produces seeds for the purpose of *reproduction*.

Activity 2-1

Design Your Own Experiment
Mealworm Behavior

You can use mealworms to show how organisms respond to changes in their environment. Mealworms aren't really worms. They are a young form of darkling beetles. Mealworms usually live in moist, dark places or where grain is stored.

PREPARE

Possible Materials
- mealworms (10)
- small cardboard box with lid, such as a shoe box
- eyedropper
- cotton-tipped swabs
- penlight or small flashlight
- dry bran flakes (15 mL)
- chopped apple (15 mL)
- chopped banana (15 mL)
- vinegar (5 mL)
- ice cubes
- plastic straw
- hot water (not boiling)
- tap water at room temperature

What You'll Investigate
How do mealworms respond to changes in their environment?

Form a Hypothesis

What types of things might or might not attract mealworms? Will they be attracted to a flashlight or to food? Dry food or moist food? Hot food or cold food? **Make a hypothesis** about how mealworms will respond to changes in these things.

Goals

Observe the traits of mealworms.

Predict how mealworms will respond to different things.

Safety Precautions

Remember that mealworms are living organisms. Be sure to handle them gently and in a responsible manner. Return them to your teacher at the end of your experiment.

PLAN

1. Working in a group, **make a list** of some possible things that you can put in front of the mealworms.
2. As a group, **make hypotheses** about how the mealworms will respond to each thing.
3. As a group, **decide** how the different things will be shown to the mealworms. Will you show each item alone, or will you use more than one item at a time?

4. How many mealworms will you use?
5. **Prepare** a data table in your Science Journal.
6. With your partners, **write** a list of all the steps you will take in your experiment.
7. **Read** over your experiment to make sure that all steps are in logical order.

DO

1. **Read** over your plan to make sure that it will test your hypotheses.
2. Make sure your teacher approves your plan before you do it.
3. Carry out the experiment as planned.
4. While the experiment is going on, **record** all observations that you make and fill in the data table in your Science Journal.

CONCLUDE AND APPLY

1. Which things attracted the mealworms? Which things did the mealworms avoid?
2. Did the results of your experiment agree with your hypotheses?
3. What did the results of your experiment tell you about mealworm behavior?
4. **APPLY** The behaviors shown by mealworms are called instincts. Instincts are behaviors that organisms are born with. Why are instincts important to organisms? How do instincts help mealworms?

Basic Needs of Living Organisms

If you did the previous activity, you may have learned that food was one thing that attracted mealworms. It shouldn't be surprising to see that mealworms are attracted to food. Aren't you sometimes willing to go out of your way for a snack when you're hungry? Mealworms need food for the same reason you need food—for energy! What other things do living things need?

Living Things Need Energy

None of life's important activities would be possible without some form of energy. Growth, movement, and reproduction could not take place without it. Energy is an important need of living things. How do organisms

USING TECHNOLOGY

A New Chess Champion?

The final game of their six-game championship chess tournament ended much sooner than people had expected. Garry Kasparov, the reigning world champion, quit the game after just 19 moves. His opponent showed no emotion, and no one was surprised. Kasparov's opponent was a computer. The 1997 tournament made headlines because it marked the first time ever that a computer had beaten a human champion at chess. The computer that beat Kasparov was a huge computer named Deep Blue. Kasparov had defeated Deep Blue a year earlier. But in their second matchup, Deep Blue was more prepared.

GARRY KASPAROV　　DEEP BLUE

Deep Blue was programmed to allow it to choose the best move it could play. However, scientists are far from making a robot or computer that can think the same way people think. A computer can only make a choice from options that it has been given.

SCIENCE Online

What is artificial intelligence? Check Glencoe Science Online, *science.glencoe.com*, for a link to a site with several definitions of computer intelligence.

obtain the energy they need? Most living things get their energy from the sun. Plants use the sun's energy and carbon dioxide from the air to make food. This important activity takes place inside some plant cells. Plants use the food energy they make for growth and other plant activities.

Most other organisms depend on this food-making ability of plants. People obtain energy when they eat plants. They can also obtain energy when they eat animals that have eaten plants. **Figure 2-6** shows a girl obtaining energy from food.

Living Things Need Water, Oxygen, and Minerals

Have you ever forgotten to water a plant? If a plant goes too long without water, it will probably die. You can't live long without water, either. Organisms are made up mostly of water. Your body, in fact, is about 70 percent water. Cells that make up organisms need water to carry out all important life activities. So, life on Earth wouldn't be possible without water.

Whether food comes from plant or animal sources, it has energy in it. Many organisms need the chemical oxygen to release the energy stored in their food. The oxygen you use for this purpose is found in air. As shown in **Figure 2-7,** organisms take in this oxygen in different ways.

FIGURE 2-6

Plants are the source of energy for most organisms. This girl is obtaining energy by eating both plants— tomatoes, lettuce, onions, wheat—and plant-eating organisms—beef cattle.

FIGURE 2-7

Some organisms, such as earthworms, absorb oxygen through their moist skin. Fish have gills that get oxygen that is in the water around them. *How do deer obtain water, oxygen, and minerals?*

Besides water and oxygen, organisms need other chemicals called minerals to live. Minerals are chemicals, such as sodium and chlorine, found in the air, soil, and water. When plants take in water through their roots, they absorb minerals that are in the water at the same time. **Figure 2-8** illustrates water absorption by a plant. Animals obtain these important minerals when they eat plants or when they eat plant-eating animals.

In this section, you've learned that there are five life activities that are common to all living things. To be considered alive, an organism must be able to respond to its surroundings, reproduce, grow and develop, and move. Organisms must also be organized and have cells to carry out these necessary life activities. But not all organisms are organized in the same way. And not all carry out these activities in the same way. Earth is filled with many different living things. In the next section, you'll discover how scientists study and learn about the different organisms on Earth.

FIGURE 2-8

Water and minerals used by a plant enter by way of its roots.

Section Wrap-up

1. List the five traits of living things.

2. Describe the needs of organisms.

3. **Think Critically:** Is fire a living organism? Explain your answer.

4. *Skill Builder*
   Making and Using Tables Make a table of the five traits of living things. In one column, list the five traits. In the second column, give an example of each of the five traits using an organism with which you are familiar. If you need help, refer to Making and Using Tables on page 546 in the **Skill Handbook**.

Science Journal

Imagine discovering a mysterious green blob in an unexplored part of a rain forest. In your *Science Journal*, describe some tests that you might perform to find out whether the green blob is alive.

Classifying Life

Classification of Living Things

Imagine walking into a video store to rent a movie and discovering that someone has rearranged all the movies on the shelves in no particular order. Would this lack of order make it hard for you to find what you're looking for? Fortunately, most stores, including grocery stores and clothing stores, are organized to help customers. Items in the store are usually grouped according to their similarities and placed together in their own sections or aisles. In a similar way, scientists organize, or classify, living things into groups. **Classification** (klas if uh KAY shun) is the grouping of objects or information based on common traits.

How are organisms classified?

When classifying organisms, scientists often use similarities in body parts to group organisms. Eastern gray squirrels and tufted-eared squirrels, shown in **Figure 2-9,** have a lot of features in common. Both squirrels are medium-sized animals with long, bushy tails and strong back legs. So, scientists have classified them in a group of organisms that share these common traits. However, the eastern gray squirrels and tufted-eared squirrels also look different. As a result, scientists call each different kind of organism a species. A **species** is a group of organisms that can mate with one another. Therefore, members of the eastern gray group can mate with one another, and members of the tufted-eared squirrels can also mate with one another.

What YOU'LL LEARN

- How to classify things
- The six kingdoms of living organisms
  Science Words:
  classification
  species
  kingdom

Why IT'S IMPORTANT

Classification is a tool that you use every day.

FIGURE 2-9

The tufted-eared squirrel and the eastern gray squirrel are two different species, even though they have many body parts in common.

Mini LAB

Classifying Animals with Bones

1. Scientists recognize five major groups of animals with bones: fish, amphibians, reptiles, birds, and mammals. Choose one of these groups to work with.
2. Obtain photographs of four different animals from the group you chose.
3. Study and compare the animals in the photographs. Make observations about the similar traits among the different species of your group.

Analysis

1. Which group of animals did you study—fish, amphibians, reptiles, birds, or mammals?
2. List the traits shared by the animals within your group.

An Example of Classification

Animals are classified into groups based on shared traits. For example, some animals have bones and others don't. There are five groups of animals with bones. Fish live in water. They have gills for breathing. Most fish have fins as well for moving through the water easily. Amphibians, such as frogs, live both in water and on land. Reptiles live on land and include animals such as turtles, alligators, and snakes. They have dry skin and most lay eggs. Animals that have feathers are placed in the bird group. Animals that have hair and give birth to live offspring are classified as mammals. Female mammals can nurse their young.

There is more to classification than just common body parts. Living organisms also can be classified according to similarities in the materials that make up their bodies. **Figure 2-10** shows one example of scientific classification based on chemicals found inside the organisms' bodies.

FIGURE 2-10

Scientists once agreed that guinea pigs should be classified with hamsters because they shared similar body parts. But by studying an important chemical in their cells, they decided that guinea pigs should really be put in a group by themselves.

Kingdoms of Life

All organisms on Earth have been separated into six large groups called kingdoms. A **kingdom** is a large group of organisms that share certain features. In the next activity, you could say that you will classify objects within the button kingdom.

The six kingdoms of living things are Archaebacteria, Eubacteria, Protists, Fungi, Plant, and Animal. Organisms can be placed into a kingdom based on the four features shown below.

- what their cells are like
- how many cells they are made up of
- whether or not they can move from place to place
- how they obtain energy

Problem Solving

Classifying Animals

Scientists don't just classify living organisms to make things easier. They also group living things to show the relationships between the different organisms. It seems logical that if two organisms look similar, they are related. But that's not always true. Think about birds, bats, and insects. All these organisms have wings and all fly, but are they related to each other? Do you see why scientists can't classify organisms just because of the way they look or act?

Solve the Problem:
Study the illustrations of the shark and the dolphin. Observe the traits that sharks and dolphins share.

Think Critically:
Sharks and dolphins look alike in many ways. But that doesn't mean they are closely related. What traits do you think could be used to tell sharks and dolphins apart?

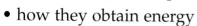
Spinner dolphin

Sandbar shark

Button Classification

Scientists classify living things so they can talk and write about organisms more easily. Classification systems also show how different species are related. To be related, the different species must share some traits in common. How do scientists make up classification systems?

What You'll Investigate
How do scientists classify organisms?

Procedure
1. **Obtain** ten to 15 different buttons.
2. **Study** the different types of buttons. **Choose** one feature that will allow you to separate the buttons into two smaller groups. Size, shape, color, number of holes, and how they feel (texture) are some possible features. **Separate** the buttons into groups based on the feature you chose.
3. Begin to make a chart of your classification system as shown on this page. **Write** down the feature you use to separate your buttons into two groups.
4. **Repeat** step 3 for each of the smaller groups you form. Keep track of your separations in your chart.
5. Continue following step 3 until there is only one button in each group.
6. **Complete** your classification chart and compare it with the results from other lab groups.

Goals
- Classify a group of buttons.

Materials
- assorted buttons (10 to 15 different types)
- pencil
- paper

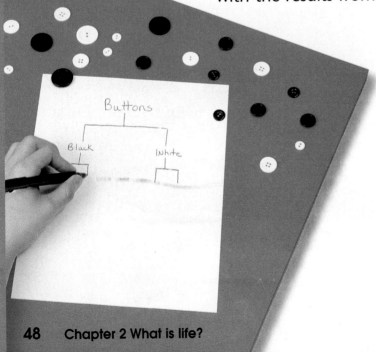

Conclude and Apply
1. Why did some groups develop different classification systems for the buttons?
2. **Compare** the way you **classified** your buttons with the way scientists classify living things.
3. Why do you end up with only one button in each group?

Bacteria

Bacteria is a term that has been used to describe members of the eubacteria (YEW bak teer ee uh) kingdom and members of the archaebacteria (ar kee back TEER ee uh) kingdom. Bacteria in the eubacteria kingdom are known as true bacteria. Archaebacteria are bacteria that live in extreme environments. All eubacteria are single-celled organisms. They include many bacteria that cause diseases in other organisms. There are also bacteria that are helpful. Bacteria used to make yogurt and some cheeses belong to the eubacteria kingdom.

There are thousands of known species of bacteria, but some scientists predict that millions more may exist. Some bacteria species take in food for energy and move around as many animals do. Others make their own food using sunlight, just as plants do. These bacteria contain the same chemical that plants use in making food.

Archaebacteria live in some of Earth's most difficult environments, such as hot sulfur springs and extremely salty lakes. Archaebacteria don't get energy from eating food as animals do. They also don't produce food by using energy from sunlight as plants do. Instead, archaebacteria get energy by taking in chemicals from their surroundings. One group, for instance, makes energy from the chemical called carbon dioxide found in the soil where they live. Another type makes energy from another chemical—sulfur. These archaebacteria grow in sulfur springs where it is extremely hot.

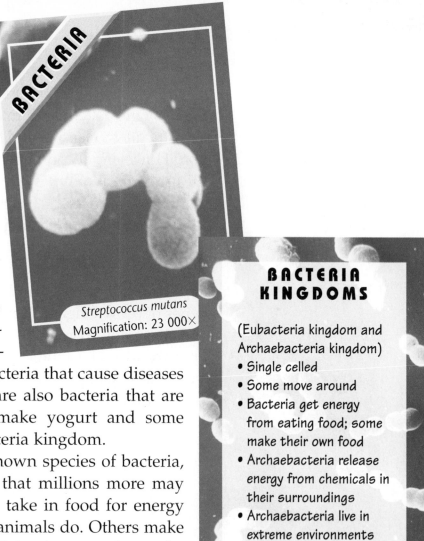

BACTERIA

Streptococcus mutans
Magnification: 23 000×

BACTERIA KINGDOMS

(Eubacteria kingdom and Archaebacteria kingdom)
- Single celled
- Some move around
- Bacteria get energy from eating food; some make their own food
- Archaebacteria release energy from chemicals in their surroundings
- Archaebacteria live in extreme environments

BACTERIA

Escherichia coli
Magnification: 14 000×

PROTIST KINGDOM

- Single-celled and many-celled species
- Some move around
- Some obtain energy from eating food; some make their own food

PROTISTS

Diatoms
Magnification: 250×

SCIENCE *Online*

Visit Glencoe Science Online, *science.glencoe.com*, to learn more about mushrooms, which are a type of fungi.

Protists

The Protist (PROH tust) kingdom includes many different single-celled and many-celled organisms. They are difficult to classify and seem to have little in common. However, all protists are classified based on their cell structure. Two types of protists include plantlike and animal-like protists.

If you have seen green algae (AL jee) growing in a fish tank, on the surface of a pond, or on the sides of a swimming pool, you're already familiar with plantlike protists. You may also be familiar with seaweeds, which are types of brown algae. Green algae and brown algae are called plantlike protists. Plantlike protists use the sun's energy to make food. Some seaweeds, such as kelp, even look like plants. Kelp is a long brown algae that can often be found on ocean beaches.

The protists also include many animal-like forms. These single-celled protists move around and take in food from the environment just like animals do. Some protists also cause diseases. One type of animal-like protist causes a disease called malaria. Malaria kills millions of people every year.

Fungi

If you like pizza with mushrooms, you are a fungus eater! Mushrooms are a type of fungi (FUN ji). You've seen other types of fungi if you've ever noticed fuzzy, black mold growing on old bread or green mold on a piece of fruit.

FUNGI KINGDOM

- Single-celled and many-celled species
- Do not move around
- Obtain energy by feeding on dead or decaying tissue

FUNGI

Scarlet waxy cap mushroom

In some ways, fungi are like plants. In other ways, they are similar to animals. You may have thought that mushrooms were plants because they grow in the ground and don't move about. Mushrooms, however, don't make food using sunlight as plants do. Instead, mushrooms and other fungi get their nutrients from the substances they grow on. As a fungus grows, it digests the material around it. When fungus grows on bread or fruit, it's actually eating the food!

The Fungi kingdom includes single-celled and many-celled forms. People have found many uses for fungi. Besides as a food, people use fungi in making medicines.

Plants

How many different kinds of plants have you seen today? They're all around—in gardens, in lakes and streams, and even between the cracks in sidewalks. They are also found in houses and office buildings and in your salad bowl at lunchtime. Grass, trees, bushes, and ferns are all members of the Plant kingdom.

All plants are many-celled organisms. Most contain the green chemical chlorophyll (KLOR uh fihl), which enables them to make food. As you've learned, plants capture energy from sunlight and store it as food. Food made by plants is the source of energy for many organisms. Every species of animal depends in one way or another on the plants in its environment.

PLANT KINGDOM

- Many celled
- Do not move around
- Make their own food

PLANT

Viola

PLANT

Alaskan Cedar

ANIMAL KINGDOM

- Many celled
- Most species move around
- Get energy by eating food

Red-faced macaque monkeys

Tunicates

Animals

Pigeons and squirrels in the park, penguins in Antarctica, rhinoceroses in Africa—Earth is home to more than a million different species of animals. Some, such as cats and dogs in your neighborhood, are familiar to you. Others, such as the rare birds, insects, monkeys, and snakes that live in tropical rain forests and other faraway places, are less well known. Even though there is a great number of animals, they are surprisingly similar in many ways.

It's easy to recognize animals. They're the only organisms you can see moving around from place to place. Moving around in their environment helps animals find food, shelter, and mates, and allows them to escape from enemies.

The animal kingdom is a group made up of a lot of different organisms. All animals are many-celled creatures. Some animals do not have bones. Sponges, worms, crabs, and insects are all boneless. Earlier, you learned that birds, fish, and reptiles are animals with bones.

Tropical butterfly
Euphaedra uganda

Macaws

In this section, you've learned that scientists classify living things into six large groups of organisms. Later in the book, you'll learn more about the different kinds of life on Earth and how living things carry out their life activities.

Section Wrap-up

1. What is a species? Give an example of two different species.

2. What's the difference between an animal and a plant? Be specific.

3. **Think Critically:** Briefly describe how you might go about classifying a collection of CDs into two groups.

4. *Skill Builder*
   Developing Multimedia Presentations
   Give a multimedia presentation that compares and contrasts the six kingdoms of living things. Be sure to include the name of the kingdom, how its members get energy, whether or not they can move around, and whether they are made up of one or many cells. If you need help, refer to Developing Multimedia Presentations on page 564 in the **Technology Skill Handbook.**

Science Journal

In your Science Journal, write a one-page summary about an organism from one of the six kingdoms. In your summary, include where it lives, details about its appearance and behavior, and why it interests you.

2•3 Viruses

What YOU'LL LEARN

- How the AIDS virus and the cold virus both affect humans
- How to use outlining as a tool for remembering the big ideas of an article

Science Words:
virus
vaccine

Why IT'S IMPORTANT

You'll be able to use outlining to organize information.

FIGURE 2-11

This computer-enhanced image is of a virus that causes influenza in humans. Influenza is often accompanied by fever and body aches.

What is a virus?

A **virus** (VI rus) is a particle that has things in common with both living and nonliving things. Viruses are like living things in that they are able to reproduce. But they can reproduce only inside a living cell. Viruses are like nonliving things because they don't grow, eat, or respond to their environments.

Viruses need living things to survive. Once a virus attaches to a cell, it invades the cell and begins to make copies of itself. Eventually, the cell occupied by the virus explodes, releasing many new viruses. These new viruses then enter other cells.

Examples of Viruses

What do measles, AIDS, colds, and chicken pox all have in common? Viruses cause all of these diseases. While viruses cannot be classified into any of the six kingdoms, they do infect all kinds of living things. Viruses infect bacteria as well as plants and animals. Some plant viruses destroy food crops such as potatoes and tomatoes. Other types, such as the one pictured in **Figure 2-11**, can cause flu and others can cause cancers in house cats and humans.

Viruses also cause cold sores and chicken pox in humans. Humans can be infected with either of these viruses by touching someone who already has one of these viruses. Hepatitis B and HIV are viruses that can be passed from person to person by blood and other body fluids and on needles carrying the virus. Hepatitis B damages the liver. HIV can lead to AIDS. AIDS is a disease that eventually destroys the body's ability to fight off diseases that can be life threatening.

Treating Viruses

Most viruses, like the one that causes the common cold, can't be treated. Other viruses that affect humans, however, can be prevented with vaccines. **Vaccines** are made from either dead or weak viruses and are given by mouth or by injection, as shown in **Figure 2-12**. Vaccines cause the body to make substances that resist particular viruses. Many states require vaccines for polio, measles, and mumps. These diseases can affect children and adults.

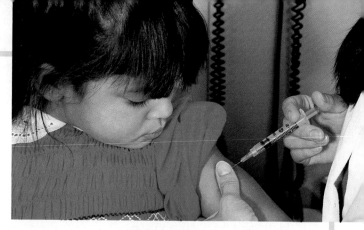

FIGURE 2-12
Many states require that children receive vaccinations before starting school.

Skill Builder: Outlining

LEARNING the SKILL

1. Carefully read the material to be outlined.

2. Choose at least two or three main ideas from the passage that summarize each section or paragraph within a section. Label each main idea with a separate Roman numeral (I, II, III, etc.).

3. Now pick out key words in each section or paragraph that support each main idea. These supporting details are written separately under each main idea. Write a capital letter (A, B, C, etc.) before each.

4. Arrange the key phrases under each key word with a number before each phrase (1, 2, etc.).

5. List additional details under the numbered key phrases with a lowercase letter (a, b, c, etc.) before the details.

6. Check the accuracy of your outline by rereading the passage with your outline in hand. Make any necessary changes.

PRACTICING the SKILL

1. Make an outline using the main ideas in this feature. Use the questions that follow to help you organize your outline.

2. How many main ideas do you have?

3. What key terms and phrases summarize the definition of a *virus?*

4. List examples of viruses and the types of organisms they affect.

5. In the final part of your outline, summarize how diseases caused by viruses are prevented.

APPLYING the SKILL

Use references and what you have learned about viruses to make an outline that compares and contrasts viruses that affect plants with those that affect animals and bacteria.

Hatchet
by Gary Paulsen

Brian opened his eyes and screamed.

For seconds he did not know where he was, only that the crash was still happening and he was going to die, and he screamed until his breath was gone.

Then silence, filled with sobs as he pulled in air, half crying. How could it be so quiet? Moments ago there was nothing but noise, crashing and tearing, screaming, now quiet.

Some birds were singing.

How could birds be singing?

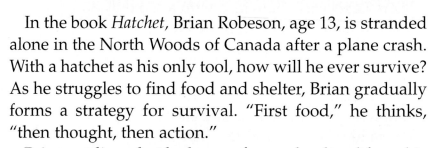

Science Journal

When do you use scientific skills and methods outside of school? Think it over. Then, in your Journal, write a short play about your experiences. That's thought, then action. What's missing? Brian would want you to have a sandwich first.

In the book *Hatchet,* Brian Robeson, age 13, is stranded alone in the North Woods of Canada after a plane crash. With a hatchet as his only tool, how will he ever survive? As he struggles to find food and shelter, Brian gradually forms a strategy for survival. "First food," he thinks, "then thought, then action."

Brian applies what he knows from school and from his parents and friends to make sense out of what his senses tell him. After he carries out a plan of action, he stops to think about it. What does he need to change to get a better result? Brian makes a guess about what kind of wood he needs to make a bow and how it should be shaped. The first result? Failure! When Brian pulls back on the shoelace he's used for a bow string, the bow snaps in half. He thinks carefully about the failure and makes a new plan, a new guess, and this time, it works. Brian thinks like a scientist, not to get an answer to a school question or to make dramatic discoveries, but to survive.

Reviewing Main Ideas

Read the statements below that review the most important ideas in the chapter. Using what you have learned, answer each question in your Science Journal.

1. Five life traits define living things: response to the environment, organization, movement, reproduction, and growth and development. *How would you decide whether or not something is alive?*

2. Living things need energy, oxygen, water, and minerals to carry out their life activities. *How do you obtain energy?*

3. Scientists classify living things in order to keep track of the millions of species on Earth. *Name three different organisms that live in your neighborhood.*

4. Organisms on Earth are classified into major groups called kingdoms. *Which kingdom of living things do you belong to?*

Using Key Science Words

cell	organism
classification	species
development	trait
environment	vaccine
kingdom	virus

Match each phrase with the correct term from the list of Key Science Words.

1. group of organisms that can breed with one another
2. grouping objects based on common traits
3. a living thing
4. is made from either dead or weak viruses
5. an organism's surroundings

Checking Concepts

Choose the word or phrase that completes the sentence.

6. A particle that has some things in common with both living and non-living things is called a(n) _____.
   - a. cell
   - b. vaccine
   - c. virus
   - d. organism

7. Archaebacteria, Eubacteria, and Protist are names of different _____ of living things.
   - a. traits
   - b. species
   - c. kingdoms
   - d. environments

8. The smallest unit of life in living things is the _____.
   - a. species
   - b. organism
   - c. kingdom
   - d. cell

9. The changes that take place when a puppy grows into a dog are known as _____.
   - a. organisms
   - b. traits
   - c. homeostasis
   - d. development

10. The spots on a dalmatian dog are an example of a(n) _____.
   - a. cell
   - b. trait
   - c. environment
   - d. classification

Thinking Critically

Answer the following questions in your Science Journal using complete sentences.

11. Think about an inflated balloon. A sharp object can easily get a response from an inflated balloon. Are inflated balloons living things? Why or why not?

12. Give an example of when you reacted to something in your environment today.

13. People organize and classify many things in their homes. Briefly describe one example of a classification system in your home.

14. Describe how the classification of living things is similar to the way videotapes are organized in a store.

15. If you have a cold, how can you help prevent your cold from spreading to other people in your class?

Developing Skills

If you need help, refer to the description of each skill in the Skill Handbook.

16. Observing and Inferring: On the way to school, you find a mysterious-looking object on the ground. The object appears to be dead. You put the object under a microscope and discover that it is organized into tiny, bricklike units. Could the object be a living thing? Explain your answer.

17. Making and Using Tables: The table below shows kingdoms of living things. Complete the table by filling in the missing information.

Kingdoms of Life			
	How many cells does it have?	Does it move?	How does it obtain energy?
Bacteria	single celled		absorb chemicals from their surroundings
Protists	single celled and many celled	some move	
Fungi		most do not move	eat food
Plants	many celled	do not move	
Animals		most move	

18. Outlining: Outline the five ways to decide when something is alive. These ways were discussed in Section 2-1 of this chapter.

19. Observing and Inferring: When opening a can of food for lunch, you notice that your cat comes running into the kitchen. Explain why this behavior might be an important instinct for a cat in the wild.

20. Comparing and Contrasting: Organisms in the protist kingdom are often called plantlike or animal-like organisms. Compare and contrast these two types of protists.

Performance Assessment

1. Poster: Choose an organism in the animal or plant kingdom. Research information about the organism. Make a poster showing the classification of this organism. What other organisms are closely related?

2. Making and Using a Classification System: Select a collection of seeds, cards, or toys. Create a classification system using these objects. Figure out a way to display the objects. Indicate the traits you used to classify the objects.

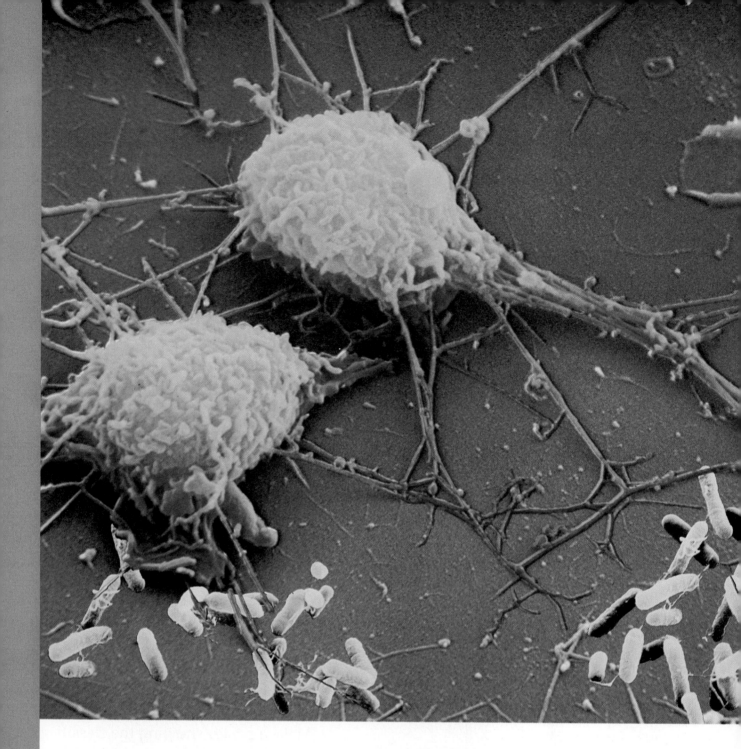

Chapter Preview

Skills Preview

▶ **Skill Builders**
  • make and use
    a table
  • sequence

▶ **MiniLABs**
  • model
  • compare

▶ **Activities**
  • compare
  • interpret
  • diagram
  • hypothesize
  • infer

Magnification:
5400×

Chapter 3

Cells

Y ou are about to enter an active world that exists in all of us, and in all other living things. It is a world so important that life couldn't exist without it. Yet it is a world that, most of the time, you can't see with just your eyes. Microscopes are windows into this world. The photograph on the left shows two white blood cells attacking bacteria that have invaded a human body. Welcome to the microscopic (mi kruh SKAH pihk) world of cells—the building blocks of life.

EXPLORE ACTIVITY

Observe Onion Cells

1. Cut a 2-cm hole in the middle of an index card.
2. Tape a piece of plastic wrap over the hole.
3. Turn down the two shorter ends of the card about 1 cm.
4. Place the hole over a piece of newspaper with the letter *e* printed on it. Draw what you see.
5. Place the hole over a piece of onion skin.
6. Put a drop of water on the plastic wrap.
7. Look through the water drop and observe the piece of onion. Draw what you see.

Science Journal

In your Science Journal, describe how the newspaper *e* and the onion skin looked when viewed with your magnifier.

3•1 Cells—Building Blocks of Life

What YOU'LL LEARN

- The parts of animal and plant cells
- The purpose of different cell parts

Science Words:
cell membrane
nucleus
mitochondrion
photosynthesis
chloroplast

Why IT'S IMPORTANT

You will better understand how your cells carry out the activities of life.

The World of Cells

Different cells have different jobs to do in living things. Cells found in onion stems help move water and other substances throughout the onion plant. White blood cells, found in humans and many other animals, help protect you from disease. White blood cells, onion cells, and all other cells are alike in many ways. A cell is the smallest unit of life in all living things. Cells are important because the activities of life take place in and around cells. They help living things carry on all of the important activities of life discussed in Chapter 2—movement, growth, and reproduction.

FIGURE 3-1

All living things are made up of cells.

A Animals have groups of specialized cells that make it possible for them to do different life activities. This tiger shark and the kangaroo and her joey have similar types of cells that are specialized to allow them to live in different environments.

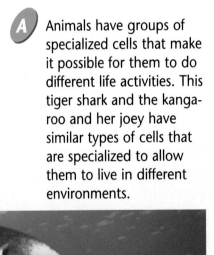

The Microscopic Cell

All the living things pictured in **Figure 3-1** are made up of cells. The smallest organisms on Earth are bacteria. Bacteria are single-celled organisms, which means they are made up of only one cell. Larger organisms are made of many cells. These cells work together to complete all of the organisms' life activities. The living things that you see every day without using a microscope—trees, dogs, frogs, people—are many-celled organisms. Your body alone contains about 10 trillion cells. Ten trillion is written as 10 000 000 000 000.

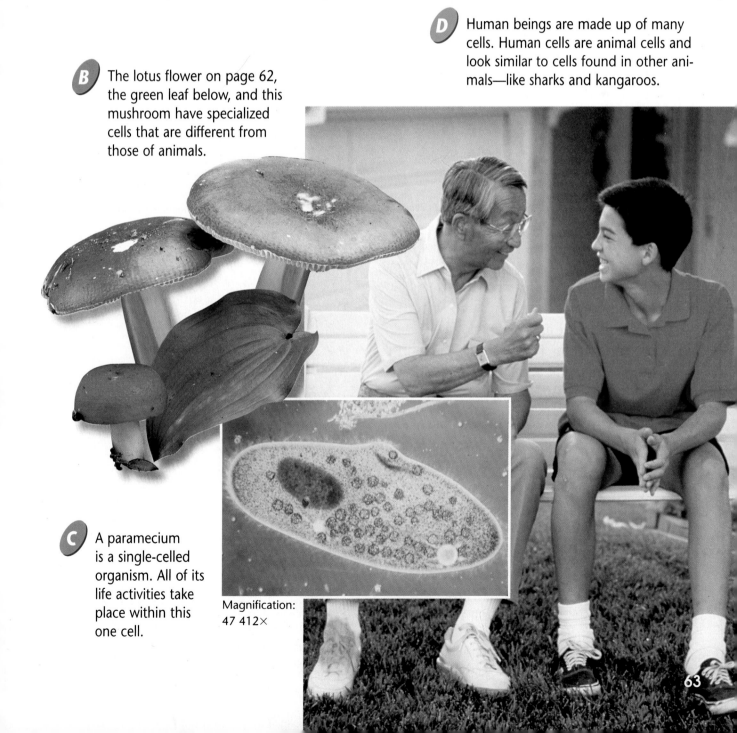

D Human beings are made up of many cells. Human cells are animal cells and look similar to cells found in other animals—like sharks and kangaroos.

B The lotus flower on page 62, the green leaf below, and this mushroom have specialized cells that are different from those of animals.

C A paramecium is a single-celled organism. All of its life activities take place within this one cell.

Magnification: 47 412×

63

Magnification: 55 000×

What makes up a cell?

As small as most cells are, as shown in **Figure 3-2**, they are made up of even smaller parts that do many different jobs. A single cell can be compared to a pizza shop, as illustrated in

FIGURE 3-2

How small are cells? If red blood cells were placed end to end in a line, about 150 cells could fit in the thickness of a penny.

USING TECHNOLOGY

Microscopes

Scientists have been able to see and study cells for only about 300 years. In that time, they have learned a lot about cells and unlocked many of their mysteries through the use of better and better microscopes. Some modern microscopes allow scientists to study the small details inside cells.

Your Microscope

The microscope used in most classrooms is called a compound light microscope. In this type of microscope, light passes through the object you are looking at and then through two or more lenses. The lenses make and enlarge an image of the object. Using the low-power lens, your microscope can let you see an object 40 times larger than its actual size. Using a high-power lens, the microscope will enlarge the image about 400 times.

Magnification: 40×

Magnification: 400×

SCIENCE *Online*

Microscopes have changed throughout the years. For links about the history of the light microscope, visit Glencoe Science Online. *science.glencoe.com*

Figure 3-3. A pizza shop needs a building. The space inside is needed to perform the business activities. The different pizza ingredients—dough, cheese, tomato sauce, and toppings—must be assembled, cooked, and served. Storage for supplies and products is also needed. Electricity runs the ovens, powers the lights, and heats the shop.

A manager is in charge of the entire operation. The manager makes a plan for every employee of the shop. He or she also has a plan for every step of the process of making and selling the pizzas.

A living cell operates in a similar way. A cell has a boundary inside which its life activities take place. Smaller parts inside the cell act as storage areas and power sources. It also has parts that use ingredients—oxygen, water, minerals, and other nutrients—to release energy and make substances that are necessary for maintaining life. **Figure 3-4** on page 66 and **Figure 3-5** on page 67 show some of the important parts of the cell and the life activities that these parts perform.

Figure 3-4 on page 66 and **Figure 3-5** on page 67

FIGURE 3-3

A cell can be compared to a pizza shop. *How is a cell like a pizza shop?*

FIGURE 3-4

Cells perform the activities necessary for life. Each part of the animal cell has a special job to do.

Animal cell

A Cell membrane: This flexible structure helps control what enters and leaves the cell.

B Cytoplasm: Cytoplasm is a gel-like substance in which cell parts move and where many of the cell's life activities take place.

C Mitochondrion: In a mitochondrion, the energy a cell needs to carry out its life activities is released from food.

D Nucleus: The nucleus controls most of the cell's activities.

E Vacuole: A vacuole is a storage area for food, water, minerals, and wastes.

F Chromosome: A chromosome contains DNA. DNA is a chemical that controls what traits an organism will have.

FIGURE 3-5

In addition to many of the cell parts found in an animal cell, a plant cell has two special parts—a cell wall and chloroplasts.

Plant cell

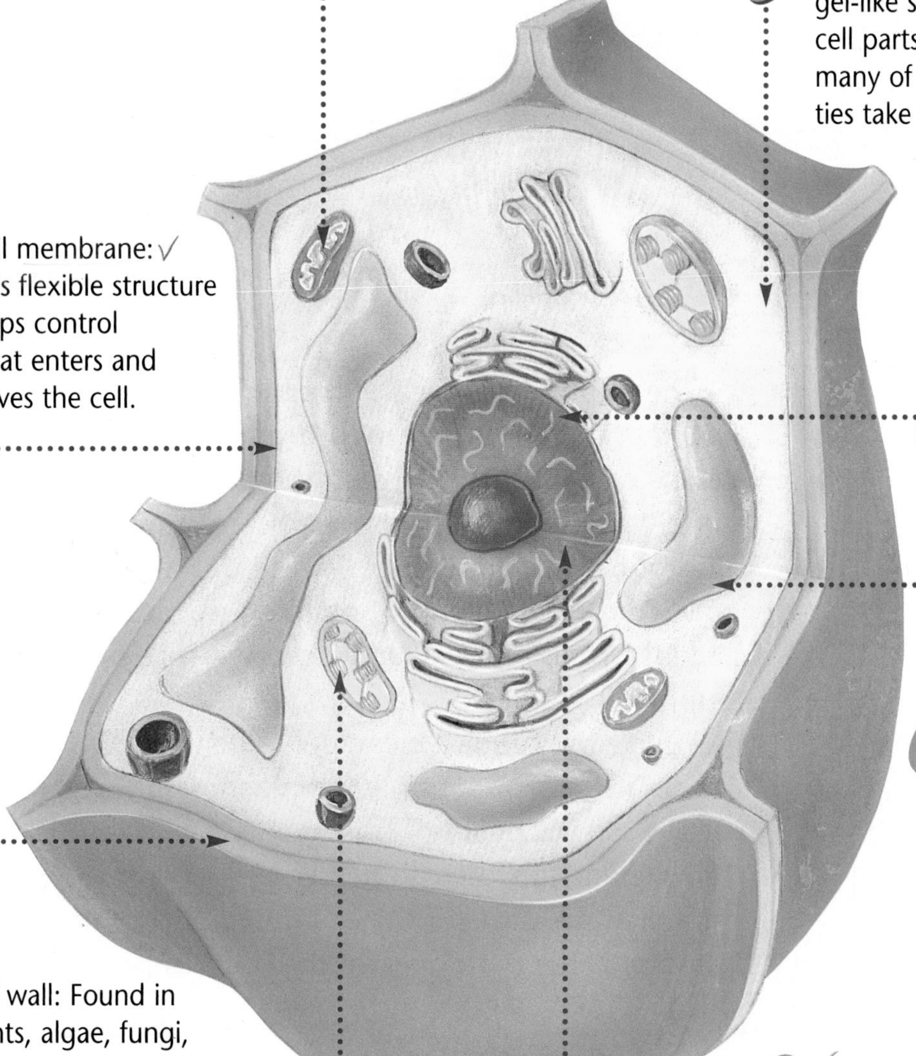

B √ Mitochondrion: In a mitochondrion, the energy a cell needs to carry out its activities is released from food.

C Cytoplasm: Cytoplasm is a gel-like substance in which cell parts move and where many of the cell's life activities take place.

A Cell membrane: √ This flexible structure helps control what enters and leaves the cell.

D √ Nucleus: The nucleus controls most of the cell's activities.

E √ Vacuole: A vacuole is a storage area for food, water, minerals, and wastes.

H √ Cell wall: Found in plants, algae, fungi, and some bacteria, the cell wall is a thick outer covering located outside the cell membrane. *Why is the cell wall important?*

F √ Chromosome: A chromosome contains DNA. DNA is a chemical that controls what traits an organism will have.

G √ Chloroplast: A chloroplast contains chlorophyll. Chlorophyll is a chemical that traps energy from the sun.

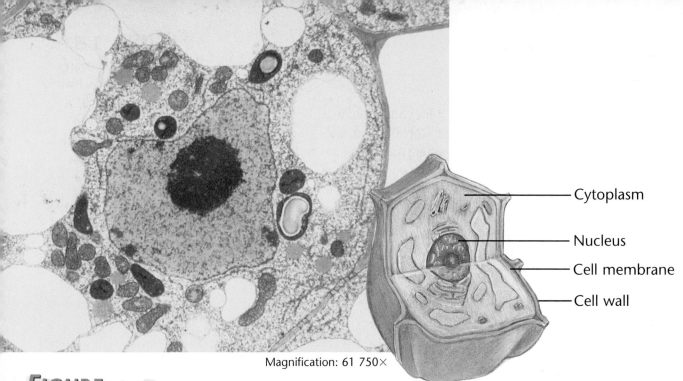

Cytoplasm

Nucleus

Cell membrane

Cell wall

Magnification: 61 750×

FIGURE 3-6
This plant cell has a cell wall and a cell membrane. Animal cells only have a cell membrane.

Cell Structure

The cell membrane is a flexible structure that holds the cell together. The **cell membrane,** shown in **Figure 3-6,** forms a boundary between the plant or animal cell and its environment and also helps control what goes into and out of the cell. Plants, algae, fungi, and many types of bacteria have an additional tough, outer structure outside the membrane called a cell wall. The cell wall helps support the plant.

Filling the cell is a clear, gel-like substance called cytoplasm (SI toh plaz um). Cytoplasm is like the work area inside the pizza shop. It is mostly water, but it also contains chemicals the cell needs. Cytoplasm is the material in which the parts of the cell move. It provides the space and holds the materials for the cell's important activities.

Inside the Cell

Pizza shops have employees who perform the shop's activities. They also have equipment for specific purposes. Some cells contain organelles (or guh NELZ), as illustrated on pages 66 and 67. Organelles are specialized cell parts that perform the cell's activities. Examples of organelles are the nucleus, chloroplasts, mitochondria, and vacuoles. All of the cell's organelles move within the gel-like cytoplasm. You could think of these organelles as the employees of the cell because each type of organelle does a different job.

The manager of the cell is the nucleus (NEW klee us). The **nucleus** controls most of the cell's activities. Inside the nucleus are the chromosomes (KROH muh zohmz), which can be likened to the business plan that the manager follows in running the shop. A business plan describes how the business operates. This could include such plans as what kinds of food and drink the business will serve. Just as the pizza shop has a business plan, chromosomes contain a plan for the cells. Chromosomes contain an important chemical called DNA that determines what kinds of traits an organism will have, such as height, hair color, and eye color. In **Figure 3-7,** you can see the nucleus and chromosomes of an animal cell.

Mini LAB

Modeling a Cell

1. Using **Figure 3-4** or **Figure 3-5** as a guide, make a model of a cell using common food items and gelatin that has not yet set firmly. Candies, pasta, and fruit and vegetable pieces are food items you could use.
2. Carefully put 125 mL of unset gelatin and pieces of food into a resealable bag. Reseal the bag.
3. Identify the cell parts by taping masking tape labels to the outside of the bag, and make a labeled sketch of your cell in your Science Journal.
4. Gently poke the center of the bag with your fingers.

Analysis

1. How is your model like a cell?
2. Can you change the shape of the cell? Does the shape stay changed? What helps keep the shape of the bag? How does the bag relate to a real cell?
3. What type of cell did you model?

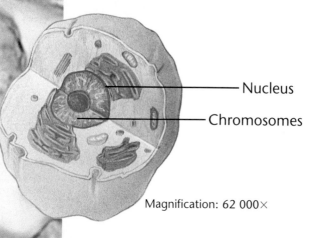

FIGURE 3-7

The cell's chromosomes are the more darkly stained objects located on the opposite sides of the nucleus.

Nucleus

Chromosomes

Magnification: 62 000×

Pantries, closets, refrigerators, and freezers store food and other materials that a pizza business needs. Trash cans hold garbage until it can be picked up. In cells, food, water, and other substances are stored in bag-like structures in the cytoplasm called vacuoles (VAK yew ohlz). Some plant cells have large vacuoles to store the food the plant produces. Other vacuoles store wastes until the cell is ready to get rid of them.

Energy and the Cell

Power plants and gas companies supply the energy needed to turn on the lights and run the ovens at a pizza shop. Cells need energy, too. Cells contain organelles called **mitochondria** (mi tuh KAHN dree uh) that provide power for all of the cell's activities. Mitochondria are the sausage-shaped power plants of both plant and animal cells. This is where energy that the cell needs is released from food. *Mitochondrion* is the singular of *mitochondria.* As shown in **Figure 3-8,** an important chemical reaction called cellular respiration (SEL yuh lur • res puh RAY shun) takes place inside a mitochondrion. Cellular respiration is the process in which food and oxygen combine to make carbon dioxide and water, and release energy. Both animal and plant cells use the energy from cellular respiration to do all of their work.

FIGURE 3-8

Cellular respiration takes place in the power plant of the cell—the mitochondrion.

Mitochondrion

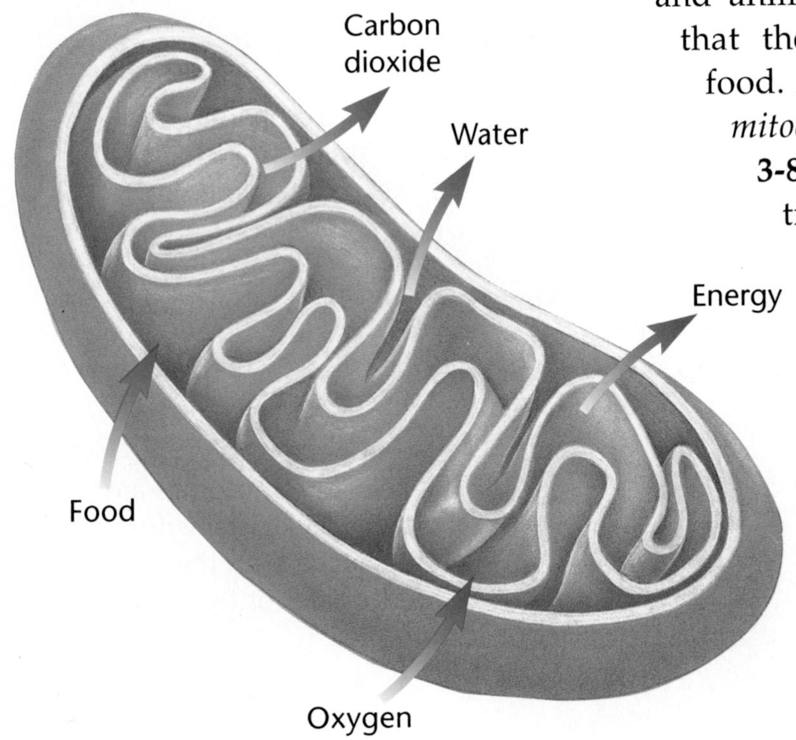

Carbon dioxide

Water

Energy

Food

Oxygen

Cellular Respiration

food and oxygen $\xrightarrow[\text{into}]{\text{is changed}}$ carbon dioxide and water and energy

Comparing Cells

Animals must obtain food from their surroundings. Plants, such as *Elodea*, make their own food using energy from the sun. You can compare animal and plant cells to find out how their structures differ.

What You'll Investigate

How do animal and plant cells differ?

Procedure

1. Observe a slide of live *Elodea* leaf cells first under low power. **Focus** the microscope on a single cell.
2. With the help of your teacher, find the green chloroplasts floating in the cytoplasm. Find a nucleus.
3. In your Science Journal, **draw** what you see. **Label** the cell wall, nucleus, cytoplasm, and chloroplasts.
4. **Place** a prepared slide of human cheek cells on the microscope and observe these cells under low power.
5. **Draw** what you see. **Label** the cytoplasm, nucleus, and cell membrane.
6. Wash your hands.

Conclude and Apply

1. **Describe** the two types of cells that you observed.
2. **Compare** the sizes and shapes of the animal and plant cells. How are they similar? How are they different?
3. In what ways are animal and plant cells the same? Why do you think they are different in some ways?

Goals

- **Compare** animal and plant cells.
- **Infer** why there are differences between animal and plant cells.

Materials

- microscope
- slide of *Elodea* leaf cells
- prepared slide of human cheek cells

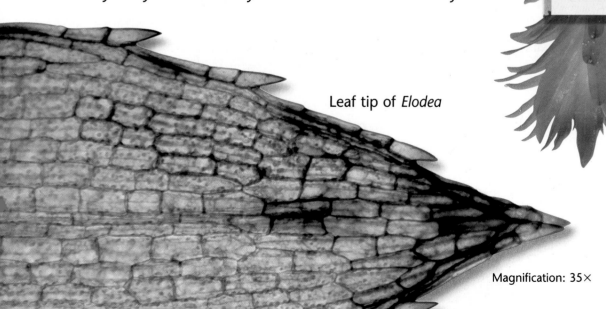

Leaf tip of *Elodea*

Magnification: 35×

71

FIGURE 3-9

Photosynthesis takes place inside the chloroplasts of these heads of lettuce. Chloroplasts are located inside the cells of green leaves.

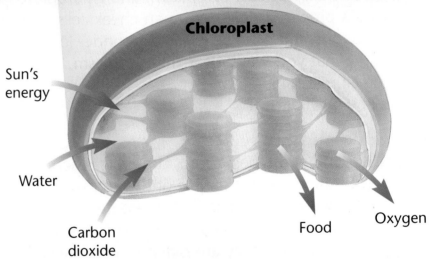

Chloroplast

Sun's energy

Water

Carbon dioxide

Food

Oxygen

Photosynthesis

sun's energy and water and carbon dioxide $\xrightarrow[\text{into}]{\text{is changed}}$ food and oxygen

Plants—Nature's Solar Energy Factories

If you've ever watched a cow grazing in a pasture, a bird pecking at worms, or a dog eating from its bowl, you know that animals get food from their surroundings. But unless you've observed a Venus's-flytrap, you've probably never seen a plant trap anything. How do plants get the food they need for energy? Plants, green algae, and many types of bacteria make their own food through a process called photosynthesis (foh toh SIHN thuh sus).

During **photosynthesis**, as shown in **Figure 3-9**, green plants trap energy from the sun. They then convert this energy into food needed by the plant. This food-making activity occurs inside the cell's chloroplasts (KLOR uh plasts). **Chloroplasts** are green organelles that trap energy from sunlight and turn it into food. This food is needed by the plant to stay alive. As the plant needs energy, its mitochondria release the food's energy. Most photosynthesis occurs in a plant's leaves. Leaves are green because leaf cells contain many of these green chloroplasts.

Organelles perform special jobs in cells. But all cells are not identical. In this section, you've learned about the different parts of animal and plant cells. In the next section, you'll learn that large, many-celled organisms like you are made up of many different kinds of cells that perform different functions.

SCIENCE *Online*

How is photosynthesis related to leaves changing their colors in autumn? For answers to these and other questions about leaves, visit Glencoe Science Online. *science.glencoe.com*

Section Wrap-up

1. How is a living cell similar to a business?

2. Why is the nucleus so important to the living cell?

3. How do cells get the energy they need to carry on their activities?

4. **Think Critically:** Suppose your teacher gave you a slide of an unknown cell. How would you tell whether the cell was from an animal or from a plant?

5.  *Skill Builder*
   Making and Using Tables Make a table listing the parts of animal and plant cells and the jobs they do in the cells. If you need help, refer to Making and Using Tables on page 546 in the **Skill Handbook.**

Using Computers

Internet Visit Glencoe Science Online at *science.glencoe.com* to learn more about microscopes. Draw a diagram of a microscope and label the different parts.

3•2 The Different Jobs of Cells

What YOU'LL LEARN

- How a cell's shape is important for doing its job
- The differences among tissues, organs, and organ systems

Science Words:
tissue
organ
organ system

Why IT'S IMPORTANT

Understanding how cells work and are organized can help you keep yourself healthy.

Special Cells for Special Jobs

Have you ever heard the phrase "choose the right tool for the right job"? It's a common expression that means that the best tool for a job is the one that has been designed for that job. For instance, you wouldn't use a hammer to saw a board in half, and you wouldn't use a saw to pound in a nail. Having the right tool applies to the world of cells, too. Inside your body, many different types of cells keep you healthy.

The large numbers of cells making up both plants and animals are classified into groups. Each group has a different job to do within the organism. This section of the chapter will give you examples of these different types of cells.

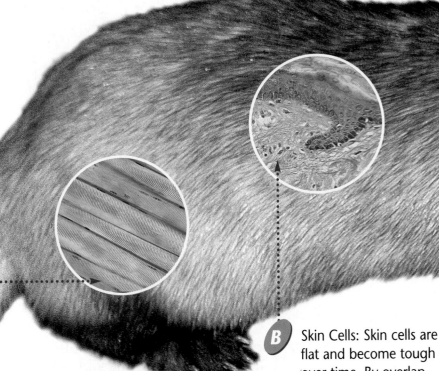

A Muscle Cells: Muscle cells are long and thin and can contract, or become shorter. Muscle cells are active and use up large amounts of energy. *Why do muscle cells have lots of mitochondria?*

Magnification: 900×

B Skin Cells: Skin cells are flat and become tough over time. By overlapping with other skin cells, they form a protective layer around an animal's body.

Magnification: 5000×

Different Animal Cells

Your body is made up of about 200 different types of cells. The same is true for other animals. **Figure 3-10** shows some different cell types in a ferret. Notice the variety of sizes and shapes. A cell's shape and size can be related to its function in the body. The MiniLAB on this page will help you see how the shapes of human cells are related to their jobs.

Mini LAB

Analyzing Cells

Analyze different animal cells. Determine how an animal cell's shape is related to its job.

1. Examine slides of human blood cells, muscle cells, skin cells, and nerve cells under low power of a microscope.
2. Draw each type of cell in your Science Journal. Label cell parts that you can see.

Analysis

1. In what ways were the cells that you observed similar? How were they different?
2. How are the cells' shapes related to their jobs?

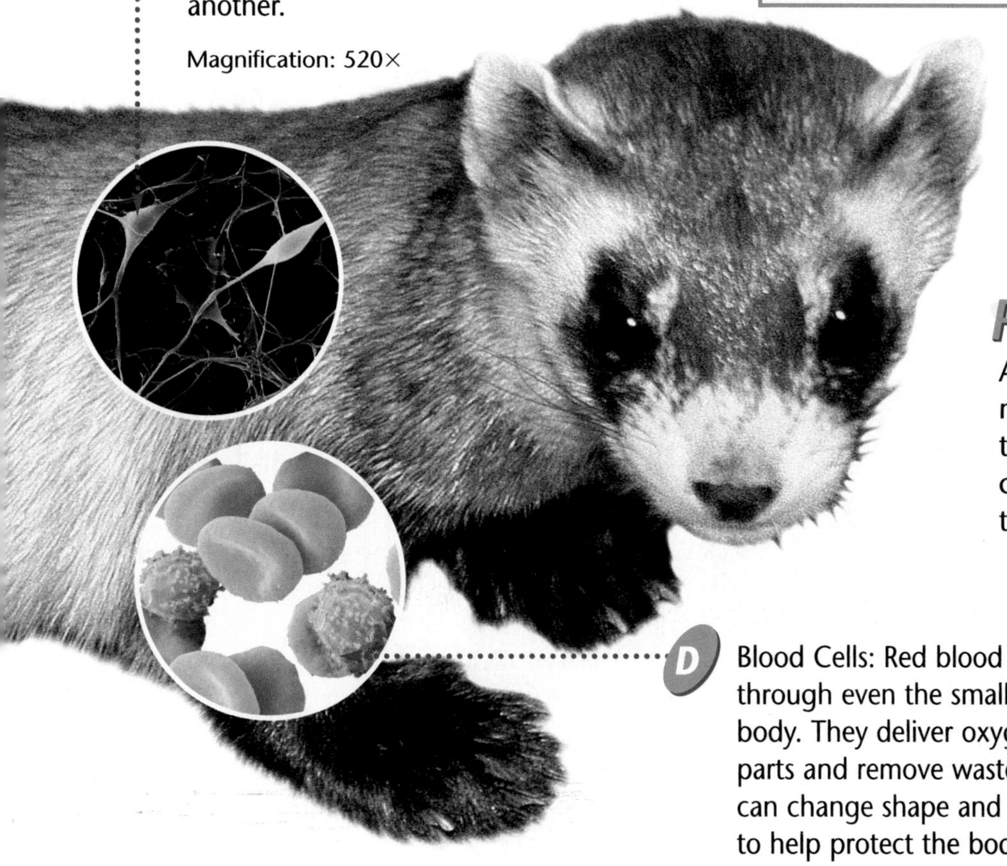

C Nerve Cells: Nerve cells are long and shaped like wires, which allows them to deliver nerve messages quickly from one part of the body to another.

Magnification: 520×

FIGURE 3-10

Animals such as this ferret are made from many different types of cells. These cells come in shapes and sizes that are related to their jobs.

D Blood Cells: Red blood cells move easily through even the smallest blood vessels in the body. They deliver oxygen to different body parts and remove wastes. White blood cells can change shape and leave the blood vessels to help protect the body from disease.

Magnification: 9000×

Problem Solving

Calculating Numbers of Blood Cells

Benito wanted to solve a problem. In science class, he learned that about 200 different types of cells can be found in the human body. There are also billions of cells of each type. Benito wanted to find out how many red blood cells the human body has. After a visit to the library and a few questions to his teacher, Benito found out that every milliliter of blood contains 5 million red blood cells, and an average person has about 3.5 liters of blood.

Solve the Problem:
One piece of data Benito discovered is expressed in milliliters. The other data are expressed in liters. Find out how many milliliters there are in a liter. Refer to Appendix B on page 526 if you need help.

Think Critically:
Using the information Benito discovered, calculate the number of red blood cells in a person with 3.5 liters of blood. List the steps you would take to answer this problem.

Different Plant Cells

Like animals, plants are also made of several different cell types, as shown in **Figure 3-11.** Plants, for instance, have different cells in their leaves, roots, and stems. Each type of cell has a specific job. Some cells in plant stems are long and tube shaped. Together, they form a system through which water, food, and other materials move in the plant. Other cells, like those that cover the outside of the stem, are smaller or thicker. They provide strength to the stem.

Cells that make up many-celled organisms are specialized. Unlike single-celled organisms, such as bacteria, specialized cells found in larger organisms cannot survive outside of the organism. Single-celled organisms carry out all life activities. But specialized cells work as a team to meet the life activities of every cell inside a many-celled organism. Cells are the building blocks of life.

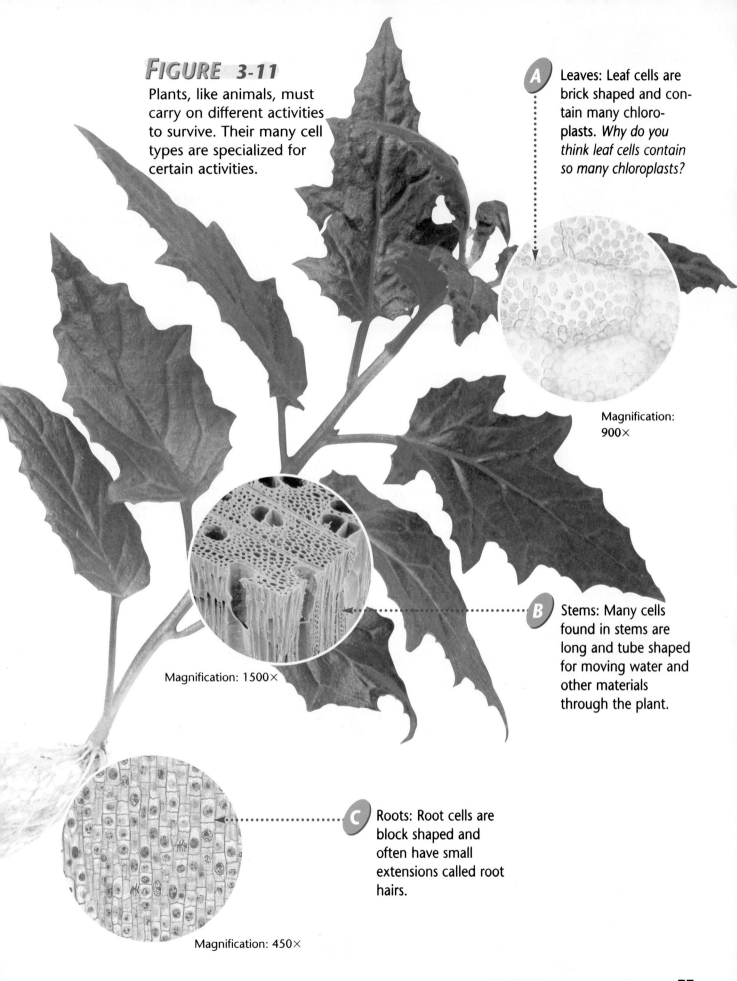

FIGURE 3-11

Plants, like animals, must carry on different activities to survive. Their many cell types are specialized for certain activities.

A Leaves: Leaf cells are brick shaped and contain many chloroplasts. *Why do you think leaf cells contain so many chloroplasts?*

Magnification: 900×

Magnification: 1500×

B Stems: Many cells found in stems are long and tube shaped for moving water and other materials through the plant.

C Roots: Root cells are block shaped and often have small extensions called root hairs.

Magnification: 450×

Design Your Own Experiment
Where does the water go?

When you are thirsty, you may sip water from a glass or drink from a fountain. Plants must get their water in other ways. Their cells help with the task. In most plants, water is absorbed from the soil by cells in the roots. How does this water go into other parts of the plant?

Possible Materials

- fresh stalk of celery with leaves
- clear drinking glass
- scissors
- red food coloring
- water

PREPARE

What You'll Investigate
Where does water travel in a plant?

Form a Hypothesis

Think about what you already know about how a plant functions. In your group, **make a hypothesis** about where you think water travels in a plant. Remember that a hypothesis is a statement that can be tested. Write down your hypothesis in your Science Journal.

Goals

Demonstrate how some cells move water through a plant.

Safety Precautions

Use care when handling sharp objects such as scissors. Avoid getting red food coloring on your clothing.

PLAN

1. **Decide** on a way to test your group's hypothesis. How will you use the materials to show that water moves through the celery?
2. How will you carry out your experiment? **Make a list** of the different steps you will take during the experiment.
3. How long will your experiment take? At what time intervals will you make your observations? At the end of class? The next day?
4. **Prepare** a data table to record your observations.
5. What kinds of observations will you make? Will you use drawings?

DO

1. **Read** over your entire experimental plan to make sure that it is designed to test your hypothesis.
2. Make sure your teacher approves your plan before you proceed.
3. When you cut the celery stalk, put the celery under water and carefully use the scissors to **make a sharp cut** on the bottom of the stalk, or ask your teacher to **cut** the stalk with a sharp knife. Place your materials where they cannot be knocked over.
4. Carry out your plan. Remember to **record** the time you set up the experiment and the time you **record** your observations.
5. While doing the experiment, be sure to **record** all of your observations in your Science Journal.

CONCLUDE AND APPLY

1. **Compare** the color of the celery stalk before, during, and after the experiment.
2. **Make a drawing of** the stalk material after it is cut. Label your drawing. Why do you suppose only some of the stalk material has turned red?
3. Was your hypothesis supported?

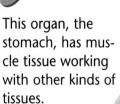

C This organ, the stomach, has muscle tissue working with other kinds of tissues.

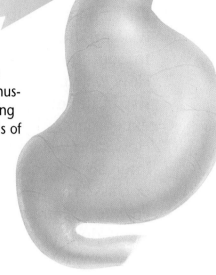

B Muscle cells working together make up muscle tissues.

A The cell is the simplest level of organization. This muscle cell is the smallest unit in the body to do "muscle work."

FIGURE 3-12

In many-celled organisms such as an alligator, the cells are arranged into different levels of organization.

Cell Organization

How well do you think your body would work if all the different cell types were just mixed together in no particular pattern? Could you walk if your leg muscle cells were scattered here and there, each doing its own thing, instead of being grouped together in your legs? How could you think if your brain cells weren't close enough together to communicate? Many-celled organisms are not just mixed-up collections of different types of cells.

Figure 3-12 shows the different levels of organization in an alligator. Cells that are alike are organized into tissues (TIH shews). **Tissues** are groups of similar cells that all do the same sort of work. For example, in animals with muscles, there is muscle tissue made up of just muscle cells. Bone tissue is made up of just bone cells, and nerve tissue is made up of just nerve cells. There is also blood, a liquid tissue, that is made up of just blood cells.

As important as individual tissues are, they do not work all by themselves. Instead, organisms are made up of many organs (OR gunz). An **organ** is a structure made up of two or more different types of tissue that work together. For example, the stomach is an organ made up of muscle tissue, nerve tissue, and blood tissue. Other organs include the heart and the kidneys.

The stomach is an organ in the alligator's body that digests food. Other organs in its body are also involved in digesting food—a mouth, intestines, and a liver. A group of organs that work together to do a certain job is called an **organ system.** Other organ systems found in your body include the respiratory system, the nervous system, and the reproductive system.

D Organ systems are the next level of organization. Systems are made up of different organs working together. In the alligator's digestive system, the mouth, stomach, intestines, and liver are the organs working together.

Section Wrap-up

1. How is a cell's shape important to its job? Give an example.

2. What is the difference between a tissue and an organ?

3. **Think Critically:** Give an example of a human organ system, and name some of the parts that make up the organ system.

4. *Skill Builder*
   Sequencing Make a list that sequences the different levels of cell organization. Start with the cell and list all levels of organization. End with an organ system. For each level of organization, provide an example from the human body. If you need help, refer to Sequencing on page 543 of the **Skill Handbook.**

Science Journal

In terms of organization, organisms can be compared to a school band. In your *Science Journal,* write a short paragraph explaining how an organism is like a band.

3•3 Test-Tube Tissues

What YOU'LL LEARN

- Many human tissues can now be grown in labs
- How to analyze news sources

Why IT'S IMPORTANT

Learning to analyze news media will help you make better decisions.

Tissue Transplants

Transplanting tissues and organs from one person to another is done often in hospitals. Burn victims and patients with one or more organs that are not working properly can live normal lives with transplants. Too many times, however, donors aren't available. In other cases, tissues and organs are available, but the person's body cannot accept the new tissue or organ.

Growing Tissues

Because of these problems of tissue not being available and bodies not accepting new tissue, scientists have found ways that allow them to grow some kinds of tissue from tissue that already exists. For example, patches of skin tissue smaller in size than quarters can be used to make more skin. Special nutrients are added to the skin cells that make up the tissue. The cells then multiply to make flat sheets of new skin tissue, as shown in **Figure 3-13.** Pieces of skin as large as postcards are being made to treat some of the 13 000 burn victims who need skin grafts in the United States each year.

Parts of a magazine article that was in *Science World* on March 7, 1996, are shown on the next page. Read these paragraphs carefully and then answer the questions that follow.

FIGURE 3-13

New skin tissue is grown in laboratories from the skin cells of the patient.

Grow Your Own Body Parts

Suddenly, fire filled the room and Emerson was engulfed in flames. When doctors saw the burns covering more than 85 percent of his body—so deep that the skin would never grow back—they thought Emerson would die.

But by the next day, Emerson's skin was growing. A miracle? Emerson's skin was growing in a laboratory 900 miles from his Kalamazoo, Michigan, hospital bed. The idea was to grow enough skin in the lab so Emerson could receive a transplant of his own skin.

Scientists experimenting with this technique, called tissue engineering, say that one day it may be used to replace other body parts that can't grow back on their own—like ears, livers, and heart valves.

"For burn patients, the most urgent need is to replace the protective, moisture-sealing layer—even if it means doing without some of the skin's other functions. . . ." says Dr. Howard Green.

Eight months after he was burned, Emerson returned to work and began playing softball again—all thanks to tissue-engineering technology. "If I'd been burned just 10 years earlier," he said, "I don't think I would have survived."

 ## Skill Builder: Analyzing News Media

LEARNING the SKILL

1. To get an accurate view of current events and technologies, you must think critically about the news items you hear or read. First, think about the audience. For example, if the item is a magazine article, is the magazine written for scientists or for everyone? If the item is on the radio or television, does the reporter state needed details or does he or she only give a brief summary?

2. Next, try to determine whether the article is true. Were any experts in the field interviewed? Was the item simply reported or was the item also analyzed and interpreted?

3. Finally, ask yourself whether the item presents both sides of the issue. Does the author or reporter of the news item have a bias?

PRACTICING the SKILL

1. From the excerpt given above, what general point is the author of the article trying to make?

2. Were any experts cited or referenced?

3. Were both sides of the issue presented? If not, where might you go to find information about the other side?

APPLYING the SKILL

Find out more about tissue engineering using many different sources. Other books, magazines, newscasts, and even the Internet can be checked for more information. Interview a doctor who does tissue transplants or someone who has had a transplant done.

People & Science

Dr. Khristine Lindo, Surgeon

Q Dr. Lindo, have you always been interested in science?

A I became interested in science in seventh grade when I read a book called *The Making of a Woman Surgeon* by Elizabeth Morgan. That book really inspired me. I realized I wanted to become a physician—a surgeon—and science was the way to reach that goal.

Q What areas of science did you like best?

A Well, in high school, I took every science class I could—biology, chemistry, and physics. I liked biochemistry because it focused on what was happening inside cells.

Q Why is understanding cells important for a surgeon?

A Cells truly are the building blocks of living things. And just like you can't build a house without laying one brick at a time, you also can't manage disease, illness, and especially surgical wounds without knowing a lot about cells. When you know what's going on at the cellular level, you can better understand why patients have certain symptoms, what different medicines do, and how healing occurs.

Q What steps do you take to reduce damage to cells during surgery?

A In any surgery, it's important to be precise in order to limit the damage done to surrounding tissues and cells as much as possible. And it's important to prevent infection before, during, and after surgery. Infection is destructive to cells.

Q What is the most rewarding part of your job?

A Probably when a patient doesn't ever have to see me again! Or when I see someone ten years later who remembers me and is beaming and smiling and well, and I can't remember he or she was ever a patient. Yes, when former patients have fully recovered and are enjoying life—that's the best!

Career Connection

There are many different kinds of medical doctors. Interview a doctor whose area of interest isn't surgery. Find out what areas of medicine he or she is interested in, and then write your own "People & Science" feature.

Read the statements below that review the most important ideas in the chapter. Using what you have learned, answer each question in your Science Journal.

1. All living things are made up of one or more cells. *What are cells?*

2. Cells produce energy, proteins, and other substances necessary for life. *How is a cell like a fast-food business?*

3. Many-celled organisms are made up of different kinds of cells that perform different tasks. *Name two different kinds of cells found in animals.*

4. Many-celled organisms are organized into tissues, organs, and organ systems that perform specific jobs to keep an organism alive. *How is the organization of cells into tissues similar to the organization of a sports team?*

Using Key Science Words

cell membrane
chloroplast
mitochondrion
nucleus
organ
organ system
photosynthesis
tissue

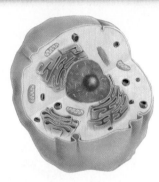

Match each phrase with the correct term from the list of Key Science Words.

1. cell part that traps energy from sunlight
2. food-making process in plants
3. controls most of the cell's activities
4. provides power for the cell's activities
5. group of similar cells working together

Checking Concepts

Choose the word or phrase that completes the sentence.

6. The _____ is the cell part that controls what enters and exits the cell.
   a. mitochondrion   c. vacuole
   b. cell membrane   d. nucleus

7. _____ are found inside the nucleus of the cell.
   a. Vacuoles   c. Chloroplasts
   b. Chromosomes   d. Mitochondria

8. _____ is the gel-like substance that fills the cell.
   a. Cytoplasm   c. Chromosome
   b. Tissue   d. Mitochondrion

9. Groups of different tissues working together form an _____ , such as the stomach.
   a. organelle
   b. organ system
   c. organ
   d. organism

10. Photosynthesis is a process that makes _____ for cells.
    a. food   c. energy
    b. organs   d. tissues

Thinking Critically

Answer the following questions in your Science Journal using complete sentences.

11. What would happen to a cell if the cell membrane were solid and watertight?

12. What might happen to a plant cell if all its chloroplasts were removed? Explain.

13. Why are cells called the "building blocks of life"?

14. What cells in an animal may not have a lot of mitochondria present?

15. Identify the cell shown to the right as an animal cell or a plant cell. Explain.

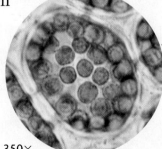

Magnification: 350×

Developing Skills

If you need help, refer to the description of each skill in the Skill Handbook.

16. **Comparing and Contrasting:** Compare and contrast photosynthesis and cellular respiration.

17. **Making and Using Tables:** Make a table that lists the functions of each cell part below.

Cell Part	Function
Nucleus	Controls most of the cell's activities
Cell membrane	Helps control what enters and leaves the cell
Mitochondrion	Releases the energy the cell needs to carry out its activities
Chloroplast	Traps energy from the sun

18. **Observing and Inferring:** Why is the bricklike shape of some plant cells important?

19. **Designing an Experiment:** Describe an experiment you would do to determine how different temperatures affect how water is moved through a plant.

20. **Making and Using Graphs:** Sunlight is necessary for most plants to make food. Using the graph below, determine which plant produced the most food. How much sunlight was needed every day to produce the most food?

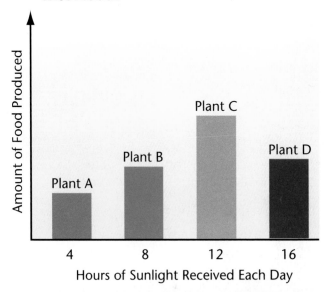

Performance Assessment

1. **Skit:** Working with three or four classmates, develop a short skit about how a living cell works. Have each group member play the role of a different cell part.

2. **Model:** Using assorted household materials such as clay, cardboard boxes, Styrofoam, pipe cleaners, yarn, buttons, dry macaroni, or other objects, make a three-dimensional model of an animal or plant cell large enough that you can walk into. On a separate sheet of paper, make a key to your model cell. On your key, indicate which materials represent the different cell parts.

Chapter Preview

Skills Preview

▶ **Skill Builders**
- sequence
- map concepts
- use a CD-ROM

▶ **MiniLABs**
- observe
- predict

▶ **Activities**
- observe
- compare and contrast
- hypothesize
- collect data

Heredity and Reproduction

Dandelions! Those plants seem to pop out of thin air every spring, showing up in lawns, along roadsides, and even in sidewalk cracks. In just a few days, a fresh, green patch of grass can become blanketed with yellow dandelion flowers. Soon, the flowers turn to fluffy puffballs that bear the plants' seeds—thousands of future dandelions. For dandelions—and all living things—reproduction (ree proh DUK shun) is an important process. For life to continue, living things must produce more living things like themselves.

EXPLORE ACTIVITY

Comparing Types of Seeds

1. Obtain seeds from your teacher, such as pinto beans, dandelion seeds, corn, burrs, beans, and acorns.
2. Using a hand lens, study each seed. Make a list of the characteristics of each seed in your Science Journal.

Science Journal

Draw and label a picture of each type of seed in your Science Journal. Why do you think the seeds are shaped differently?

4•1 Continuing Life

What YOU'LL LEARN

- How cells divide
- The importance of reproduction for living things
- The differences between sexual and asexual reproduction

Science Words:
mitosis
asexual reproduction
sexual reproduction
sex cells
meiosis
fertilization

Why IT'S IMPORTANT

You'll understand how a living thing can inherit characteristics from its parents.

Cell Division

Have you ever seen frog or toad eggs developing in a pond? Frogs and toads lay eggs by the hundreds in gooey clumps, as in **Figure 4-1.** Some other kinds of organisms, including humans, usually produce only one offspring at a time.

Millions of different kinds of living things inhabit Earth. New organisms are the result of the process of reproduction. Reproduction is important to all living things. Without reproduction, a particular species could not continue. During reproduction, information is passed on from parent to offspring. This information is contained in a chemical called DNA found inside the nucleus of cells. DNA controls what the new individuals will look like and how their bodies work.

How did you grow from a baby to the size you are now? Most of the cells that make up organisms form new cells by dividing into two. All organisms are made up of cells, and all cells come from other cells. As you developed from an infant to a preschooler and then to the size you are now, you grew because your cells divided. Before a cell divides into two, all the information contained in the chromosomes of that cell is copied.

FIGURE 4-1

These frog eggs will hatch into free-swimming tadpoles about six days after fertilization.

Before mitosis begins

Nucleus

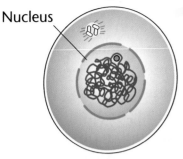

Chromosomes in nucleus are duplicated

Step 1

Chromosomes become visible

Step 2

Chromosome pairs line up along the middle of the cell

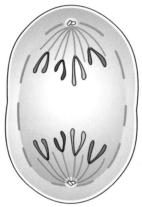

Step 3

Chromosome pairs split and individual chromosomes are pulled to opposite ends of the cell

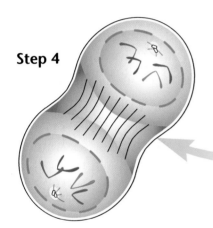

Step 4

The rest of the cell divides

FIGURE **4-2**

In mitosis, cells go through several steps to produce two cells with identical nuclei.

This way, each new cell has exactly the same information as the cell it came from.

First, the chromosomes, which are located in the nucleus (NEW klee us), are duplicated. Then the nucleus divides into two. Each new nucleus receives a copy of the same chromosomes. This division of the nucleus is called mitosis (mi TOH sus). **Mitosis** is the process that results in two nuclei with the exact same information. You can see the process in **Figure 4-2.** Once the nucleus divides, the rest of the cell divides into two cells of equal size. Almost all the cells in any plant or animal divide using this process. Cells divide for growth and to replace aging or injured cells.

Mini LAB

Yeast Budding

1. Fill a small jar or beaker halfway with very warm water.
2. Add a pinch of yeast—the amount you can pinch between your thumb and index finger—to the water. Add the same amount of sugar. Stir. Wash your hands thoroughly.
3. Use a dropper to place a drop of the yeast mixture onto a microscope slide. Add a coverslip.
4. Examine the slide under low power, then high power. Make a new slide after five minutes. Record your observations in your Science Journal.

Analysis

1. What did you observe on the first microscope slide? The second?
2. Why do you think it was necessary to add sugar?

Reproduction from One Parent

Have you ever seen shoots develop from the eyes of a potato as shown in **Figure 4-3?** The method of reproduction in which one organism can produce a new organism is called **asexual reproduction.** In asexual (ay SEK shul) reproduction, all the DNA in the new organism comes from one parent organism. In the case of the potato, the DNA of the growing eye is the same as the DNA in the cells of the parent potato.

Let's Split!

There are several different types of asexual reproduction. Some organisms, such as bacteria and other single-celled organisms, divide in two, forming two cells. The DNA in a bacterium cell is copied. When the cell divides, each new cell gets an exact copy of that DNA. The two new cells are exactly alike. The original cell no longer exists.

Budding Out and Breaking Up

Microscopic organisms, such as bacteria, are not the only species that reproduce asexually. Many species of mushrooms, plants, and even a few animals have some type of asexual reproduction. **Figure 4-4A** shows asexual reproduction in *Hydra,* a relative of jellyfish and corals. When *Hydra* reproduces asexually, a new individual develops from the parent by a process called budding. As you can see, the *Hydra* bud has the same shape and characteristics as the original *Hydra.* The bud matures and will eventually break away and live on its own.

FIGURE 4-3

Many plants in addition to potatoes reproduce asexually. Some of these are cattails, strawberries, and creosote bushes.

In some species of organisms, a whole individual can grow from just a part of a parent organism, as you can also see in **Figure 4-4.** Cuttings taken from plants such as spider plants and ivy develop into whole new plants. Gardeners take advantage of this when they take cuttings from the stems, leaves, or roots of plants. They can grow many plants from one.

In a process called regeneration (ree jen uh RAY shun), some organisms are able to replace body parts that have been lost because of an injury. Lizards, such as chameleons, grow a new tail if theirs is broken off. In sea stars, regeneration is a form of asexual reproduction. If an arm is removed from a sea star, the sea star will grow a new arm. But the arm that was removed can often develop into a whole new sea star, too.

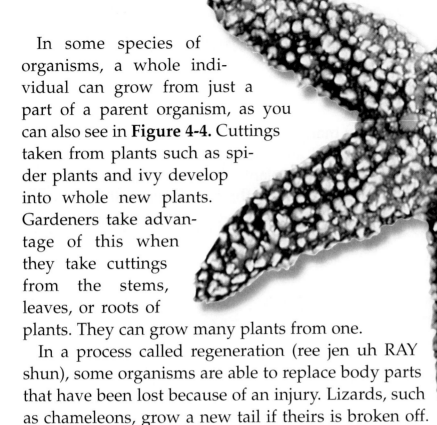

FIGURE 4-4

There are several ways organisms can reproduce asexually.

C Sea stars can regrow or regenerate missing parts or a whole new organism.

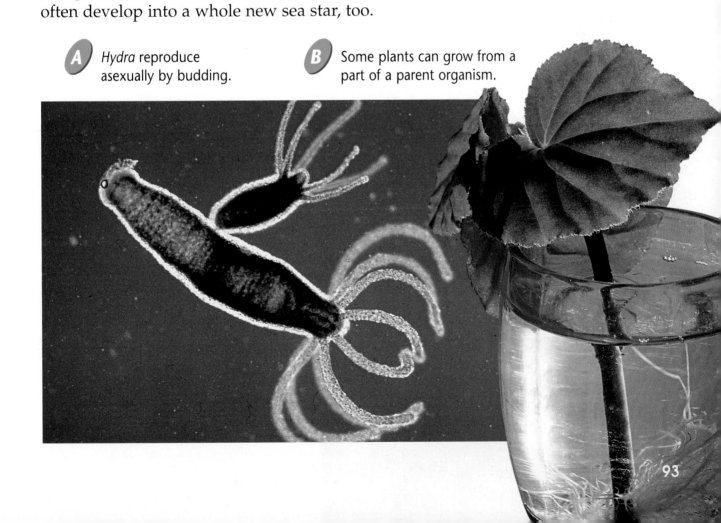

A *Hydra* reproduce asexually by budding.

B Some plants can grow from a part of a parent organism.

Activity 4-1

Design Your Own Experiment
Plant Cuttings

You can grow many familiar houseplants by cutting off parts of the plant and rooting them. Because the cuttings come from a single plant, all of the cells in the plant have DNA that is identical to the parent plant. If there are differences in the way the plants grow from the cuttings, the differences will be due to the conditions under which they were grown.

Possible Materials
- plant trays or paper cups
- soil
- water
- metric ruler
- spoon
- watering can
- masking tape
- marker
- goggles
- stem cuttings from common fast-rooting houseplants

PREPARE

What You'll Investigate
What conditions do plants need to grow from cuttings?

Form a Hypothesis

Think about the things plants need to grow, such as light, water, and a certain temperature range. In your Science Journal, **make a hypothesis** about how changing the amounts of any one of these factors could affect plant growth.

Goals

Demonstrate that a plant can reproduce asexually from plant cuttings.

Analyze the effects that different conditions have on plant growth.

Safety Precautions

Be sure to wash your hands thoroughly after handling soil and plant cuttings. Wear goggles and an apron to protect yourself from any spills.

PLAN

1. As a group, agree upon and write a hypothesis.
2. What steps will you take to **test** your hypothesis? With your group, brainstorm possible experiments.
3. **Choose** the best possible way to test your hypothesis. **Make a list** of the steps you will take for the experiment your group will perform. Be specific and describe exactly what you will do at each step.
4. Which environmental factor will you study? Be sure you choose only one factor at a time to study. You might test the number of hours a day the cutting is in light or different types of water—distilled or tap.
5. What kinds of data will you collect? Will you take measurements? How often will you collect data?
6. **Design a data table** that clearly organizes the data from your experiment.
7. Be sure all the cuttings are from the same plant.
8. As a group, review your experimental plan to make sure that all steps are in a logical order.
9. How will you present the results of your experiment? Will you use charts, graphs, photos, or drawings?

DO

1. Make sure your teacher approves your plan.
2. Carry out your plan.
3. As you do the experiment, be sure to write down all observations in your data table.

CONCLUDE AND APPLY

1. Which environmental factor did you study?
2. **Compare** any differences in the plants grown. What differences in the conditions affected the growth of the cuttings?
3. What type of reproduction did these plants have?
4. **APPLY** How would the owner of a greenhouse use information about plant cuttings and the conditions needed to grow them?

FIGURE 4-5

Sexual Reproduction

Does a new baby look exactly like its father or its mother? Most likely, the baby has features of both of its parents, or it may even look a lot like one of its grandparents. The baby might have her dad's nose shape and hair color, and her mom's eyes and chin. But the baby probably doesn't look exactly like either of her parents. Each person is an individual. That's because humans, as well as many other organisms, are a product of sexual (SEK shul) reproduction. In **sexual reproduction,** a new cell is produced when DNA from both parents combines. This cell multiples and becomes a new individual with its own DNA.

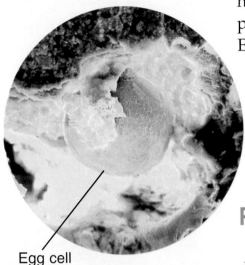

Egg cell

Magnification: 1850×

Production of Sex Cells

In Chapter 3, you learned that your body is made up of different types of cells. Specialized cells called **sex cells** are involved in reproduction. They are shown in **Figure 4-5.** Female sex cells are called eggs and male sex cells are called sperm.

Each human body cell, such as a skin cell or muscle cell, has 23 pairs of chromosomes in its nucleus for a total of 46 chromosomes. The chromosomes in each pair are similar but are not the same. When a body cell divides, two new cells are formed. The new cells have identical sets of chromosomes and they are identical to the chromosomes of the original cell.

Sex cells are different from body cells. A sex cell cannot divide. Also, its nucleus does not have pairs of chromosomes like a body cell's nucleus does. That's because sex cells are not formed by mitosis but are formed by a process called **meiosis** (mi OH sus). Meiosis only happens in body cells called reproductive cells. Before meiosis, the chromosomes in a reproductive cell's nucleus are copied. Then, during meiosis, the reproductive cell's nucleus divides twice. This means that each sex cell formed only receives one chromosome from

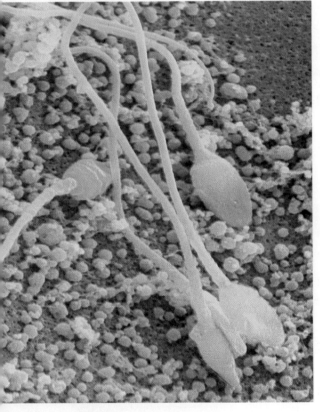

Magnification: 12 000×

each pair of chromosomes instead of pairs of chromosomes. The nucleus of a human sex cell, a sperm or an egg, has 23 chromosomes not 23 pairs (46 chromosomes). That way, when a sperm fertilizes an egg, the new individual receives 46 chromosomes, 23 from each sex cell, making 23 pairs of chromosomes. **Figure 4-6** shows how **fertilization** happens.

FIGURE 4-6

Many animals and plants produce offspring through sexual reproduction.

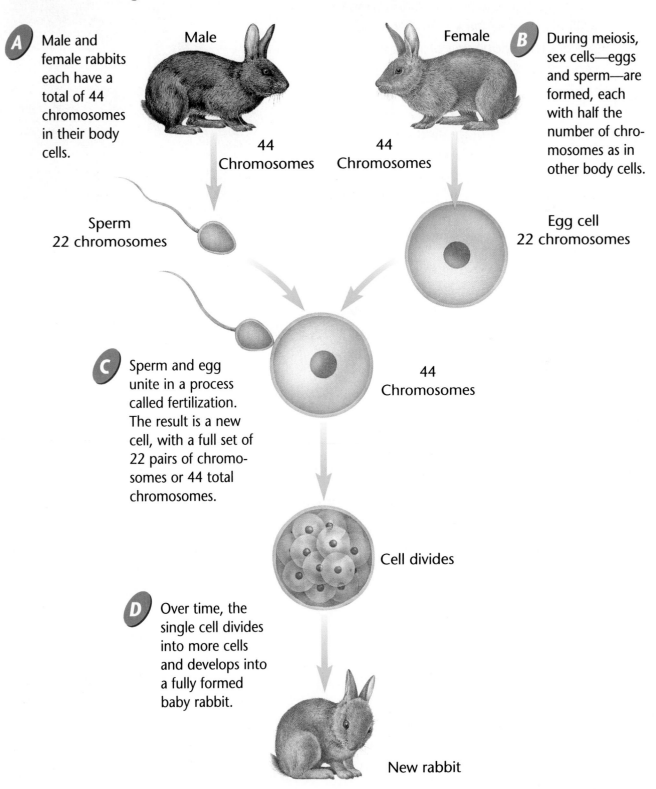

A Male and female rabbits each have a total of 44 chromosomes in their body cells.

Male

Female

B During meiosis, sex cells—eggs and sperm—are formed, each with half the number of chromosomes as in other body cells.

44 Chromosomes

44 Chromosomes

Sperm
22 chromosomes

Egg cell
22 chromosomes

C Sperm and egg unite in a process called fertilization. The result is a new cell, with a full set of 22 pairs of chromosomes or 44 total chromosomes.

44 Chromosomes

Cell divides

D Over time, the single cell divides into more cells and develops into a fully formed baby rabbit.

New rabbit

4-1 Continuing Life   **97**

USING TECHNOLOGY

The Millennium Seed Bank

In 1996, researchers at the Royal Botanic Gardens near London, England, began the Millenium Seed Bank. The Seed Bank is a project they hope will protect the future of more than 24 000 species of plants around the world. The seed bank building was completed in the year 2000, and is the world's largest storage and research facility for seeds. In a seed bank, large amounts of plant seed are stored in cold underground vaults, where they can last for hundreds of years.

Why save seeds?

Within 50 years, one quarter of the world's plants could disappear from the planet due to loss of habitat, pollution, or for other reasons. If this happens, other organisms that depend on the plants may disappear as well.

SCIENCE Online

Visit Glencoe Science Online, *science.glencoe.com*, for a link to further information about the Millenium Seed Bank. On what areas of the world are scientists concentrating their efforts? Where will they keep all the seeds?

Seed Production

It may seem that flowers are just for decoration. But flowers contain the reproductive structures in many plants. Many flowers have male and female parts. Male flower parts produce pollen, which contains sperm cells. Female flower parts produce eggs. When a sperm and an egg join, a new cell forms. As the new cell develops, it becomes enclosed in a seed, which protects the developing cell and helps keep it alive.

Soon after fertilization, rapid changes begin to take place in the flower. The petals and most other parts fall off. The flower's ovary (OH vuh ree), where the eggs are, swells and develops into a fruit which contains seeds, as shown in **Figure 4-7.**

FIGURE 4-7

An apple blossom, when fertilized, will develop into an apple containing seeds.

In this section, you have learned about how cells divide. You have also learned that organisms reproduce in different ways. Asexual and sexual reproduction are different, but in both, DNA is passed on to the new organisms. Later in this chapter, you'll learn that in organisms that have sexual reproduction, the information contained in DNA is passed on in specific patterns.

Section Wrap-up

1. Explain the process of mitosis.

2. Compare the outcome of the process of mitosis with the outcome of the process of meiosis.

3. Explain why offspring produced by asexual reproduction are identical to the parent that produced them.

4. **Think Critically:** Describe the differences between asexual reproduction and sexual reproduction.

5. *Skill Builder*
   Using a CD-ROM Use a CD-ROM encyclopedia to research information on the process of reproduction in a flowering plant. Briefly outline the steps of sexual reproduction in one type of plant. If you need help, refer to Using a CD-ROM on page 563 in the **Technology Skill Handbook.**

USING MATH

A female bullfrog produces 350 eggs. All of the eggs are fertilized and hatch in one season. Assume that half of the tadpoles are male and half are female. If all the female tadpoles survive and produce 350 eggs each one year later, how many eggs would be produced?

4•2 Cloning

What YOU'LL LEARN

- How cloning is done and possible benefits of cloning research
- How to identify main ideas in science articles

Science Words:
cloning
embryo

Why IT'S IMPORTANT

Knowing how to identify the main idea will make you a better reader.

Wood frog, 4-cell stage

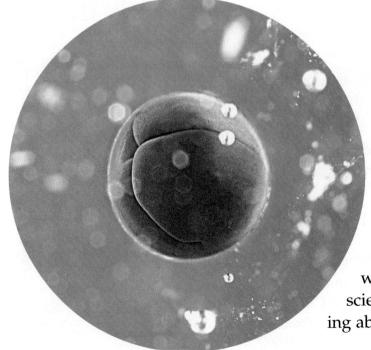

What is cloning?

Cloning means creating an organism that has exactly the same DNA as another organism. Some scientists use the word *cloning* to describe a procedure that could more accurately be called embryo splitting. An **embryo** is a fertilized egg that has begun dividing into more cells. In embryo splitting, a method that has been used for decades by animal breeders, an embryo is divided into several portions. Each portion can develop into a whole organism. For example, if frog eggs are used, each of the portions can grow into a normal frog. All of the frogs produced from one embryo are genetically identical to each other. But these frogs are not true clones. They grew from a fertilized egg cell that received half of its genetic information from the mother and half from the father. A clone receives all of its DNA from just one parent. Until now, this process was not possible except in plants.

The Science of Cloning

In 1997, a Scottish scientist, Dr. Ian Wilmut, announced that he had successfully cloned an adult Finn Dorset sheep he named Dolly. Dolly was the first successfully cloned mammal. She grew from a single cell taken from another sheep. She did not have two parents. Dr. Wilmut said the real value of cloning research is not that Dolly was created in a laboratory, but that scientists now have a better understanding about how cells work.

Through Dolly, scientists learned more about the details of how a single cell can develop into an organism with skin cells, bone cells, muscle cells, and other body cells. Dr. Wilmut hopes other female animals can be changed genetically to produce medicines in their milk. Other scientists think that the new technology may make it possible to grow different types of cells—such as nerve and muscle cells—in the laboratory.

As you read about new developments in science, you come across many details. These details make more sense when you can connect them to one main idea. Understanding the main idea allows you to grasp the big picture. Follow the steps below to identify the main idea and the supporting details in what you just read.

Dolly

 ## Skill Builder: Identifying the Main Idea

LEARNING the SKILL

1. Before reading the material, review related background material from your text, notes, or from class. In this case, review information about cells, cell division, and DNA.

2. Study any photographs or illustrations that accompany the material.

3. What points in the passage are stated most strongly? What is the main idea in each paragraph?

4. Identify the main idea of the article.

5. Identify details that support the main idea.

PRACTICING the SKILL

1. What is the main idea of this article?

2. Briefly describe the difference between a clone and an embryo that has been split.

3. What points in the passage are stated the most strongly?

4. What details are given that support the main idea?

APPLYING the SKILL

Many people have different opinions about cloning. Bring news articles showing these differing opinions on cloning to class. Identify the main idea and supporting details. Discuss different viewpoints with your class.

4•3 Genetics—The Study of Inheritance

What YOU'LL LEARN

- How traits are inherited
- The structure and function of DNA
- How mutations add variation to a population

Science Words:
genetics
gene
DNA
variation
mutation

Why IT'S IMPORTANT

You will understand why you have the traits you have.

Heredity

When you go to a family reunion or browse through family pictures, like the one in **Figure 4-8,** you can't help but notice similarities and differences among your relatives. You notice that your mother's eyes look just like your grandmother's, and one uncle is tall while his brothers are short. Scientists also think about these similarities and differences. Heredity (huh RED uh tee) is the passing on of traits from parents to offspring. Cracking the mystery of heredity has been one of the great success stories of biology.

Look around for a moment at the students in your classroom. What makes each person a unique individual? Is it hair or eye color? Is it the shape of a nose or the arch in a person's eyebrows? Eye color, hair color, skin color, nose shape, and many other features are types of traits that are inherited from a person's parents. A trait is a physical characteristic of an organism. Every organism is a collection of traits. The study of how traits are passed from parent to offspring is called **genetics** (juh NET ihks).

Genes on Chromosomes

All the traits that you have are inherited. Earlier in the chapter, you learned that you're a sort of combination of your parents. Half of the information in your chromosomes came from your father, and half came from your mother. This information was contained in the chromosomes of the sex cells that joined and eventually became you.

FIGURE 4-8

Family members often share similar physical features.

All chromosomes contain genes (JEENZ). A **gene** is a small section of a chromosome that determines a trait. Genes are arranged on a chromosome, one next to another. Humans have about 30 000 different genes arranged on the 23 pairs of chromosomes shown in **Figure 4-9**. Genes control all of the traits of organisms, even traits that can't be seen, such as the size and shape of your stomach. Genes provide all the information needed for the growth and life of that species.

Life's Master Molecule—DNA

You've probably seen or heard about science fiction movies in which DNA is used to bring prehistoric animals back to life. Perhaps you've heard about using DNA to solve crimes. What is DNA? How does it work?

DNA is the material that chromosomes are made of. It is a chemical found in the nuclei of almost all cells. All of the information in the DNA in your chromosomes is called your genetic information. You can think of DNA as a genetic blueprint that contains all the instructions for how an organism looks and functions. Your DNA controls the texture of your hair, the shape of your ears, and even how you digest your food. The same is true for other organisms—all of an organism's traits are affected by the DNA in its cells.

Chromosome Pair

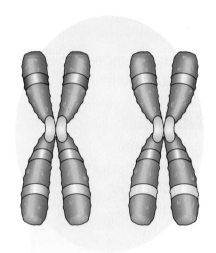

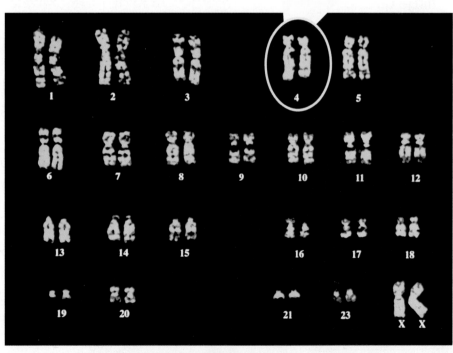

FIGURE 4-9

Each chromosome in a human body is made of DNA. The 23 pairs of chromosomes of one person are shown in this photograph. This person is a female.

Getting DNA from Onion Cells

Throughout the life of an organism, DNA provides instructions for the millions of cell processes that occur daily. It is found in the cells of all living things. In this activity, you will be able to see the actual material that contains all the instructions for a living thing—an onion.

Goals

- Remove DNA from onion cells.
- Practice laboratory skills.

Materials

- prepared onion mixture
- toothpicks
- rubbing alcohol
- measuring cup (1 cup)
- large glass or other glass container
- magnifying glass or microscope

What You'll Investigate

How is DNA taken out of cells?

Procedure

1. **CAUTION:** *Be sure to wear an apron and goggles throughout this activity. Avoid getting the rubbing alcohol in your eyes or on your clothing.*
2. **Obtain** a cup of prepared onion mixture from your teacher.
3. Slowly **pour** an equal amount of rubbing alcohol down the side of the container into the mixture. The alcohol should form a separate layer on top of the onion mixture.
4. **Observe** the gooey strings of DNA floating to the top.
5. Use a toothpick to gently **stir** the alcohol layer. Use another toothpick to remove the slimy DNA strings.
6. **Observe** DNA with a magnifying glass or a microscope. Record your observations in your Science Journal.
7. When you're finished, **pour** all liquids into containers provided by your teacher.

Conclude and Apply

1. Based on what you know about DNA, **predict** whether DNA removed from other plants would look different from the DNA you obtained.
2. **Infer** whether this method of taking DNA out of cells could be used to compare the amount of DNA between different organisms. **Describe** an experiment you might use to find out which types of plants have the most DNA.

Life's Code

If you could look at DNA in detail, you would see that it is shaped like a ladder. This structure, shown in **Figure 4-10,** is the key to how DNA works. The two uprights of the ladder form the backbone of the DNA molecule. The uprights support the rungs of the ladder. It is the rungs that hold all the genetic information. Each rung of the ladder is made up of a pair of chemicals called bases. The secret of DNA has to do with how these bases are put in order. A DNA ladder has billions of rungs, and the bases are arranged in thousands of different orders. The order or sequence of bases along the DNA ladder forms a sort of code. When the cell's machinery reads the code, the cell gets instructions about what to do and how to do it. The largest DNA molecule in a human is approximately 4 cm in length.

FIGURE 4-10

The sequence of bases that form the "rungs" of the DNA molecule forms a code. This code gives the instructions for running the body.

Problem Solving

Analyzing a DNA Fingerprint

A DNA fingerprint looks something like the bar codes on items you buy at the grocery store. DNA fingerprinting is a method of distinguishing individuals based on their DNA. This method produces a pattern of a person's DNA. Each person's DNA is unique.

DNA fingerprinting is an important, accurate tool for investigating crimes in which biological clues such as blood or hair samples are left behind. By comparing the DNA pattern of hair found at a crime scene to a suspect's DNA patterns, investigators can determine whether the hair is likely to have come from the suspect. DNA fingerprinting is also useful for determining whether two people are related.

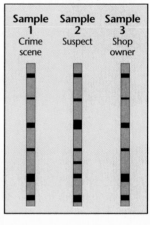

DNA Fingerprints

| Sample 1 Crime scene | Sample 2 Suspect | Sample 3 Shop owner |

Solve the Problem:
The police wanted to find out whether a particular person was involved in the robbery of a shop. These DNA fingerprints show hair samples that were taken from the suspect compared with those left at the scene of the crime.

Think Critically:
After looking at the DNA fingerprints, is it possible that the suspect committed the crime?

What determines traits?

FIGURE 4-11

Pea flowers can be purple or white. The chromosome pair from a pea plant shows that both have an allele for the trait of flower color. *What color will the flowers produced by a plant with this chromosome pair be?*

Remember that in body cells, such as skin cells or muscle cells, chromosomes come in pairs. The genes on those chromosomes are in pairs, too. You have learned that genes control traits. A single pair of chromosomes may control many different traits.

The genes that make up a gene pair may or may not be the same—they may come in different forms. For example, the genes for the trait of flower color in pea plants may be of the purple variety or the white variety. When genes that control a trait come in different forms, those forms are called alleles (uh LEELZ), as shown in **Figure 4-11.**

The combination of alleles in a gene pair determines how a trait will be shown. If a pea plant has two copies of the purple-flower allele, it will have purple flowers. If a pea plant has two copies of the white-flower allele, it will have white flowers.

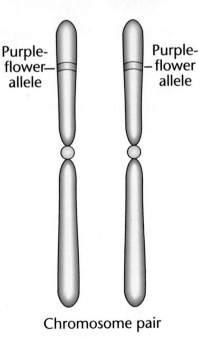

Purple-flower—allele   Purple-flower allele

Chromosome pair

What if the plant has one purple-flower allele and one white-flower allele? Will the flowers be purple or white? That depends on something called dominance (DAW muh nunts). Dominance means that one trait covers over or masks another form of the trait. Some alleles dominate others, as **Table 4-1** shows. For instance, if a pea plant has one purple-flower allele and one white-flower allele, its flowers will all be purple, just as if the plant had two purple-flower alleles. In pea plants, purple is the dominant flower color. White is what's known as the recessive flower color. Recessive means that the trait is hidden or masked if the dominant form of the trait is present. When an organism has two identical alleles for a trait, it's called pure. If it contains different alleles for a trait, it's called a hybrid (HI brud).

Table 4-1

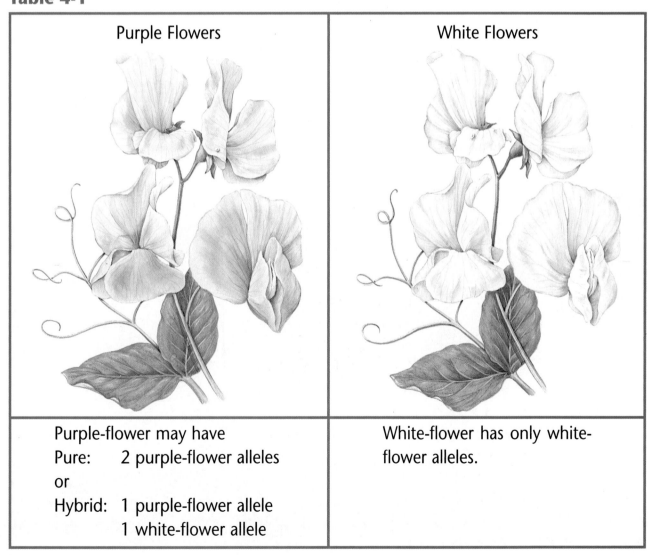

Purple Flowers	White Flowers
Purple-flower may have Pure: 2 purple-flower alleles or Hybrid: 1 purple-flower allele 1 white-flower allele	White-flower has only white-flower alleles.

Passing Traits to Offspring

How are traits passed from parents to offspring? Let's use the trait of flower color in pea plants as an example. Suppose a hybrid purple-flowered pea plant (one with two different alleles for flower color) is mated with a white-flowered pea plant. To mate one pea plant with another, pollen from one plant is placed on the pistil of a flower on another plant. What color flowers will the offspring have?

The traits that a new pea plant will inherit depend upon which genes are carried in the parents' sex cells. Remember that sex cells are produced during meiosis. In meiosis, pairs of chromosomes separate as sex cells form. Pairs of genes therefore also separate from one another. As a result, each sex cell contains one allele for each trait. Because it is a hybrid, the purple-flowered plant in **Figure 4-13** produces half of its sex cells with the purple-flower allele and half with the white-flower allele. On the other hand, the white-flowered plant is pure. All of the sex cells that it makes contain only the white-flower allele.

FIGURE 4-12
The colors of pea flowers are a result of the alleles for flower color the plant receives from its parent plants.

In fertilization, one sperm will join with one egg. Which egg and sperm will join? We don't know. Many events, such as flipping a coin and getting either heads or tails, are a matter of chance. In the same way, chance is involved in heredity. In the case of the pea plants, there was an equal chance that the new pea plant would receive either the purple-flower allele or the white-flower allele from the sperm cell.

FIGURE 4-13
The traits an organism has depend upon the combinations of genes that were carried in the parents' sex cells. This diagram shows how the trait of flower color is passed on in pea plants.

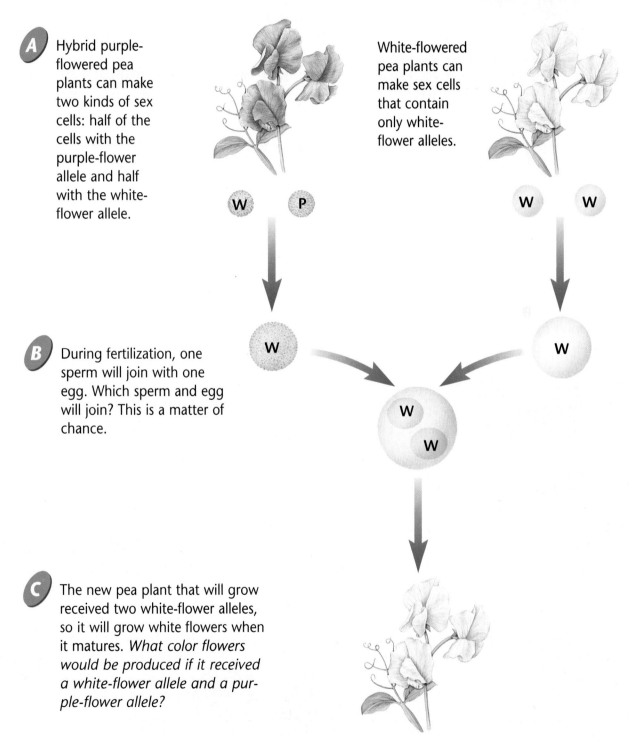

A Hybrid purple-flowered pea plants can make two kinds of sex cells: half of the cells with the purple-flower allele and half with the white-flower allele.

White-flowered pea plants can make sex cells that contain only white-flower alleles.

B During fertilization, one sperm will join with one egg. Which sperm and egg will join? This is a matter of chance.

C The new pea plant that will grow received two white-flower alleles, so it will grow white flowers when it matures. *What color flowers would be produced if it received a white-flower allele and a purple-flower allele?*

FIGURE 4-14

Height is a trait that has many variations.

Mini LAB

Probability

Flip a coin to understand how events happen by chance.

1. Flip a coin ten times. Count the number of heads and the number of tails.
2. Now flip the coin 20 times. Count the number of heads and tails.
3. Record your data in your Science Journal.

Analysis

1. What were your results when you flipped the coin ten times? Was this what you expected?
2. What were your results when you flipped the coin 20 times? Were your observed results closer to your expected results when you flipped the coin more times?
3. How is the flipping of a coin similar to the joining of egg and sperm at fertilization?

Differences in Organisms

Now you know why a new baby can have characteristics of either of its parents. The genes he or she inherited from the parents determined hair color, skin color, eye color, and other traits. But what accounts for the differences, or variations (vayr ee AY shuns), in a family? **Variations** are the different ways a certain trait appears. An example of variations in height is shown in **Figure 4-14.**

Multiple Genes and Multiple Alleles

Earlier, you learned how the trait of flower color in pea plants is passed from parent to offspring. Flower color in pea plants shows a simple pattern of inheritance. Sometimes, though, the pattern of inheritance of a trait is not so simple. Many traits in organisms are controlled by more than two or multiple alleles. In humans, multiple alleles control blood types A, B, or O.

Traits can also be controlled by more than one gene. In humans, for example, height, weight, eye color, skin color, and hair color are traits that are controlled by several genes. This type of inheritance provides for differences or variations in a species. Humans, for instance, can be tall, short, and every height in between.

Mutations: The Source of New Variation

When you hear the words *mutation* (myew TAY shun) or *mutated*, what comes to mind? Maybe you think of a horrible creature straight out of a science fiction movie or a cartoon character with supernatural powers. But if you've ever hunted through a patch of clover until you found one with four leaves instead of three, you've come face-to-face with a real mutation. A four-leaf clover is the result of a mutation. Actually, the word *mutate* simply means "to change." In genetics, a **mutation** is a change in a gene or chromosome due to an error in meiosis or mitosis or due to an environmental factor. Many mutations happen by chance, but some mutations are caused by outside influences, such as X rays or dangerous chemicals in the environment.

What are the effects of mutations? Sometimes mutations affect the way cells grow, repair, and maintain themselves. Some mutations are harmful to organisms. Others are beneficial. But many, such as the four-leaf clover example, have no effect at all. Whether a mutation is beneficial, harmful, or neutral, all mutations add variation to the genes of a species. **Figure 4-15** shows a mutation called albinism that changes the way skin and other body cells work.

FIGURE 4-15

Mutations happen in all organisms. Some cause visible changes such as in the clover and the squirrel. Others are not seen. Some are harmful, but most are harmless.

Section Wrap-up

1. What is heredity?

2. Describe the function of genes.

3. **Think Critically:** Why is DNA sometimes called the "blueprint" for an organism?

4. *Skill Builder*
   Concept Mapping Make a concept map that shows the relationships between the following concepts: genetics, genes, chromosomes, DNA, variation, and mutation. If you need help, refer to Concept Mapping on page 544 in the **Skill Handbook.**

Science Journal

Find out what a transgenic organism is, then go to the library and find books or magazine articles on these organisms. In your Science Journal, write a brief summary of your findings.

Science & *History*

SCIENCE *Online*

Visit Glencoe Science Online at *science. glencoe.com* for a link to more information on different uses of horses through history.

A Horse for Every Job

No one is sure exactly when the first wild horses were tamed, or domesticated, by people. But people in Asia were riding horses as long ago as 3000 B.C. Once horses began to play a role in human culture, people bred them to do specific jobs. Large horses were bred with strong ones to produce breeds suited for hard work. Smaller, sleeker horses were bred for riding and racing. Today, there are hundreds of different breeds, each with its own unique characteristics. However, all horses can be grouped into several basic categories.

Draft horses are large, strong breeds that were bred for pulling plows and hauling heavy loads. Draft breeds include Percherons, Clydesdales, and shires. Shires are the largest and strongest horses in the world. They stand 180 cm (72 inches) high at the withers (the ridge between a horse's shoulders) and weigh at least 1200 kg (2640 pounds). Shires can pull five times their own weight!

Today, the saddle horse breeds show the most variety. These smaller, lighter breeds were developed for such jobs as riding, herding, hunting, and racing. The high-stepping American saddlebred, the thoroughbred, and the American quarter horse are examples of saddle horses.

Reviewing Main Ideas

Read the statements below that review the most important ideas in the chapter. Using what you have learned, answer each question in your Science Journal.

1. Organisms can reproduce sexually or asexually. *Why does sexual reproduction provide more variety in a species than asexual reproduction?*

2. A clone is an organism that is genetically identical to another organism. *What are some possible benefits of cloning technology?*

3. Genetics is the study of how traits are passed from parent to offspring. *How is it possible that you have traits of both of your parents?*

4. Mutations are changes in a gene or chromosome. *How do mutations provide new variation in a population?*

Using Key Science Words

asexual reproduction
cloning
DNA
embryo
fertilization
gene
genetics

meiosis
mitosis
mutation
sex cells
sexual
  reproduction
variation

Match each phrase with the correct term from the list of Key Science Words.

1. chemical that chromosomes are made of
2. reproduction with one parent
3. a change in a gene or chromosome
4. reproduction involving two parents
5. the process of nuclear division

Checking Concepts

Choose the word or phrase that completes the sentence.

6. The process of producing an organism that is genetically identical to another organism is called _____.
   a. fertilization
   b. sexual reproduction
   c. cloning
   d. mutation

7. Sperm and eggs are types of cells called _____.
   a. embryos   c. mutations
   b. sex cells   d. genes

8. During meiosis, _____ for a trait separate.
   a. cells   c. clones
   b. sex cells   d. genes

9. Albinism is a type of _____.
   a. mutation
   b. sexual reproduction
   c. gene
   d. embryo

10. A(n) _____ is a feature or characteristic of an organism.
    a. sex cell   c. trait
    b. embryo   d. gene

Thinking Critically

Answer the following questions in your Science Journal using complete sentences.

11. Explain the relationship among DNA, genes, and chromosomes using the pea as an example.

12. Two brown-eyed parents have a baby with blue eyes. Explain how this could have happened using what you know about heredity.

13. The photo shows a picture of a spider plant. Why is this plant an example of asexual reproduction? How could the plant reproduce through sexual reproduction?

14. How is the process of meiosis important in sexual reproduction?

15. Explain how a mutation in a gene could be harmful or beneficial to an organism.

Developing Skills

If you need help, refer to the description of each skill in the Skill Handbook.

16. Measuring in SI: Gather 50 sunflower seeds or pinto beans. Using a metric ruler, measure the total length of each seed in millimeters. Record the results in your Science Journal.

17. Making and Using Graphs: Make a bar graph from the seed data you collected in question 16. Discuss the variation in seed length for this collection.

18. Designing an Experiment: Design an experiment to test whether or not plants produced by asexual reproduction grow faster or taller than plants produced by sexual reproduction.

19. Recognizing Cause and Effect: Stomach cells produce chemicals that help digest certain foods. What effect might a mutation in a stomach cell gene have on the life of an organism?

20. Predicting: A pure, purple-flowered pea plant is crossed with a pure, white-flowered pea plant. What color of flowers will the resulting pea plant be able to produce?

Performance Assessment

1. Scientific Drawing: Use your imagination and make vocabulary illustrations for each of the following science words: *asexual reproduction*, *genetics*, and *mutation*.

2. Newspaper Article: Many scientists have reported that it's possible to get DNA from prehistoric creatures. Go to the library and find a newspaper article that describes the discovery of ancient DNA. Write a summary of the article in your Science Journal.

Chapter Preview

Skills Preview

▶ **Skill Builders**
- observe
- infer
- make an events chain
- recognize cause and effect

▶ **MiniLABs**
- hypothesize
- infer
- compare and contrast
- predict

▶ **Activities**
- measure
- use tables
- infer
- conclude

Chapter 5

Diversity and Adaptations

Hiking through the rain forest, you come across the scene pictured in the photo. At first, you see nothing unusual. Then your guide whispers, "Look at that cool insect! Can you spot it?" This animal from the tropical rain forest of South America is called a stick insect. The insect blends into its surroundings. Its body shape and color make it almost invisible among the branches and leaves of the plants. Stick insects are found in many parts of the world including North America. Stick insects are just one of the many species that can be found in a tropical rain forest.

EXPLORE ACTIVITY

Identify Different Species

1. Obtain a species category card from your teacher.
2. Look at the card. How many different types of organisms can you name that fit this category?
3. In your Science Journal, list as many different examples of organisms in your category as you can.

Science Journal

Find a picture of one of the living things you listed in your Science Journal. Observe the features of this living thing. What features help it survive in its environment?

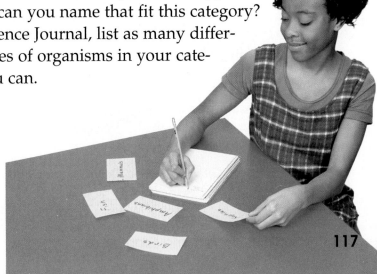

5•1 Diversity of Life

What YOU'LL LEARN

- The term *biodiversity*
- Why some environments have high biodiversity
- How organisms are adapted to their environments

Science Words:
biodiversity
adaptation

Why IT'S IMPORTANT

Learning the connection between environments and the organisms that live there will help you make better decisions about Earth's resources.

Life's Endless Variety

Earth is filled with an enormous number of different living things. Life can be found almost everywhere on Earth—in the air, on water, on land, underground, and in the soil. Why do you think so many different forms of life exist on Earth?

How many different types of animals and plants did you name in the Explore activity? Perhaps you named people, dogs, cats, birds, squirrels, bees, and other familiar species in your environment. No matter how many species you named, they will be only a small portion of the many species known.

How do scientists keep track of all these species? All living things on the planet can be classified into categories called kingdoms. Within these kingdoms, approximately 1.4 million species have been identified and named. Scientists estimate that many more millions of species are yet to be discovered. Some scientists estimate that the total of all life on Earth may be somewhere between 10 and 100 million species. The graph in **Figure 5-1** shows that plants, insects, and other animals show the most *diversity*, or variety, of living organisms that we know. Are you surprised to learn that insects represent more than half of all known species?

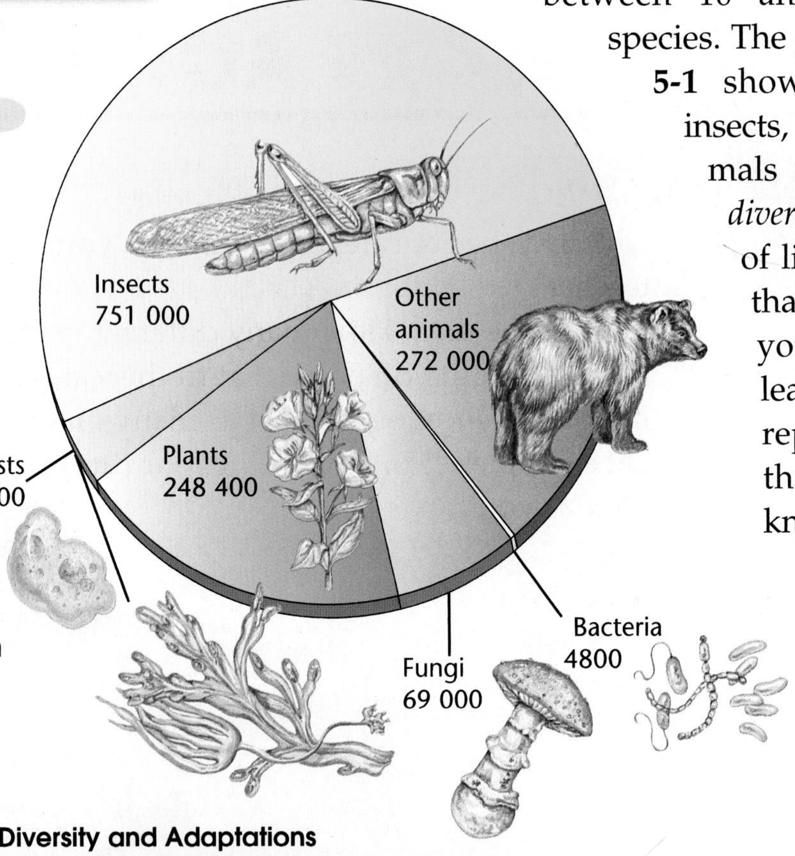

FIGURE 5-1

Which group of organisms has the most species? The fewest?

Number of Species Currently Known

Insects
751 000

Other animals
272 000

Plants
248 400

Protists
57 700

Fungi
69 000

Bacteria
4800

Diversity in the Tropical Rain Forest

You're back in the tropical rain forest now, still staring at the amazing stick insect. The forest is thick with vines, mosses, orchids, and trees, trees, trees. Insects buzz all around. Brightly colored birds flit from branch to branch. Other animals feed and rest in the treetops.

Tropical rain forests are full of life! These environments have a high number of species. Biological diversity or **biodiversity** (bi oh duh VUR suh tee) is the measure of the number of different species in an area. Tropical rain forests contain more different types of living things than any other places on the planet. They have a high biodiversity.

Why do tropical rain forests have such high biodiversity? To answer this question, think about an environment that contains fewer species, such as a desert. In the desert, temperatures are high and very little rain falls. Some desert landscapes have little plant life. What kinds of animals might be able to survive in this environment? Where would they live? What would they eat?

In tropical areas of the world, the climate provides more resources for organisms. Temperatures remain warm and steady all year, and rainfall is high. These conditions are perfect for the growth of plants. In the rain forest, you might find more than 200 different kinds of plants growing in an area the size of a football field. Living among these plants are many insects, birds, mammals, reptiles, and the amphibian shown in **Figure 5-2.** These animals use the plants for food and shelter. In short, tropical environments have high biodiversity because food, water, and shelter are plentiful.

FIGURE 5-2

Animals such as the red-eyed tree frog (middle) find many places to live and many sources of food in the tropical rain forest (top). Deserts (bottom) have fewer sources of food and a harsh environment.

Prickly pear cactus

Namib viper

Diversity and Adaptation

From the stick insect in the rain forest, to a deep-sea fish, to a lizard in the desert, Earth is filled with living things that seem to fit into their environments. Their colors, shapes, sizes, and behaviors allow them to live in their surroundings. Any body shape, body process, or behavior that allows an organism to survive in its environment and carry out its life processes is called an **adaptation.**

As you can see in **Figure 5-3,** species of organisms have many different adaptations to help them survive. Some adaptations involve physical features, such as the body shape and color of a stick insect. These features cause the insect to blend into its environment, hidden from predators. The sharp spines of a cactus are an adaptation for protection. Some animals that have coloration that allows them to blend into the environment may also have behaviors that help protect them. The bobwhite quail and viper remain absolutely still in times of danger. As a result, these animals also stay hidden from predators.

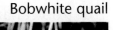

Bobwhite quail

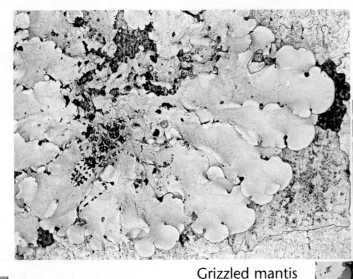

Grizzled mantis

FIGURE 5-3

Plants and animals show a variety of adaptations for survival. *What adaptations do you see in these organisms?*

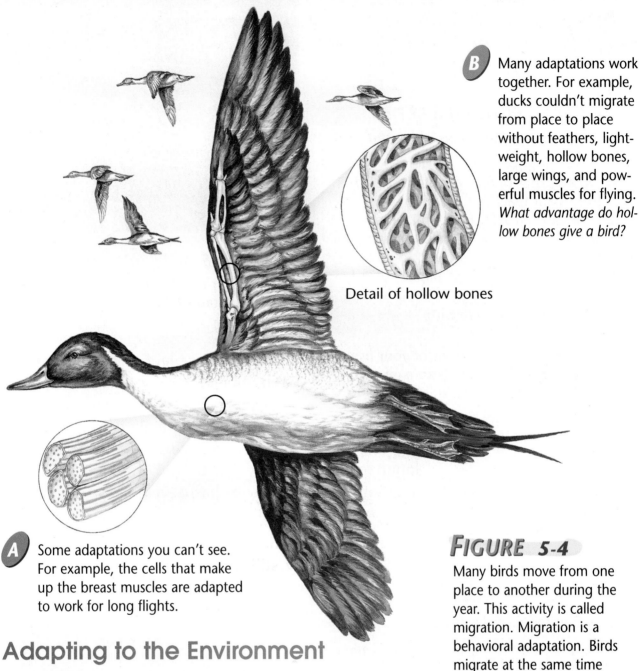

B Many adaptations work together. For example, ducks couldn't migrate from place to place without feathers, lightweight, hollow bones, large wings, and powerful muscles for flying. *What advantage do hollow bones give a bird?*

Detail of hollow bones

A Some adaptations you can't see. For example, the cells that make up the breast muscles are adapted to work for long flights.

FIGURE 5-4

Many birds move from one place to another during the year. This activity is called migration. Migration is a behavioral adaptation. Birds migrate at the same time every year. Some ducks travel thousands of miles to a suitable environment. *What are some other behavioral adaptations of birds?*

Adapting to the Environment

How do you think an organism living in the frozen Arctic would be different from one that lives in the rain forest or the desert? Why would it have different adaptations? An organism's environment includes all of the living things in its surroundings. It also includes the nonliving things such as water, sunlight, and soil. For example, the thick white fur of a polar bear protects it from the cold and helps it blend into its icy, snowy surroundings. These are adaptations the polar bear has to its particular surroundings. Some organisms such as the bird in **Figure 5-4** can move to find a suitable environment.

Mini LAB

Identifying Adaptations of Fish
What kinds of adaptations are useful for fish?

1. Obtain a set of four adaptation cards from your teacher.
2. Look at the cards. On a small piece of paper, draw a picture of a fish that matches the features given to you.
3. On a separate piece of paper, make a drawing of the type of environment your fish could survive in.

Analysis
1. Describe the adaptations of your fish.
2. Explain how the adaptations of your fish make it well suited to its environment.

FIGURE 5-5
A waxy outer layer and sharp spines are protective adaptations for many cacti.

Figure 5-5 shows up close an adaptation of a cactus plant. Cacti live in desert environments where water is in short supply. Thick, waxy coatings on the stems of cactus plants help prevent the cactus from drying up. This is an adaptation for conserving water. Try the MiniLab to see what adaptations a fish species has developed for its watery environment.

Section Wrap-up

1. Define the term *biodiversity*.
2. Explain why tropical rain forests have high biodiversity.
3. What are adaptations? Give an example of one.
4. **Think Critically:** What adaptations would an animal living in a desert need to survive?
5. **Skill Builder**
   Observing and Inferring Think of an organism, not a pet, that is familiar to you. Make a list of its traits. Next to each trait, describe how it helps the organism survive in its environment. If you need help, refer to Observing and Inferring on page 550 in the **Skill Handbook.**

USING MATH

Scientists have identified and named approximately 1 023 000 different species of animals. If 751 000 of these are insects, what percentage do insects make up in the total animal population? Show all your work.

RED-HEADED WOODPECKER
Melanerpes erythrocephalus

Description: — 10" (25 cm) *Whole head red,* wings and tail bluish black, with *large white patch on each wing;* white underparts; white rump, conspicuous in flight. Immature resembles adult, but has gray head, 2 dark bars on white wing patch.
Voice: — A loud *churr-churr* and *yarrow-yarrow-yarrow.*
Habitat: — Open country, farms, rural roads, open parklike woodlands, and golf courses.
Nesting: — 5 white eggs placed without nest lining in a cavity in a tree, telephone pole, or fence post.
Range: — Saskatchewan, Manitoba, and Quebec south to Florida and the Gulf Coast. Scarce in northeastern states. Winters in southern part of range.

Using Field Guides to Explore Diversity

A bird with a red head and black wings lands on a tree. It hammers into the bark with its beak. You're pretty sure it's a woodpecker, but what kind?

One place to find out is in a field guide, such as the *National Audubon Society Field Guide to North American Birds.* Field guides are handbooks for identifying living things, from mushrooms to mammals. They contain descriptions and photographs or illustrations, as well as information about where organisms live. Field guides are useful tools for making sense of the often-bewildering diversity of life on Earth.

Some field guides are organized around easily identifiable characteristics, such as shape or color. A field guide to flowers might group yellow flowers in one section, red in another, and so on. Other guides group together species that may look different, but that all belong to the same scientific family.

Using field guides is a bit like detective work. You follow clues, gradually narrowing your search until you can finally answer the question: "What's that?"

Science Journal

Spend several hours observing nature. Describe five plants or animals you saw. Make sketches and note colors, shapes, and sizes. Then, using field guides, try to identify the organisms on your list. If you have a beetle on your list, good luck—there are nearly 300 000 different kinds!

5•2 Unity of Life

What YOU'LL LEARN

- How species change and new species form through natural selection
- About the evidence that supports the idea that organisms change through time

Science Words:
natural selection
common ancestor
fossil

Why IT'S IMPORTANT

If you know how organisms are adapted to their environments, you can better understand how the environment can affect organisms.

Why are they like that?

You probably would never confuse an eagle with a hummingbird. Both have wings, beaks, feathers, and other familiar features of birds, but they are different in size, shape, and where they live. On the other hand, you might have trouble recognizing different kinds of birds called warblers unless you were an experienced bird-watcher. Many species of these small birds look alike. Why are some species more similar than others? And how do new species form? Through the study of living and once-living organisms, scientists try to find the answers.

Adaptation Through Natural Selection

In the last section, you learned that organisms are adapted to the environments in which they live. How do these adaptations happen? Recall from Chapter 4 that individual organisms are not identical to each other, even if they belong to the same species. In every species, variations, or differences, occur in the traits of that species. In a population of gray squirrels, for example, individuals may have slightly different colors of fur. Most will have brown-gray fur color. A few will have dark fur, and some will have light fur.

FIGURE 5-6

This illustration shows how a population of squirrels can change through natural selection.

A In all species, individual organisms have differences, or variations.

B In this population, some squirrels may be alert and hide more quickly when predators come near.

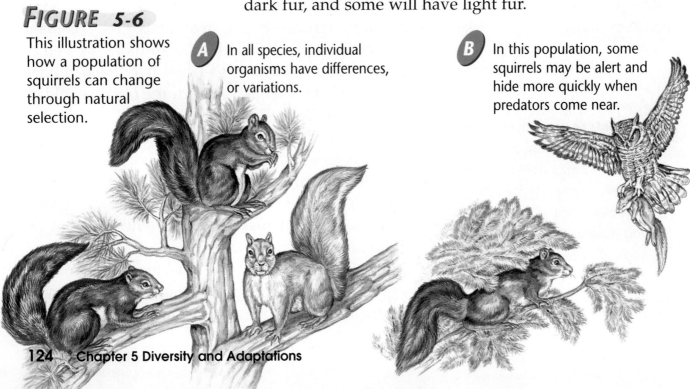

Now, imagine this same population of squirrels in an environment that changed. Suppose the environment changed so that squirrels with dark fur are better able to survive than squirrels with a lighter fur color because predators are not able to see squirrels with darker fur as easily. Dark squirrels would be better off in this environment. They would be more likely to survive and have offspring, as shown in **Figure 5-6.** This process, in which organisms with characteristics best suited for the environment survive, reproduce, and pass these traits to their offspring, is called **natural selection.**

Now, think about the process of natural selection happening over many, many years. With each generation, more and more dark squirrels will survive and pass their coat color on to their offspring. Squirrels with lighter fur don't survive and so produce fewer offspring. The population changes. Every generation, the population consists of more dark squirrels than the one before it. This process of natural selection is the main way changes happen in a population.

Natural selection can explain why polar bears are white, why some other bears are brown, and why some insects that live in the leaves of a tree are green. Think of how the stick insect at the beginning of the chapter came to look the way it does.

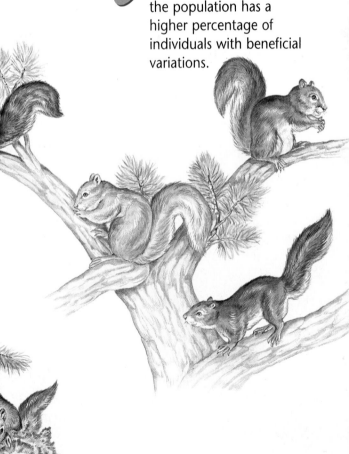

D There is still variation, but the population has a higher percentage of individuals with beneficial variations.

C Organisms that survive reproduce and pass on their traits.

Activity 5-1

Design Your Own Experiment
Simulating Selection

Natural selection causes a population to change. In this activity, you will design an experiment to discover how camouflage adaptations—those adaptations that allow organisms to blend into the environment—happen through natural selection.

PREPARE

Possible Materials
- bag of small-sized jelly beans in assorted colors (approximately 100 beans)
- meterstick
- paper
- pencil
- Astroturf rug or another solid colored rug
- watch with a second hand

What You'll Investigate
How does natural selection work?

Form a Hypothesis

Any body shape, structure, or coloration of an organism that helps it blend in with its surroundings is a camouflage adaptation. Make a hypothesis about how natural selection can explain camouflage adaptations.

Goals

Model natural selection in a population of insects.

Explain how natural selection produces camouflage adaptations.

Safety Precautions

Do not eat any jelly beans.

PLAN

1. With a partner, **discuss** the process of natural selection. How does natural selection cause species to change?
2. In this activity, one student will play the role of the bird that eats insects (jelly beans) that live in grass. With your partner, think about how you can model natural selection using some of the materials suggested.

3. With your partner, **make a list** of the different steps you might take to model natural selection.

4. You might start the experiment with a particular insect population. How many insects (jelly beans) will be in your starting population? How will you show variation in your starting population?

5. Think about how many generations of insects you will use in the experiment.

6. What data will you collect? Will you need a data table for this experiment? If so, **design a table** in your Science Journal for recording data.

7. What happens to individual organisms if they have favorable variations for a particular environment? What happens if an organism has unfavorable variations? Make sure you think about these questions in designing your experiment.

DO

1. **Review** your list of steps to make sure that the experiment makes sense and that all of the steps are in logical order.

2. Make sure that your teacher has approved your plan before you continue.

3. While doing the experiment, **record** your observations and **complete** your data table in your Science Journal.

4. Carry out the experiment.

CONCLUDE AND APPLY

1. **Observe** which types of insects the bird in your experiment was able to locate most quickly. Why?

2. Did your population of insects change over time? **Explain** your answer.

3. **APPLY** How can your experiment be used to **explain** camouflage adaptations in organisms?

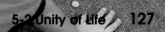

A The members of this population of lizards all look alike and are able to interbreed.

FIGURE 5-7

One way new species form is for a population to be separated.

The Origin of New Species

If you've ever watched nature programs showing lions and tigers in the wild, you've seen the animals hunting, washing themselves, and caring for their young. You may have seen these same behaviors in a house cat. Maybe you've also noticed that all types of cats—lions, tigers, bobcats, and others—have similar features. How did the various species of cats form, and why do they share so many features?

Remember that scientists define a species as a group of organisms that look alike and reproduce. One way a new species can form is if a large population of organisms becomes separated into smaller populations that no longer breed with others of its kind. A barrier, such as a mountain range or river, can cause this separation. **Figure 5-7** shows an example.

A New Species Develops

Let's look at a population of lizards. Suppose a river divides the whole population into two smaller populations. Lizards in one group can no longer mate with lizards in the other groups. They can mate only among themselves. Over time, each small population will adapt to its own environment. Because each environment is a little bit different, each small population will develop different adaptations. Over time, the populations may become so different from each other that offspring couldn't be produced even if the lizards could again mate with one another. Each small population of lizards

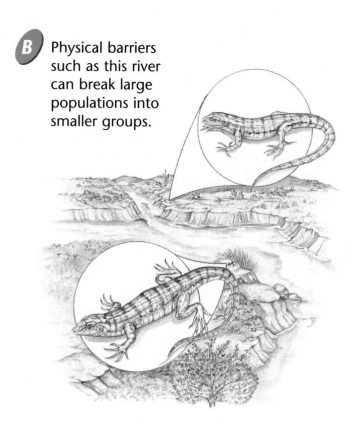

B Physical barriers such as this river can break large populations into smaller groups.

C New species can form when a population is separated and individuals in one population no longer mate with individuals on the other side of the barrier.

has become a new species. Because the different species of lizards in this example all arose from one population, we can say that they share a **common ancestor** (AN ses tur). Each of the new lizard species now has its own specific traits. But all the species share some traits that they inherited from their common ancestor.

Evidence for Change Over Time

You might say seeing is believing. But we can't always see everything we'd like to study. Objects may be too tiny or too far away for us to see. Processes may happen too quickly or too slowly for us to observe. That's the problem with natural selection. It occurs in all species, but it usually occurs so slowly that it's usually not possible to see the process in action. So how can scientists be sure that species have changed over time? They rely on several types of evidence. They compare physical traits of species that are similar. They study the DNA and proteins of species for clues to how closely they are related. They look at fossils. Let's take a closer look at each of these methods.

Mini LAB

Making a Fossil

How can you make a cast fossil from a mold?

1. A mold fossil is formed when an organism leaves an impression in clay or mud. When the mold is later filled in, a cast fossil can be formed. To make a model of a fossil, grease the inside of a pint-sized milk container with petroleum jelly.
2. Pour plaster of paris into the milk container until it is half full.
3. When the plaster begins to thicken, grease some small objects and press them into the plaster.
4. After the plaster has hardened, remove the objects. You have now made a mold.
5. Next, grease the entire layer of hard plaster. Pour another layer of colored plaster to fill in the mold.
6. After the plaster hardens, tear away the milk container and separate the two layers of plaster. The colored layer shows your cast.

Analysis

Which fossil showed the most details of the objects you used, the mold or the cast?

Fossils Show Change over Time

Although dinosaurs have been extinct for millions of years, they come to life in the movies. Do you think those giant, roaring creatures on screen are anything like real dinosaurs were? How do scientists and the movie producers know how dinosaurs looked and acted? Scientists have learned a lot about the ancient reptiles by studying their fossils. **Fossils** are the remains or traces of ancient life.

Fossils give scientists the most direct evidence that species change over time. For example, fossils of trilobites (TRI luh bites) show that change occurred in their structure over time. Trilobites are extinct relatives of animals you know today such as lobsters, crabs, and insects. Early trilobites had generalized body structures with few segments. Fossils of later trilobites show many more specializations and differences in shape. **Figure 5-8** shows examples of these fossil organisms. Through the study of fossils, scientists have shown that trilobites changed over time.

FIGURE 5-8

Trilobites are an example of an organism that changed through time.

FIGURE 5-9

Wolves and dogs are close relatives. *What physical similarities can you see between these two animals?*

DNA Shows Relationships

Appearances can trick you. Some species look a lot alike but aren't closely related. Scientists need something besides physical appearance to help them figure out how closely related species are. In Chapter 4, you learned that DNA determines the traits of organisms. The more similar the DNA of two species is, the more closely they are related. For example, in comparing the DNA of modern breeds of dogs, wolves, and foxes, the gray wolf, *Canis lupus,* turns out to be the closest relative of the domestic dog. These animals are shown in **Figure 5-9.**

Problem Solving

More Chemical Clues

You've learned that DNA can be used to show how organisms are related. Scientists also use other chemicals found in living things to show how closely related two species are. One type of chemical they study is protein. Proteins perform a variety of jobs in living things. Some are used in the building of living material, such as bones, muscles, and skin. Others help living things grow, digest foods, and fight diseases. Scientists can learn about the relationships between species by studying the structure of proteins. Each protein is made of building blocks put together in a specific order. In closely related species, similar proteins have a similar order of building blocks. In species that are more distantly related, the same proteins have different arrangements of building blocks.

Solve the Problem:

The illustration shows the structure of a type of protein used in digestion found in three unknown bird species (Species A, Species B, and

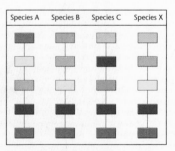

Species C) and in a known bird species (Species X). The colored blocks represent the building blocks of the digestive protein. Compare the structure of the proteins in all four species.

Think Critically:

Which bird species (A, B, C) do you think is most closely related to Species X? Why did you reach your decision?

Anatomy and Ancestry

If you were a biologist studying the many species of cichlid (SI klud) fish in Africa's Lake Victoria, you would see differences in the jaws and teeth of these fish. You would also see differences in their color. Some of these differences are shown in **Figure 5-10.** Some species have small mouths and sharp teeth for eating insects. Others have mouths adapted for scraping algae. Still others are adapted for eating the scales of other fish. However, despite the differences, all 170 species of cichlids in the lake have many more things in common.

Scientists compare similarities and differences in the body structures of all living things. This gives them clues to how species may have changed over time. After comparing the anatomy of many species, scientists conclude that all species may have come from a common ancestor.

FIGURE 5-10

The differences in the head shapes of these fish allow them to eat different things and feed in different ways. Two species may feed on insects but in different ways.

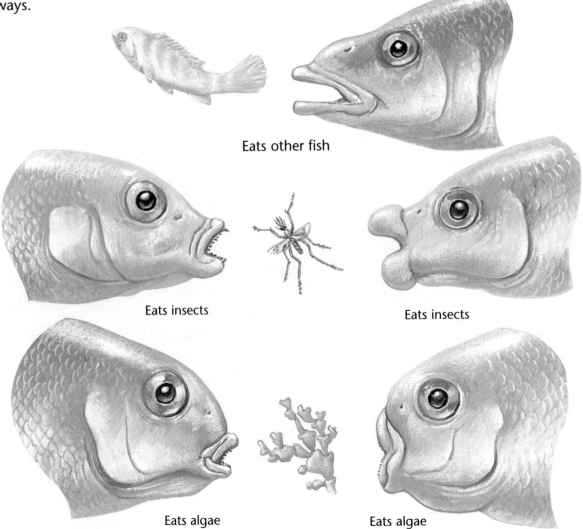

Eats other fish

Eats insects

Eats insects

Eats algae

Eats algae

Another example of similarities in anatomy between two organisms is seen in **Figure 5-11.** Do you see these similarities? If you said yes, then you made the type of observation that scientists use to find relationships among organisms. Scientists view these similarities as evidence that two organisms developed from a common ancestor.

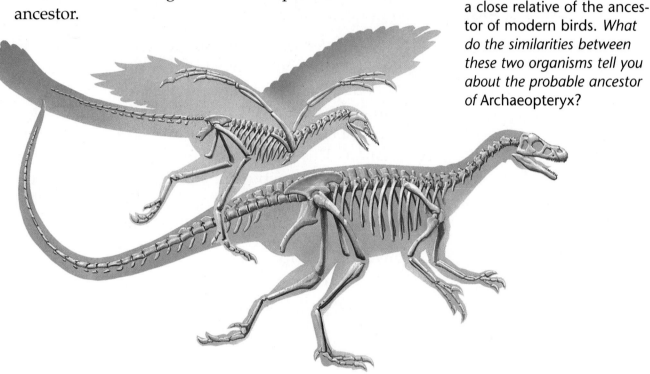

FIGURE 5-11

Archaeopteryx (above) has many similarities to *Ornitholestes* (below), a small dinosaur. *Archaeopteryx* is considered a close relative of the ancestor of modern birds. *What do the similarities between these two organisms tell you about the probable ancestor of Archaeopteryx?*

Section Wrap-up

1. Explain natural selection.

2. Briefly describe how natural selection plays a role in how species adapt to the environment.

3. **Think Critically:** Explain how fossils can show that species change over time.

4. **Skill Builder**
   Concept Mapping Make an events chain that describes how camouflage adaptations may arise in a population of birds. If you need help, refer to Concept Mapping on page 544 in the **Skill Handbook.**

Science Journal

Some scientists hypothesize that dinosaurs and birds are closely related. Imagine you are a scientist who wants to examine this hypothesis. In your Science Journal, describe methods you would use to investigate the relationship between dinosaurs and birds.

5•3 Captive-Breeding Programs

What YOU'LL LEARN

- How captive breeding programs help endangered species
- How to identify resources

Science Words:
captive breeding

Why IT'S IMPORTANT

Learning how to identify and use information resources will help improve your research skills.

Have you ever seen a Siberian tiger such as the one in **Figure 5-12** in a zoo? Zoos are great places to see and learn about animals from all over the world. But there's more to a zoo than meets the eye. In the United States, many zoos belong to the American Zoo and Aquarium Association. This group is trying to save endangered species using a variety of methods, including captive breeding programs. Endangered species are those that have only a small number of living members left.

Captive Breeding

Captive breeding is the breeding of endangered species in captivity. Usually, the species that are bred in captivity have only a few living members left in the wild. Through captive breeding, the number of individuals in the population is built up if the program is successful. Some of the animals that are bred by a zoo are released into the wild. Others are kept by the zoo or given to other zoos to breed more animals. Why do zoos trade animals for breeding? Zoos can't breed animals that are born of the same parents over and over again because this would weaken the genetic diversity and health of the species, so they trade animals with other zoos.

FIGURE 5-12

Siberian tigers once roamed Russia, China, and Korea in large numbers. It is now estimated that there are less than 5000 Siberian tigers left in the wild.

Breeding Programs

Let's look at one captive-breeding program that works with several states and zoos to help save black-footed ferrets from extinction.

In 1972, it was thought that the black-footed ferret was extinct. No one had seen one for years. But in 1981, a Wyoming rancher found a dead ferret on his land. Soon after, a small group of ferrets, like those shown in **Figure 5-13,** was discovered nearby. The last of these wild ferrets was captured in the early 1990s for captive breeding. This group has increased from 18 to more than 500.

Since then, captive-breeding programs for the ferrets have been started at zoos and wildlife centers from Virginia to Colorado. Some of the ferrets are being released into the wild; others are kept for breeding. By the year 2010, it is hoped that 1500 ferrets will be living in the wild.

Would you like to learn more about ferrets or other animals that are bred in captivity? You can do this if you know which resources to use. Resources are things that help you to learn more about a subject. Encyclopedias, magazines, the Internet, newspapers, people—all of these are resources.

FIGURE 5-13

Black-footed ferrets born in captivity in a breeding program (top) can be returned to their natural environment (bottom).

 ## Skill Builder: Using Resources

LEARNING the SKILL

1. To identify resources, first look for key words or events in the material you read. You'll use these things as guides to help you in your research.

2. With the help of your teacher or a librarian, list all the resources available for you to use.

3. Decide what type of information you're looking for. Are you looking for local events? A newspaper or an interview with a local resident might be the best resource for you to use. Are you looking for historical events? Then use an encyclopedia or the Internet.

PRACTICING the SKILL

1. What is captive breeding?

2. What keywords would you use to learn more about captive breeding?

3. Which resources would you use to find out more about captive breeding in general? What resources would you use to find out about captive-breeding programs in your area?

APPLYING the SKILL

Choose a topic that interests you to research. It could be anything from insects to stars and galaxies. Use at least three resources to research the topic. Share your research with the class.

5•4 History of Life

What YOU'LL LEARN

- The importance of the fossil record
- About the geologic time scale
- The possible causes of mass extinctions

Science Words:
fossil record
geologic time scale
mass extinction

Why IT'S IMPORTANT

Learning about the history of life on Earth will help you understand the diversity of life that is now on Earth.

A Trip Through Geologic Time

Have you ever seen a movie or read a book about time travel? What if you could travel back in time into Earth's past? What kinds of interesting creatures would you see? Time travel will always be science fiction rather than science fact. But in a way, scientists can travel back through time by studying fossils. You learned that fossils are evidence of ancient life. Fossils help scientists form a picture of the past. They provide a history for life on Earth. Let's go along on the journey through Earth's history.

The Fossil Record

Earth's rocky crust is a vast graveyard that contains the fossil remains of species that have lived throughout Earth's history. Large fossils are carefully removed from the ground as shown in **Figure 5-14.** All of the fossils that scientists have recovered from the ground make up the **fossil record.** The fossil record for life on Earth is a rich one. Fossils from almost every major group of plants and animals are part of the fossil record.

FIGURE 5-14

This scientist (right) is carefully uncovering a *Tyrannosaurus rex* fossil. Fossils allow scientists to make models of dinosaurs.

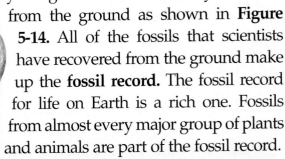

However, the fossil record doesn't show a complete history. Although many major groups of plants and animals are represented, not every species is. It is much more common for an organism to decay without ever becoming a fossil.

USING TECHNOLOGY

Blood in the Bones

In 1991, researchers at Montana State University studying a *Tyrannosaurus rex* fossil made what may turn out to be an important discovery. "I think I've found red blood cells," one of the researchers said as she focused her microscope on a slide made from the dinosaur fossil.

The Real Jurassic Park Lab

Were the mysterious structures in the dinosaur bones really red blood cells? If so, the scientists reasoned that they would find hemoglobin (HEE muh gloh bun), the blood protein that carries oxygen to body cells. To test for hemoglobin, the dinosaur tissue was hit with laser light. The tissue produced the same chemical signature as modern hemoglobin-carrying blood cells. In the second test, an extract was made from the dinosaur tissue and compared with the blood proteins from birds, crocodiles, and humans—all organisms with hemoglobin in their red blood cells. Similarities between the blood samples gave the scientists more evidence that hemoglobin, and therefore blood cells, were present in the fossil.

Scientist studying dinosaur fossils

If these cells are dinosaur blood cells, scientists may be able to get DNA from them. Scientists aren't interested in bringing *T. rex* back to life as the characters did in the movie *Jurassic Park.* But they could compare dinosaur DNA to DNA in living organisms. If so, the world would gain a better picture of the nature of these creatures.

SCIENCE *Online*

To what living organisms would the scientists compare dinosaur DNA? Check Glencoe Science Online. *science.glencoe.com*

Table 5-1

Geologic Time Scale

Era	Period	Million years ago	Major evolutionary events	Representative organisms
Cenozoic	Quaternary	1.8	First humans	
	Tertiary	65		
Mesozoic	Cretaceous	145	Large dinosaurs First flowering plants	
	Jurassic	213	First birds First mammals	
	Triassic	248		
Paleozoic	Permian	286		
	Pennsylvanian	325	First conifer trees; first reptiles and insects	
	Mississippian	360		
	Devonian	410	First amphibians and land plants; first bony fish	
	Silurian	440	First fish with jaws	
	Ordovician	505	First vertebrates, armored fish without jaws	
	Cambrian	544	Simple invertebrates	
	Precambrian		First fossilized animals and plants; protozoa, sponges, corals, and algae	
		4500	First fossil bacteria	

Geologic Time Scale

By studying the fossil record, scientists have put together a sort of diary for life on Earth called the **geologic time scale.** The geologic time scale helps scientists keep track of when a species appeared on Earth or when it disappeared from Earth.

You can see the geologic time scale in **Table 5-1**. As you can see, the geologic time scale is divided into four large intervals of time called eras, and each era is subdivided into periods. The beginning or end of each time period marks an important event in Earth's history, such as the appearance or disappearance of a group of organisms. The fossil record of life on Earth gives scientists strong evidence that life has changed over time.

How big were the dinosaurs?

Fossils can be used to estimate the size of a dinosaur. Dinosaur models also may be used.

What You'll Investigate
How can you estimate the mass of a dinosaur?

Procedure
1. **Make a data table** in your Science Journal. Include type of dinosaur, its scale, and volume of water displaced in your table.
2. Obtain a plastic dinosaur model. Fill in your data table. A good scale to use is a 1:40 scale, meaning that each dimension of the model—length, width, and height—is 1/40 of the dinosaur's actual size.
3. **Fill** the pail with water until it is almost full. Carefully place the pail in the utility pan. Fill the pail to the brim, taking care not to let any water spill over into the utility pan.
4. **Submerge** the dinosaur model into the water, allowing the water to flow out of the pail and into the utility pan.
5. Carefully remove the pail from the utility pan.
6. Pour the water from the utility pan into a graduated cylinder. The volume of the water displaced is equal to the volume of the dinosaur model.
7. **Record** your measurement in your data table.
8. To find the mass of the actual dinosaur in grams, multiply the volume of the model by the cube of the scale. If the scale is 1:40 and the model has a volume of 10 mL, you would multiply 10×40^3 ($40 \times 40 \times 40$). To find the mass in kilograms, divide by 1000.

Conclude and Apply
1. What is the estimated mass of your dinosaur in kilograms?
2. Male Indian elephants weigh about 5500 kg. What can you **infer** about the approximate size of some dinosaurs?

Goals
- Estimate the mass of a dinosaur using a model.

Materials
- plastic dinosaur model
- plastic graduated cylinder
- marking pen
- calculator
- water
- small pail
- utility pan

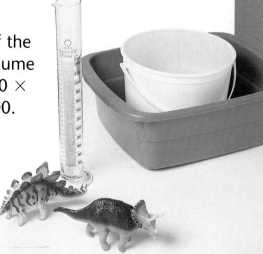

Extinction of Species

Why don't you see dinosaurs at the zoo? These large reptiles ruled Earth for more than 100 million years, but they're all gone now. Among the dinosaurs were the largest land animals, such as the plant-eating *Brachiosaurus* and the meat-eating *Tyrannosaurus rex*. Yet, despite the great success and long history of the dinosaurs, all of them became extinct about 65 million years ago. So did many other land and sea animals.

Scientists aren't surprised that dinosaurs are now extinct. That's because about 99 percent of all species that have ever existed are now extinct. Usually, extinction is a natural event. As you have learned, Earth's environments are constantly changing. When a species can't adapt to changes in its environment, it becomes extinct forever.

FIGURE 5-15

The second-largest mass extinction event in Earth's history occurred 65 million years ago at the end of the Cretaceous period. One-half of all living things—including many dinosaurs—became extinct at this time.

Mass Extinctions

The extinction event that killed the dinosaurs around 65 million years ago was a mass extinction. A **mass extinction** is a large-scale disappearance of many species within a short time.

The extinction of the dinosaurs was not the only mass extinction event in Earth's history. There were other large extinctions. The most severe mass extinction happened about 245 million years ago. Scientists estimate that nearly 96 percent of all animal species became extinct at this time. Most were animals without backbones, such as clams, jellyfish, sponges, and trilobites. Most fish and land species survived.

What causes mass extinctions? Scientists are not sure about the exact causes of all the mass extinctions. One idea that has been hypothesized for the extinction of the dinosaurs is that large asteroids slammed into Earth. According to this hypothesis, the collisions sent huge clouds of dust into the atmosphere. Over time, the clouds blocked out sunlight. The result would have been rapid cooling of the environment. Dinosaurs and other organisms probably could not have adapted quickly enough to survive, as shown in **Figure 5-15**.

SCIENCE *Online*

Visit Glencoe Science Online at *science. glencoe.com* for links to more information about dinosaurs and extinctions.

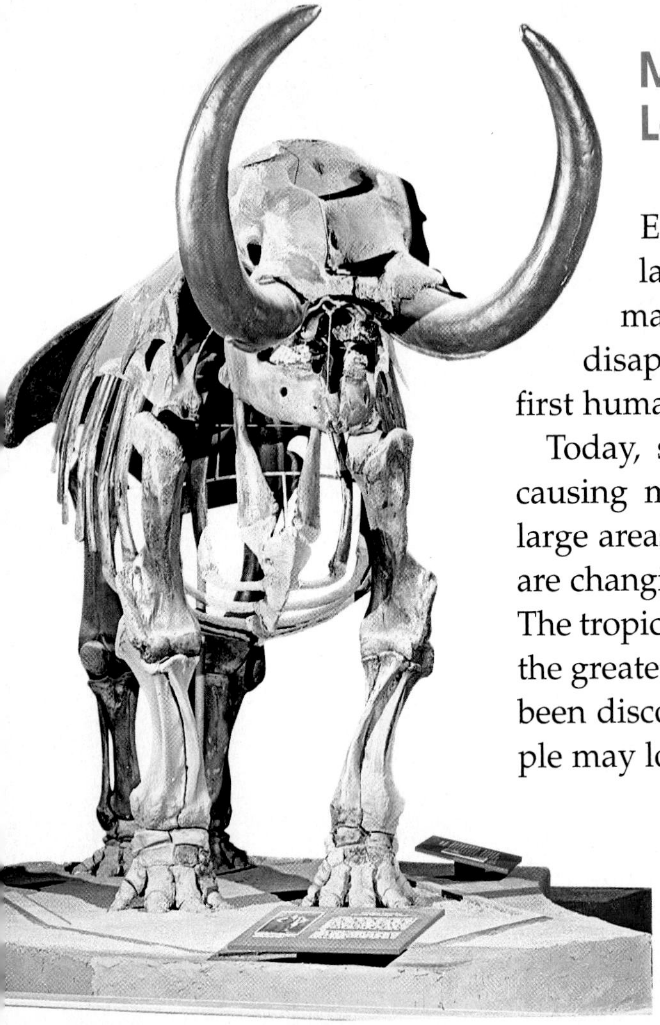

Mass Extinction and the Loss of Biodiversity

Mass extinctions are not just a part of the past. Extinctions have also been occurring within the last few thousand years. For instance, large mammals such as mammoths and mastodons disappeared from North America about the time the first humans appeared here.

Today, scientists fear that human activity is rapidly causing more species to become extinct. Humans use large areas of land for many reasons. Some of these uses are changing or limiting the habitat for many organisms. The tropical rain forests, where biodiversity is high, are in the greatest danger. Many rain forest species have not yet been discovered or named. If these species are lost, people may lose possible sources of food or new medicines.

FIGURE 5-16

Many organisms, such as the mastodon, disappear from Earth for reasons not completely known.

Section Wrap-up

1. What is the fossil record?

2. What is the geologic time scale?

3. What is a mass extinction? Give an example.

4. **Think Critically:** How might the extinction of a single plant species from a forest affect other organisms that live there?

5. **Skill Builder**
   Recognizing Cause and Effect Explain how an environmental change might cause a species to go extinct. If you need help, refer to Recognizing Cause and Effect on page 551 in the **Skill Handbook.**

USING MATH

The Precambrian represents about 4.5 billion years of Earth's 4.6-billion-year history. What percent of the total does this represent?

Read the statements below that review the most important ideas in the chapter. Using what you have learned, answer each question in your Science Journal.

1. Biodiversity refers to the variety of plants, animals, and other species in an area. *Why do tropical areas have high biodiversity?*

2. An adaptation is any structure or behavior that helps a species survive in its environment. *What are some adaptations of fish?*

3. Species adapt to the environment, and new species form through the process of natural selection. *What is natural selection?*

4. The geologic time scale shows when species appeared and disappeared during Earth's history. *When were the dinosaurs the dominant land animals on Earth?*

Using Key Science Words

adaptation
biodiversity
captive breeding
common ancestor
fossil

fossil record
geologic time scale
mass extinction
natural selection

Match each phrase with the correct term from the list of Key Science Words.

1. traces or remains of ancient life
2. numbers of different species in an area
3. mechanism for species' change over time
4. dolphin's flipper or a bird's wing
5. a large-scale disappearance of many species

Checking Concepts

Choose the word or phrase that completes the sentence.

6. A(n) _____ is the disappearance of many species in a short time.
   a. natural selection
   b. adaptation
   c. fossil
   d. mass extinction

7. Species adapt to the environment through the processes of mutation and _____.
   a. natural selection
   b. biodiversity
   c. captive breeding
   d. fossil formation

8. Dinosaur bones are examples of _____.
   a. species
   b. adaptations
   c. fossils
   d. biodiversity

9. New species can form when _____ split into small groups.
   a. fossils   c. populations
   b. adaptations   d. biodiversity

10. Tropical rain forests have high _____ compared with deserts.
    a. species   c. fossil records
    b. biodiversity   d. temperatures

Thinking Critically

Answer the following questions in your Science Journal using complete sentences.

11. Which of the following variations would be most beneficial to a bird living in a wetland: webbed feet, clawed feet, feet with toes for gripping branches? Explain.

12. If a scientist had DNA samples from four organisms, how could he find out which organisms were related?

13. What types of adaptations would be beneficial to an organism living in the desert?

14. Use the idea of natural selection to explain why polar bears are white and why some other bears are brown.

15. Horseshoe crabs are an example of an organism that has changed very little through time. What factors might have kept these organisms from changing?

Developing Skills

If you need help, refer to the description of each skill in the Skill Handbook.

16. **Observing and Inferring:** Look at the chart below. Scientists hypothesize that millions more microscopic species exist. Why do you think the number of microscopic species known to science is so low?

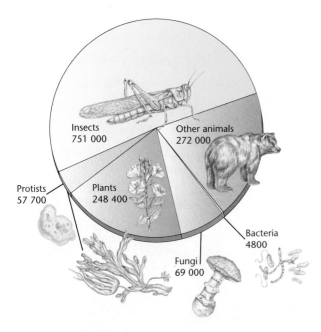

Insects
751 000

Other animals
272 000

Protists
57 700

Plants
248 400

Bacteria
4800

Fungi
69 000

17. **Hypothesizing:** How are the thorns on a rose an adaptation to the plant's environment?

18. **Recognizing Cause and Effect:** A new predator is introduced into a population of rabbits. Rabbits with longer back legs have an advantage because they can run faster. Using the concept of natural selection, explain what might happen to the rabbit population over time.

19. **Interpreting Illustrations:** Study the geologic time scale on page 138. Make a table that lists when the following groups of organisms appeared on Earth: amphibians, fish, land plants, mammals, reptiles, birds, simple animals.

20. **Hypothesizing:** Look at the fish in **Figure 5-10** on page 132. Study the mouth and head shapes and hypothesize how the shapes help the fish feed.

Performance Assessment

1. **Poster:** Make a poster that depicts the animals or plants of a particular time during Earth's history. For instance, you might make a poster about the dinosaurs of the Jurassic period, or a poster about the earliest known birds.

2. **Newspaper Article:** Write a summary of a newspaper or magazine article that discusses the discovery of a fossil species. In your summary, be sure to include information about where the fossil was found, how old it is, and its importance to science.

Chapter Preview

Skills Preview

▶ **Skill Builders**
- observe
- compare and
  contrast

▶ **MiniLABs**
- compare
- infer

▶ **Activities**
- observe
- collect data
- graph
- hypothesize
- compare
  results

146

Ecology

The day is warm and bright, perfect for a field trip to a local stream. Carefully, quietly, you push aside the cattails and kneel in the marshy grass to view an insect's progress down a water-soaked log. You lean in for a closer look when—WHAM! A sticky tongue latches onto the insect and flings it into the waiting mouth of a frog. Startled, you jump back. Your sudden movement sends the frog leaping into the water. SPLASH! You're all wet. You have just observed a system in action.

EXPLORE ACTIVITY

Observe a System

1. Choose a small area near your school to observe, such as a plot of grass or weeds. Identify the boundaries of your plot.
2. Carefully observe and record everything in your plot. Be sure to include all the parts of the plot, including air and soil.
3. Classify what you observe into two groups—things that are living and things that are not living.

Science Journal

A system is a group of things that interact with one another. In your Science Journal, describe how you think the parts of the plot you observed form a system.

6•1 What is an ecosystem?

Why IT'S IMPORTANT

Understanding interactions of an ecosystem will help you understand your role in your ecosystem.

Ecosystems

Take a walk outside and look around. What do you see? Woods? A street? A patch of weeds growing in a sidewalk crack? If you observe one of these areas closely, you may see many different organisms (OR guh nih zumz) living there. In a forest, for instance, there are birds, deer, insects, plants, mushrooms, and trees. In your backyard, you might see squirrels, birds, insects, grass, and shrubs. These organisms, along with the nonliving things in the woods or yard, such as soil, air, and light, make an ecosystem (EE koh sihs tum). An **ecosystem** is made up of organisms interacting with one another and with nonliving factors to form a working unit. **Figure 6-1** shows an example of a stream ecosystem.

FIGURE 6-1

Let's identify the living and non-living parts of this stream ecosystem.

 Rocks and water are nonliving. Pond skaters are insects that skim the surface of the water.

What does it mean to say that an organism interacts with another organism? Think back to the field trip to the stream on page 147. When the frog ate the insect, an interaction occurred between two organisms living in the same ecosystem.

What does it mean to say that an organism interacts with the nonliving parts of an ecosystem? Think about the field trip again. What did the frog do when it spotted your movement? It dove into the stream, probably for safety. The frog uses the stream for shelter. This is an example of an interaction between a living organism and a nonliving part of an ecosystem.

B Algae, fish, crayfish, and mosses covering rocks are living parts of this ecosystem. *How do these organisms interact with nonliving parts of the ecosystem?*

Activity 6-1

Ecosystem in a Bottle

You may think of ecosystems as large areas. But ecosystems can be any size. You can even make an ecosystem that fits in a plastic bottle.

What You'll Investigate
What are the parts of an ecosystem?

Procedure
1. **Rinse** out a 2-L plastic bottle with water. Using scissors, carefully **cut** off the top of the bottle.
2. **Pour** a layer of sand 5 to 10 cm deep in the bottom of the bottle.
3. **Fill** the bottle to within 5 cm of the top with water that has stood in an open container for about two days. Keep a supply of this aged water on hand to replace the water that evaporates from the bottle. The level should always be about the same.
4. **Plant** the *Elodea* and add a 2-cm layer of gravel.
5. When the water clears, **add** a guppy.
6. **Feed** the fish one or two small flakes of food every day.
7. Now you've made an ecosystem. **Observe** your ecosystem every day and **record** your observations in your Science Journal. Be sure to make observations about the living and nonliving parts of your ecosystem.

Conclude and Apply
1. **Describe** how the parts in the bottle work together to form an ecosystem.
2. What is needed to keep the ecosystem healthy?

Goals
- Make an ecosystem
- Observe an ecosystem

Materials
- 2-L bottle
- scissors
- sand
- aquarium gravel
- metric ruler
- water
- *Elodea* plants
- guppy
- fish food

The Study of Ecosystems

When you study the interactions in an ecosystem, you are studying the science of ecology (ee KAH luh jee). **Ecology** is the study of the interactions that take place among the living organisms and nonliving parts of an ecosystem. Ecologists spend a lot of time outdoors, observing their subject matter up close. Just as you knelt quietly in the cattails on your field trip, an ecologist might spend hours by a stream, watching, recording, and analyzing what goes on there. In addition, like other scientists, ecologists also conduct experiments in laboratories. For instance, they might need to analyze samples of stream water. But, most of the ecologist's work is done in the field.

The Largest Ecosystem

Ecosystems come in all sizes. Some are small, like a pile of leaves. Others are big, like a forest or the ocean. **Figure 6-2** shows the biosphere (BI oh sfeer), the largest ecosystem on Earth. The **biosphere** is the part of Earth where organisms can live. It includes the topmost layer of Earth's crust; all the oceans, rivers, and lakes; and the surrounding atmosphere. The biosphere is made up of all the ecosystems on Earth combined.

How many different ecosystems are part of the biosphere? Let's list a few. There are deserts, mountains, rivers, prairies, wetlands, forests, plains, oceans—the list can go on and on, and we haven't even gotten to smaller ecosystems yet, such as a vacant lot or a rotting tree trunk. The number of ecosystems that make up the biosphere is almost too many to count. How would you describe your ecosystem?

FIGURE 6-2

The biosphere is the part of Earth that contains all the living things on the planet. Each ecosystem that you study is part of the biosphere.

Living Parts of Ecosystems

Each of the many ecosystems in the biosphere contains many different living organisms. Think about a rotting tree trunk. It's a small ecosystem compared to a forest, but the tree trunk may be home to bacteria, bees, beetles, mosses, mushrooms, slugs, snails, snakes, wildflowers, woodpeckers, and worms. The organisms that make up the living part of an ecosystem are called **biotic factors.** An organism depends on other biotic (bi AH tihk) factors for food, shelter, protection, and reproduction. **Figure 6-3** shows some of the biotic factors in a desert ecosystem.

FIGURE 6-3

The living and nonliving components in this desert ecosystem interact in a variety of ways.

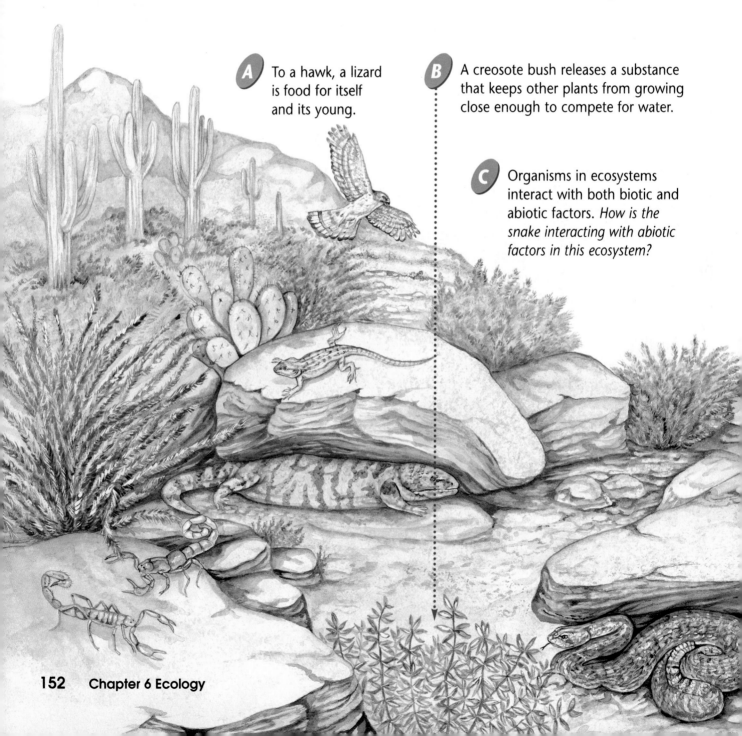

A To a hawk, a lizard is food for itself and its young.

B A creosote bush releases a substance that keeps other plants from growing close enough to compete for water.

C Organisms in ecosystems interact with both biotic and abiotic factors. *How is the snake interacting with abiotic factors in this ecosystem?*

Nonliving Parts of Ecosystems

Earlier, you listed the parts that make up an ecosystem near your school. Was your list limited to the living or biotic factors only? No. You included nonliving factors, too, such as air and soil. The nonliving things found in an ecosystem are called **abiotic factors.** Look for some abiotic (AY bi AH tihk) factors in the desert shown in **Figure 6-3.** Abiotic factors have roles in ecosystems, too. They affect the type and number of organisms living there. Let's take a closer look at some abiotic factors.

SCIENCE Online

Visit Glencoe Science Online, *science.glencoe.com*, for a link to more information about desert ecosystems.

Soil

Soil is an abiotic factor that can affect which plants and other organisms are found in an ecosystem. Soil is made up of ingredients much like a recipe. It is made up of a combination of minerals, water, air, and organic matter—the decaying parts of plants and animals. You know that salt, flour, and sugar are found in many recipes. But not all foods made from these same ingredients taste or look the same. Cakes and cookies look and taste different because different amounts of salt, flour, and sugar are used to make them. It's the same with soil. Different amounts of minerals, organic matter, water, and air make different types of soil.

E A bird seeks shelter among the spines of a cactus.

D Abiotic factors are important in all ecosystems. *What abiotic factors are important in this ecosystem?*

Mini LAB

Comparing Soils

Observe soil characteristics.

1. Fill two cups with soil. Use different amounts of the materials available to create two different "soil recipes." Pack the soil equally into each cup.
2. Pour equal amounts of water into each cup.
3. Tip the cups over to see if any water pours out.
4. Observe the characteristics of the soils you made. Record your observations in your Science Journal.

Analysis

1. What was the difference between the soil in the two cups to start with?
2. Was there a difference in how the soil in each cup held water? What could this mean for plants or other organisms living in the soil?

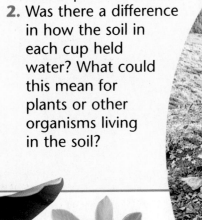

Different soils offer different materials and conditions for organisms. If you've ever visited a gardening store, you've seen all kinds of products gardeners add to their soil to make it just right for the types of plants they want to grow. The next time you dig a hole, take a close look at the soil. Is it dry? Does it have a lot of dead leaves and twigs in it? Is it tightly packed or loose and airy?

Temperature

Soil is only one of the factors that affect the organisms that live in an ecosystem. Temperature also determines which organisms live in a particular place. How do the tropical plants shown in **Figure 6-4** compare with the mountainside plants? Predict what would happen if the organisms on the mountainside were moved to a hot climate such as a tropical rain forest.

FIGURE 6-4

Plants have adaptations for their environments. The mountainside wildflowers (above) grow in clusters close to the ground, which protects them from strong winds. The tropical plants (left) have large leaves to absorb as much light as possible in the dim understory.

Water

Another important abiotic factor is water. In the field trip to the stream at the beginning of the chapter, maybe you saw a sleek trout dart through the water. Some organisms, such as fish, whales, and algae (AL jee), are adapted for life in water, not on land. But these organisms depend upon water for more than just a home. Water helps all living things carry out important life processes such as digestion and waste removal. In fact, the bodies of most organisms are made up mostly of water. Scientists estimate that two-thirds of the weight of the human body is water, as shown in **Figure 6-5.** Do you know how much you weigh? Calculate how much of your weight is made up of water.

Because water is so important to living things, it is also important to an ecosystem. The amount of water available in an ecosystem can determine how many organisms can live in a particular area. It can also serve as shelter and as a way to move from place to place.

Sunlight

The sun is the main source of energy for most organisms on Earth. Energy from the sun is used by green plants to produce food. Humans and other animals then obtain their energy by eating these plants and other organisms that have fed on the plants. When you eat food produced by a plant, you are consuming energy that started out as sunlight. You'll learn more about the transfer of energy in an ecosystem in Section 6-4.

FIGURE 6-5
Sunlight and water are two abiotic factors essential to ecosystems. Water is important to humans because 66 percent of our bodies is composed of water.

A Balanced System

Every ecosystem is made up of many different factors working together. When these factors are in balance, the system is in balance, too. Does an ecosystem ever get out of balance? Ecosystems are always changing. Many events can affect the balance of a system. A good example would be a long period of time without rain (called a drought). Predict what would happen if a drought occurred at the stream where you took your field trip. You can see a possible result in **Figure 6-6.** Some organisms, like a fish, would not survive for long periods of time without water. Some might have to find new homes. And, organisms that couldn't survive in a stream environment before might find the dried-up stream to be just the ecosystem they need.

FIGURE 6-6

Ecosystems are always changing. Some changes are small. Others, such as a stream drying up, are much larger with many more effects.

Section Wrap-up

1. What is an ecosystem?

2. How are abiotic factors important to an ecosystem?

3. **Think Critically:** If you were designing a space station on the moon, how would you use your knowledge of ecosystems to help you design it?

4. **Skill Builder**
   Observing Like all living things, people are part of an ecosystem. Describe the ecosystem you are a part of. Include both biotic and abiotic factors. If you need help, refer to Observing and Inferring on page 550 in the **Skill Handbook.**

Science Journal

Write a paragraph in your Science Journal describing the changes that you think would occur in a forest eco-system following a fire. How might the inter-actions between biotic and abiotic factors change?

Relationships Among Living Things

Organizing Ecosystems

Imagine trying to study all of the living things on Earth at once! When ecologists study living things, they usually don't start by studying the entire biosphere. Remember, the biosphere consists of all the parts of Earth where organisms can live. It's much easier to begin by studying smaller parts of the biosphere.

To separate the biosphere into smaller systems that are easier to study, ecologists find it helpful to organize living things into groups. They then study how members of a group interact with each other and their environments.

Groups of Organisms

Look at the sea horse in **Figure 6-7.** This particular sea horse lives in a coral reef in the warm, shallow waters near Australia. The sea horse uses energy, grows, reproduces, and eventually dies. The coral reef is the ecosystem the sea horse lives in. All of the sea horses that live in this particular coral reef make up a population. A **population** is a group of the same type of organisms living in the same place at the same time. Some other populations that you might find in a coral reef ecosystem are sponges, algae, sharks, and coral. What are some populations of organisms that live around your school?

FIGURE 6-7

This sea horse belongs to the population of sea horses living in a coral reef ecosystem.

What YOU'LL LEARN

- How ecologists organize living systems
- To describe relationships among living things

Science Words:
population
community
limiting factor
niche
habitat

Why IT'S IMPORTANT

Organizing living things will help you to study ecosystems and understand your role in them.

FIGURE 6-8

Many populations make up a coral reef ecosystem.

Mini LAB

Inferring

Calculate the population density of your classroom.

1. Calculate the area of your classroom in square meters by multiplying the length by the width.
2. Count the number of students in the class.
3. Divide the area of the room by the number of students to determine how much space is available for each student.

Analysis

1. What would happen to the population density if the number of students in your classroom doubled?

Groups of Populations

Many populations live in an ecosystem like the coral reef in **Figure 6-8.** All of the populations that live in an area make up a **community** (kuh MYEW nuh tee). The members of a community depend on each other for food, shelter, and other needs. For example, a shark depends on the fish populations for food. The fish populations, on the other hand, depend on coral animals to build the reef that they use to hide from the sharks.

No matter where you live, you live in and are part of a community. Make a list of as many of the populations that make up your community as you can. Compare your list with the lists of your classmates. How many populations did the class come up with?

Characteristics of Populations

Look around your classroom. Is the room big or small? How many students are in your class? Are there enough books and supplies for everyone? Ecologists ask questions like these to describe populations. They want to know the size of the population, where its members live, and how it is able to stay alive.

Population Density

Think about your classroom. A population of 25 students in a large room has plenty of space. How would the same 25 students fit into a smaller room? Ecologists determine population density (DEN suh tee) by comparing the size of a population with its area. For instance, if 100 dandelions are growing in a field that is 1 square kilometer in size, then the population density is 100 dandelions per square kilometer.

Limits to Populations

Populations cannot grow larger and larger forever. There wouldn't be enough food, water, living space, and other resources to go around. The things that limit the size of a population, such as the amount of rainfall or food, are called **limiting factors.** Think back to the stream. One biotic limiting factor in the stream ecosystem is the mosquito population. How can a mosquito population be a limiting factor? If you were a frog, you might know the reason. Frogs eat mosquitoes. If lack of rain caused the mosquito population to go down, then the frog population might not have enough food and its population size might go down. What are some other limiting factors in a stream ecosystem?

USING TECHNOLOGY

Internet Monarch

A common sign of summer is the appearance of butterflies fluttering over flower beds and fields. Some butterflies live only a short time. Others, such as monarchs, can live for years. Monarchs travel to warm climates during the winter and return to the same location year after year. This seasonal travel is called migration.

To study the monarch migration, a "monarch watcher"—often a student—carefully catches a monarch and attaches a tag to one of its wings. When someone else catches that butterfly and reads that tag, he or she can figure out how far the butterfly has flown.

Monarch watchers can use the Internet to collect and organize their data. A butterfly may be tagged in Illinois and caught several days later in South Carolina. This information is combined with similar information from other places to build a picture of the migration.

SCIENCE *Online*

Visit Glencoe Science Online, *science.glencoe.com*, for a link to information about butterfly migration. Find out how data is collected from monarch watchers and how these data are used.

Activity 6-2

Design Your Own Experiment
What's the limit?

How many blades of grass are in a park? It may seem to you like there's no limit to the number of blades of grass that can grow there. However, as you've discovered, there are many factors that organisms like the plants in the park need to live and grow. By experimenting with these factors, you can see how they limit the size of the population.

Possible Materials

- bean seeds
- small planting containers
- soil
- water
- labels
- spoons
- aluminum foil
- sunny window or other light source
- refrigerator or heater

PREPARE

What You'll Investigate
How do space, light, water, and temperature limit plant populations?

Form a Hypothesis

Think about what you already know about the needs of plants. In your group, **make a hypothesis** about how one abiotic factor may limit the number of bean plants that can grow in a single pot.

Goals

Observe how space, light, water, or temperature affect how many bean plants are able to grow in a pot.

Design an experiment that shows whether a certain abiotic factor limits a plant population using the materials listed.

Safety Precautions

Wash your hands after you handle soil and seeds.

PLAN

1. **Decide** on a way to test your group's **hypothesis**. Make a complete materials list as you plan the steps of your experiment.
2. What is the one abiotic factor you will be testing? How will you test it? What factors will you need to control? Be specific in describing how you will handle the other abiotic factors.
3. How long will you run your experiment? How many trials of your experiment will you run?
4. Decide what data you will need to collect. **Prepare a data table** in your Science Journal.
5. **Read** over your entire experiment and imagine yourself doing it. Make sure the steps are in logical order.

DO

1. Make sure your teacher has approved your plan and your data table before you proceed.
2. Carry out your plan.
3. While doing the experiment, **record** your observations and **complete** your data table in your Science Journal.

CONCLUDE AND APPLY

1. **Compare** your results with those of other groups. How did different factors affect the plants?
2. **Graph** your results using a bar graph. **Compare** the number of bean seeds that grew in the experimental containers with the number of bean seeds that grew in the control containers.
3. How did the abiotic factor you tested affect the bean plant population?
4. **APPLY Predict** what would happen to your plant population if you added another kind of plant or animal to the containers.

Interactions in Communities

Are frogs the only organisms in the stream community that eat mosquitoes? No, there are many animals that eat them, including some birds and spiders. That means that frogs must compete with birds and spiders for the same food. Feeding interactions such as those in **Figure 6-9** are the most common interactions among organisms in a community.

Imagine a large bowl of popcorn in your classroom. As long as there is enough popcorn to go around, you don't have to worry that you'll get your share. But if the bowl of popcorn were small, you would have to compete with your classmates to get some. The greater the population size of an area, the greater the competition for resources such as food. Food isn't the only resource that organisms compete for. Organisms will compete for any resource that is in limited supply. Space, water, sunlight, and shelter are all resources that may be limited in a particular ecosystem.

Eat or Be Eaten

Have you ever heard the phrase "birds of prey"? A falcon is a bird of prey, which means it captures and eats other animals. A falcon, with its razor-sharp talons, will swoop down from the sky to snatch up a field mouse. The falcon is a predator (PRED uh tur). Predation (pruh DAY shun) is the act of one organism feeding on another.

Organisms That Live Together

Predation doesn't sound like a good deal for the field mouse, does it? The falcon population, however, is limited by the size of the mouse population. There are other types of relationships among organisms. In one type of interaction, both organisms in the relationship benefit. The African tickbird, for instance, gets its food by eating insects off the skin of zebras. The tickbird gets food, while the zebra gets rid of harmful insects. In another type of relationship, only one organism benefits. The other is not harmed. A bird building a nest in a tree is an

FIGURE 6-9

One of the most common ways organisms interact in a community is by being food for another organism.

example of this. The bird gets protection from the tree, but the tree isn't harmed. In still another relationship, one organism is helped while the other is harmed. The insects on the zebra's skin, for example, benefit from the zebra. However, these insects can cause harm to the zebra. Have you ever been bitten by a mosquito? That's a firsthand experience of this type of relationship.

Where and How Organisms Live

How can a small ecosystem such as a classroom aquarium support a variety of different organisms? It's possible because each type of organism has a different role to play in the ecosystem. A typical classroom aquarium may contain snails, fish, algae, and bacteria. The role of snails is to feed on algae, which keeps the glass of the aquarium clear for light to get in. The role of the algae is to provide oxygen for the system through photosynthesis. What do you think the role of the fish might be? The role of an organism in an ecosystem is called the organism's **niche** (NIHCH). The niche of the fish might be to feed on algae to keep

Problem Solving

Graphing Populations

One way to understand more about relationships among organisms in an ecosystem is to keep track of, or monitor, and graph populations. Use the information below to make a graph of population size over time for barn owls and field mice. Then, answer the questions that follow.

Table 6-1

Monthly Population Size per Hectare (in 100s)									
Month	J	F	M	A	M	J	J	A	S
Field Mice	6	5	4	3	3	4	5	4	6
Barn Owls	2	3	4	4	2	1	4	3	4

Solve the Problem:

Set up your graph with months on the *x*-axis and numbers of organisms on the *y*-axis. Use two colors to plot your data. For more help, refer to Graphing in the Skill Handbook. Use your graph to infer how the population of field mice affects the population of barn owls. Predict how the next two months of the graph will look.

Think Critically:

Field mice eat green plants and grains. What do you think would happen to the population of barn owls if there were no rain in the area for a long time?

its population in check. All the conditions an organism needs to survive are part of its niche. How would you define your niche in your ecosystem?

The place where an organism lives out its life is called its **habitat** (HAB uh tat). Different species share a habitat. Describe the habitat of the organisms in **Figure 6-10.** Resources such as food, space, and shelter are all shared among the organisms in the habitat. This is possible because each organism has its own niche. The different ways each organism feeds and uses other resources allow all the organisms to live together.

FIGURE 6-10

Each organism in this ecosystem has its own job or niche in the ecosystem.

Section Wrap-up

1. Identify three reasons a population can't continue to grow larger and larger forever.

2. List similarities and differences in a community and a population.

3. **Think Critically:** Identify several interactions between organisms in an ecosystem. Give an example of each interaction.

4. *Skill Builder*
   Developing a Multimedia Presentation
   Choose a community such as a coral reef, grassland, wetland, tundra, or forest. Develop a multimedia presentation that describes the community and some interactions that occur within it. If you need help, refer to Developing a Multimedia Presentation on page 564 in the **Technology Skill Handbook.**

USING MATH

Find data about the population size of an organism such as bison, mountain lions, whales, or an endangered species over a period of ten years. Encyclopedias or the Internet can provide this information. Make a graph that shows how the size of the population varied over those years.

Laura Collins, Bird Curator

Q Ms. Collins, tell us about your job.

A I'm the curator of birds at Sea World Ohio. We have about 400 birds on exhibit here, representing roughly 90 species.

Q What goes into building an exhibit that re-creates a bird's natural habitat?

A It's a team effort, involving many people and a lot of research. We determine what the birds' needs are and then design an exhibit that's appropriate for them. For example, a flamingo needs a shallow pond so it can wade and filter feed, while a diving duck needs a deeper pond. Our Penguin Encounter exhibit has probably been one of our most challenging projects as far as designing a habitat for a bird.

Q Tell us about that exhibit. How do you make the penguins feel at home?

A We have Macaroni, Adélie, and emperor penguins. In the exhibit, there are mountains, rocks for the penguins that like to climb, a 45 000-gallon saltwater pool, and snow on the ground. Two snow machines on top of the exhibit make 5000 pounds of snow a day. All the water in the pool is filtered every 25 minutes, and it's chilled to around 40°F (4°C). The air is filtered and chilled, too, down to about 27°F (−3°C). Even on a 90° day, it's still below freezing in there!

Q That sounds like a huge project!

A The engineering and life-support systems, and the number of people that it takes to monitor those systems and keep them running, are phenomenal. But the real success story is that the penguins breed in the exhibit. They build nests, lay eggs, and raise chicks—it's a self-sustaining population.

Career Connection

Many careers involve working with animals. Interview a person in your community who has a career working with animals. Some examples include a keeper in a zoo, a veterinarian, a dog groomer, or an animal breeder. Present a summary of your interview to your class.

6•3 Gators at the Gate!

What YOU'LL LEARN

- How to recognize the problems of endangered and threatened species
- How to distinguish fact from opinion

Science Words:
endangered species
threatened species

Why IT'S IMPORTANT

The needs of wildlife and the needs of humans are in conflict all over the world.

FIGURE 6-11

Alligators are a common sight along the banks of rivers and canals in Florida neighborhoods.

When you think about Florida, you probably picture sandy beaches and palm trees. But do you think about alligators? Alligators are among the best-known animals that live in Florida. They can grow to be 13 feet long and weigh more than 600 pounds.

Endangered Alligators

By the 1960s, numbers of alligators were greatly reduced in Florida due to hunting and habitat loss. The numbers became so low that alligators were placed on the endangered species list. A species is considered an **endangered species** when so few of its members are living that the entire species may become extinct. In the United States, it became illegal to hunt alligators. The number of alligators went up. By 1977, they were renamed as a threatened species. A **threatened species** still needs to be protected but is not in immediate danger of becoming extinct. Now, more than a million alligators live on farms and in the wild. Good news, right? Think again. There are problems—big problems from big alligators.

As you read the following quotes from newspaper articles about alligators, try to distinguish facts from opinions. Facts can be proven. Opinions cannot; they reflect feelings and beliefs.

Alligator Problems

More and more people live, work, and play in areas where alligators live. Some people, such as Sandy Bonilla, think the alligators have to go. "I found an alligator in my swimming pool," Bonilla says. "My neighbor saw one on a golf course. I believe this whole situation is out of control. With people and gators living in the

same areas, attacks become more likely. These animals are the greatest danger we face."

But Juana Aravjo disagrees. "It seems to me that the problem is not the alligators," she says. "The problem is that there's no room for the alligators." Aravjo explains that before, when an alligator was caught in someone's swimming pool, a professional trapper released it back into a wilderness area. "Today, there are fewer wilderness areas, so when an alligator is considered a nuisance to people, it is removed by a trapper and killed," Aravjo says. "If we don't act quickly, alligators will probably become extinct."

FIGURE 6-12

Alligators eat frogs, fish, snakes, birds, and small mammals. But they've been known to eat sticks, rocks, and even aluminum cans.

 # Skill Builder: Distinguishing Fact from Opinion

LEARNING the SKILL

1. To tell the difference between facts and opinions, study the statements to see if they can be checked for accuracy and if they can be proven. Facts can be proven; opinions cannot.

2. Facts answer specific questions such as: What happened? Who did it? When and where did it happen? Why did it happen? They give specific information about an event.

3. Opinions tell someone's feelings or what they believe. They often begin with phrases such as "I think," " I believe," or "it seems to me." They often include words such as *probably, greatest, best,* or *worst.*

PRACTICING the SKILL

1. Compare and contrast an endangered species with a threatened species.

2. Examine the quotes from the newspaper. List one fact and one opinion from each speaker. Explain why the statements are facts or opinions.

3. Based on what you have read, should alligator habitats be protected? If so, how?

APPLYING the SKILL

Other species besides alligators are threatened. Bring a newspaper or magazine article to class about a threatened or endangered species. Distinguish fact from opinion in the article.

6•4 Energy Through the Ecosystem

What YOU'LL LEARN

- How organisms get the energy they need
- How energy flows through an ecosystem

Science Words:
producer
consumer
decomposer

Why IT'S IMPORTANT

Knowing about food webs and material cycles will help you understand how living things depend on each other.

It's All About Food

Think about the interactions that we've talked about so far. The frog and the mosquito, the falcon and the mouse—most of the interactions involve food. Energy moves through an ecosystem in the form of food.

Many different populations interact in a backyard ecosystem, including plants, birds, insects, squirrels, and rabbits, as shown in **Figure 6-13.** The plants in the ecosystem produce food through photosynthesis. An organism that makes its own food, like a plant, is called a **producer.** The grasshopper that nibbles on the plants is a **consumer.** A consumer eats other organisms.

Some of the consumers in an ecosystem are so small that you might not notice them. But they have an important role to play. They are the decomposers, such as bacteria and fungi. **Decomposers** are organisms that use dead organisms and the waste material of other organisms for food.

A The plants in this backyard are producers. They use energy from the sun to make food.

B A grasshopper may eat the plant.

E Decomposers such as bacteria and fungi feed on dead organisms, breaking them down.

Modeling the Flow of Energy

The food chain in **Figure 6-9** on page 162 is a simple model that shows how energy from food passes from one organism to another. Each organism is linked by an arrow. The arrows show that energy moves from one organism to another in the form of food.

Can you spot any problems with the model? How do the other organisms in the community fit into the food chain? We need a more complex model to show the true feeding interactions in the backyard ecosystem.

In an ecosystem, food chains often overlap. For instance, a bird may eat seeds and in turn be eaten by a cat. But, a cat could also eat the rabbit or the mouse. One food chain cannot model all these overlapping relationships. Scientists use a more complicated model, called a food web, to show the transfer of energy from food in an ecosystem. A food web is a series of overlapping food chains that shows all the possible feeding relationships in an ecosystem. Draw a food web for this backyard ecosystem.

FIGURE 6-13
In any community, energy flows from producers to consumers.

C) A bird may eat the grasshopper.

D) A cat may prey on a bird, rabbit, or chipmunk.

FIGURE 6-14

Horses get the materials they need to grow and maintain their bodies by eating food such as grass. *Where do you think the grass gets the materials it needs?*

Cycling of Materials

What happens when you recycle a soda can? The can is taken to a place that melts it so that the aluminum can be used again. This is an example of a simple cycle. The same aluminum can be used over and over again. Cycles are important to ecosystems. Instead of aluminum cans, however, it's the materials that make up organisms that get recycled.

The bodies of living things are made up of matter. They are composed of materials like water, and chemicals like nitrogen and carbon. To get the matter needed to build bones, muscles, and skin, you need to eat food made of the right kinds of materials, as the horse in **Figure 6-14** is doing. In an ecosystem, matter cycles through food chains. The amount of matter on Earth never changes. So matter in ecosystems is recycled, or used again and again.

Living things depend on these cycles for survival. Living things also depend on one another for food, shelter, and other needs. All the different things that make up the biosphere—from a tiny insect to a raging river—have a unique role to play.

Section Wrap-up

1. Name some organisms that are consumers. Give an example of the type of food each eats.

2. Why is less energy available in the fourth link of a food chain than at the first?

3. **Think Critically:** Imagine that you just ate a chicken sandwich. Starting with the sun, trace the flow of energy in the food chain from the beginning to you.

4. *Skill Builder*
   Comparing and Contrasting Compare and contrast a food chain and a food web. If you need help, refer to Comparing and Contrasting on page 550 in the **Skill Handbook.**

Using Computers
Model Using graphics software, make a diagram that shows how energy flows through your local ecosystem in a food web.

Read the statements below that review the most important ideas in the chapter. Using what you have learned, answer each question in your Science Journal.

1. An ecosystem is made up of organisms interacting with one another and with the nonliving things in the system. *Describe the abiotic factors and the biotic factors in a pond ecosystem.*

2. A food chain shows how food energy flows through an ecosystem. *Draw a food chain that leads to a hamburger on your plate.*

3. An organism's survival is related to how well it is adapted to the environment. Limiting factors, which may be living or nonliving, influence the survival of an organism or a species. *What adaptations might a horse need for its environment?*

4. Energy is transferred through an ecosystem in the form of food. The feeding relationships can be illustrated by a food web. *Describe what might happen in an ecosystem if several producers were removed.*

Using Key Science Words

abiotic factor   endangered species
biosphere   habitat
biotic factor   limiting factor
community   niche
consumer   population
decomposer   producer
ecology   threatened species
ecosystem

Match each phrase with the correct term from the list of Key Science Words.

1. the study of all the interactions between living and nonliving parts of an ecosystem
2. the part of Earth where an organism can live
3. any living thing in the environment
4. the place where an organism lives
5. individuals of the same species living in the same place at the same time

Checking Concepts

Choose the word or phrase that completes the sentence.

6. All are biotic factors except
_____.
   a. raccoons   c. pine trees
   b. sunlight   d. mushrooms

7. Ponds, streams, and prairies are examples of _____.
   a. niches   c. populations
   b. producers   d. ecosystems

8. A group of the same type of organism living in the same place at the same time is a _____.
   a. habitat
   b. population
   c. community
   d. ecosystem

9. A food chain is a model that shows the transfer of _____.
   a. producers   c. plants
   b. consumers   d. energy

10. An example of a producer is a
_____.
   a. grass   c. fungus
   b. horse   d. fish

Thinking Critically

Answer the following questions in your Science Journal using complete sentences.

11. Why is it correct to say that decomposers are also consumers?

12. Give examples of foods you would eat if you were eating low on the food chain.

13. Draw an ecosystem. Label biotic and abiotic factors. Describe three interactions among organisms in the ecosystem.

14. Identify three possible limiting factors for an aquarium ecosystem. Describe how each factor can limit population growth.

15. Describe your own habitat and niche.

Developing Skills

If you need help, refer to the description of each skill in the Skill Handbook.

16. **Classifying:** Make a list of ten of your favorite foods. Classify each as coming from a producer, consumer, or decomposer. Write a short explanation of your classification of each item.

17. **Observing and Inferring:** Observe the photo of the aquarium on page 164. Identify at least one food chain you might find in the aquarium ecosystem.

18. **Making and Using Graphs:** The following graph shows the changes in the size of a population of insects living on roses over the course of a year. During what month is the insect population the smallest? During what month is the population the largest?

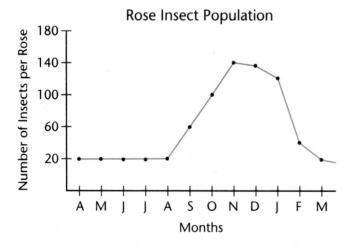

Rose Insect Population

19. **Designing an Experiment:** Design an experiment to test the hypothesis that water is a limiting factor for plants. Include a list of procedures.

20. **Predicting:** Predict what would happen to an ecosystem if its decomposers were removed.

Performance Assessment

1. **Slide Show or Photo Essay:** Find slides or photographs that show a variety of different ecosystems. Arrange a slide presentation or photo display of these images. Use titles or captions to identify each photograph or slide.

2. **Poster:** Choose an ecosystem to research. Find out what plant and animal species are found there and how they interact. Make a poster illustrating a food web in this ecosystem.

Chapter Preview

Skills Preview

▶ **Skill Builders**
- sequence
- develop multimedia presentations
- recognize cause and effect

▶ **MiniLABs**
- analyze
- model

▶ **Activities**
- make and use a table
- calculate
- diagram
- think critically

Resources

Imagine sailing across the ocean on the HMS *Rose*—the largest active wooden ship in the world. Do you see anything about it that looks unusual? Its sails are made of materials that were once plastic soda bottles and car fenders. These plastic materials were made into new things, instead of being thrown away. Using old things to make new things is one way to help the environment.

EXPLORE ACTIVITY

Classify Resources

1. In your Science Journal, make a table with five columns: *Plastic, Paper, Metal, Glass,* and *Wood.*
2. Think of things you use every day at home or in your classroom that are made of each material. List as many as you can think of in each column.

Science Journal

In your Science Journal, record how many items are listed for each category. Which column had the longest list? Where do you think these materials come from? Which category do you think you depend on most? Why?

7•1 Natural Resource Use

What YOU'LL LEARN

- How resources are used
- How resources are classified

Science Words:
natural resource
renewable resource
nonrenewable resource

Why IT'S IMPORTANT

If you understand where resources come from and how they are used, you can make better decisions about the things you buy or use.

News Flash: Trouble in the Rain Forest

For months, you've been saving your money, and today you're going shopping for a CD player. On the way to the mall, you turn on the car radio. The news reporter is saying that rain forests are being cut down at a rate of about 117 000 square kilometers per year. Look at **Figure 7-1.** That's a lot of land!

The news report goes on to explain that the rain forest problem is not a simple one. People cut down trees in the rain forests to graze animals or grow crops on the cleared land. People also use rain forest trees for fuel or to make things such as furniture or paper. Large companies from foreign countries may pay local peoples to cut down the trees to make wood products. Many of the local families need this money to buy food and supplies for their families. The news report says that everyone needs to act now to save the rain forests because these activities have a negative effect on the soil, the climate, and the organisms that normally live there. The report goes on to say that if enough people show concern and start to take positive actions, change will occur.

FIGURE 7-1

On average, an area of rain forest about the size of the state of Pennsylvania is cut down every year. Pennsylvania has an area of 116 083 square kilometers.

Things You Use Affect the Environment

You think about this news report as you reach the shopping mall. When you walk into the different stores, you can't help but notice that many products and the packages they come in are made of cardboard, a wood product. Even though this wood may not have come from a rain forest, it did come from a forest somewhere. Could these products have been packaged in a different way?

Let's take a look at the CD player you want to buy, shown in **Figure 7-2.** It is made of plastic and often comes in a package made of cardboard. Its wires, screws, and some inside parts are made of metal. Metal and plastic aren't made from trees. So where did they come from?

FIGURE 7-2

What is a CD player made of?

Natural Resources

Most of the things that you buy or use are made of materials that come from natural resources. **Natural resources** are things found in nature that living things use. **Figure 7-3** shows some other examples of natural resources. Living things use natural resources to meet their needs. Vegetables that you eat are natural resources. They fill your need for nutrients. The trees and the minerals that were used to make the lumber, plastic, and metal in your house are natural resources. They fill your need for a place to live. Natural resources are also used to make other things in our lives, such as CD players.

FIGURE 7-3

Cotton, gold, trees, and water are examples of natural resources.

What goes into making a CD player?

You already know that the cardboard box the CD player came in was made from trees, and that trees are a natural resource. What about the plastic used to make the CD player? Where did it come from? Plastic is made from oil, a resource that's found underground. Oil is a thick, dark liquid. Deep holes have to be drilled in the ground to reach oil. Products made from this oil are then taken to a factory to be made into plastic and other products, such as gasoline or plastic soda bottles. What about the metal? Where did the metal come from to make the screws that hold the CD player together? **Figure 7-4** answers this question.

FIGURE 7-4

The screws used in your CD player are usually made of steel. Iron ore is the natural resource used to make this steel.

B Railroad cars carry iron ore to factories where the ore is melted down and the impurities are removed.

C The iron is then trucked to a factory where various iron and steel products are made, including the screws used in CD players.

A Iron ore is a mineral that contains iron. It is found below Earth's surface and must be mined.

More Natural Resources

Trees, oil, minerals—were any other natural resources used to make the CD player? Cutting down trees, drilling for oil, mining, and getting natural resources to factories all require energy. Once the natural resources are at the factories, it takes energy to make them into plastic, cardboard packaging, and metal wires and screws. Where does all this energy come from?

If you guessed natural resources, you're right. Trucks that take the natural resources to the factories use gasoline, and gasoline is made from oil. In many parts of the country, the electricity used to power machines that make natural resources into materials for CD player parts may come from burning coal. Like oil, coal is a natural resource that forms underground. It takes energy to mine the coal and this energy comes from natural resources.

D Different parts used in making electronic equipment are transported to a factory where the parts are assembled into products.

E The final product—the CD player—is packaged and then trucked to a store, where it is sold.

Mini LAB

Analyzing Gift Wrap

Can you think of a better way to wrap a gift?

1. Your teacher will give you an object to wrap. Discuss different methods and unusual materials for wrapping this object with your group.
2. Think about how different ways of wrapping could make the best use of the materials you have. Does the wrap waste material? Could it be used again? Is it easy to dispose of?
3. Wrap the object. In your Science Journal, write down all the resources that would have been used to wrap it in gift wrap.

Analysis

1. What problems did you have with your method of wrapping?
2. Why do you think your choice of wrapping material is a good one?

You're beginning to see that it takes a variety of natural resources to make one CD player, doesn't it? Think of all the natural resources that are used to make something large like a house or a school building. Now think of all the houses and apartments in the world. Are there enough natural resources to meet everybody's need for a place to live?

Maybe. But people use natural resources to meet other needs, too. In fact, all living things on Earth use natural resources. Plants and animals use natural resources for food and shelter. Will we ever run out of natural resources? That may depend on the particular resource.

Availability of Resources

Imagine that you're riding your bike on a warm spring day. Your destination: the city park, such as the one shown in **Figure 7-5.** When you get to the park, you head straight for the pond. You hop off your bike, take an apple out of your backpack, then lean against a tree. Later, you might take a hike around the pond. But for now, you're content just to watch the sunlight sparkling on the water.

Sunlight, water, trees, apples . . . These are all natural resources. They have something else in common, too. They will likely be around for a long time. Why? Because they are all renewable (ree NEW uh bul).

FIGURE 7-5

Natural resources are found everywhere—in national parks and in the middle of large cities like Boston.

Renewable Resources

Resources that can be replaced by natural processes in 100 years or less are called **renewable resources.** Look at **Figure 7-6.** Energy from the sun is a renewable resource because the sun gives off light energy every day, and it will continue to do so every day for millions of years. Trees are renewable resources because most trees will grow back and be cut again in less than 100 years. Water is another renewable resource because we use the same water over and over again.

FIGURE 7-6
Sunlight and trees are both renewable resources.

USING TECHNOLOGY

From Plastic Bottles to Play Towers

What happens to all of those clear, 2-L plastic soft-drink bottles people put into recycling bins? What is recycling, and what objects can be made from recycled plastic?

Recyclable plastic objects are first taken to a recycling plant where the different types of plastic are sorted. The objects are chopped into small pieces and washed. After the plastic is dried, it is melted and pushed through a screen. This step forms plastic pellets that will be used to make new products.

Recycled Products

Many recycled plastic pellets are re-formed into plastic packaging for different products. Pellets are also used to make stuffing for toys and beanbag chairs. Trash cans, pet food bowls, flowerpots, park benches, and plastic lumber are other products made from recycled plastic pellets. The play tower pictured here is made of recycled materials. It is made from 15 765 recycled plastic containers, 20 232 recycled aluminum cans, and 6180 recycled soup cans.

SCIENCE *Online*

Check out the link at Glencoe Science Online at *science.glencoe.com* to find out about other products that are made from recycled materials.

40% Oil
22% Coal
23% Natural Gas
7% Nuclear
8% Other

FIGURE 7-7

As you can see from this circle graph, 40 percent of the world's energy needs are met by oil and 22 percent are met by coal. Scientists estimate that we have enough coal to last another 200 years. But if we continue using oil at present rates, some scientists estimate that we will run out of this natural resource in 30 or 40 years. *What energy sources do you use in your home?*

Nonrenewable Resources

Do you see coal or oil among Earth's energy sources in **Figure 7-7?** Coal, natural gas, and oil take millions of years to form inside Earth. They are examples of nonrenewable (NAHN ree new uh bul) resources. **Nonrenewable resources** are resources that cannot be replaced by natural processes within 100 years. If all the coal and oil that we can recover is used up during our lifetimes, there won't be any more available for use for millions of years. So, because nonrenewable resources form slowly over long periods of time, they need to be used carefully. Also, scientists are finding other energy sources. **Figure 7-7** shows how the world's energy needs are being met. Sections 7-2 and 7-3 will discuss ways to reduce the amount of nonrenewable resources used.

Using Water

Water is an important resource that we use every day. You wash dishes and clothes and you wash yourself and brush your teeth. You cook meals. All of these activities need water.

What You'll Investigate
Calculate how much water your family uses in three days.

Procedure
1. Use the table below to **calculate** how much water your family uses.
2. For three days, have the people who live in your house **keep a record** of when they do the activities listed in the table.

Conclude and Apply
1. The numbers in the table describe approximately how many liters one person uses in a single day for the activity listed. **Multiply** these numbers by the number of people who did these activities.
2. **Add** up the totals for each day and the final sum will be the total amount of water used for these activities in three days.
3. **List** ways in which your family could control the amount of water used.

Goals
- Calculate how much water the people in your household use in three days.
- Make a plan to control the amount of water used.

Materials
- calculator

Activity	Conditions	Amount of Water Used
Washing dishes by hand	Water is running all the time	113 L/person/day
Washing dishes by hand	Sink is filled with water	19 L/person/day
Washing clothes in machine	Small load with high water setting	68 L/person/day
Washing clothes in machine	Full load with high water setting	45 L/person/day
Taking a shower	10 minutes long	150 L/person/day
Taking a bath	Bathtub is full of water	113 L/person/day
Flushing the toilet	Water-saving toilet	23 L/person/day
Brushing teeth	Water is running all the time	17 L/person/day

Natural resources make up the environment that we live in. These resources are classified into nonrenewable and renewable resources. Examples of renewable and nonrenewable resources are listed in **Table 7-1.** In Section 7-2, you will learn how people affect these natural resources.

Table 7-1

Natural Resources	
Renewable Resources	**Nonrenewable Resources**
Plants	Coal, oil, and gas
Sunlight	Some ores and metals
Water	Land

Section Wrap-up

1. What is a natural resource? Give some examples of how a squirrel uses natural resources.

2. List three examples of nonrenewable resources and three examples of renewable resources. Explain why your choices fit into each category.

3. Think Critically: Is land a renewable or a nonrenewable resource? Explain your answer.

4. **Skill Builder**
   Sequencing Make an events chain showing, in sequence, the steps that are taken to make a tree into a baseball bat displayed in a store. If you need help, refer to Sequencing on page 543 in the **Skill Handbook.**

USING MATH

A regular showerhead puts out 15 L of water per minute. A water-saving showerhead puts out only 9.5 L of water per minute. If you take a five-minute shower every day, how much water would you save in one week by using a water-saving showerhead?

People and the Environment

Exploring Environmental Problems

Look at **Figure 7-8.** Have you ever seen a construction site for a new highway? Sometimes, hillsides have to be dynamited to make room for the highway. The trees and plants that grew on the hillside are destroyed. The animals that lived on the hillside depended on the trees and plants for food and shelter. Some might die if their food source is destroyed, but most will survive and find new habitats to live in. Construction companies now have to restore land that they dynamite so that plants and animals will be able to continue to live there.

What if there isn't another place to live? Many plants and animals lose their natural habitats because people use land for growing crops, grazing animals, or building homes. That's what's happening in the rain forest. Because large areas of rain forest are being destroyed at a rapid rate, certain species in the rain forest have no habitat to fill their needs for food, shelter, and other things. They are threatened with extinction.

Frequently, human activities affect some of our most precious natural resources: land, water, and air. Let's see how this happens.

What YOU'LL LEARN

- How people affect the environment
- The different types of pollution

  Science Words:
  landfill
  pollutant
  acid rain

Why IT'S IMPORTANT

Knowing how your actions affect the environment will help you to make choices that could help reduce environmental problems.

FIGURE 7-8

Construction destroys some parts of the environment. Laws are now in place to reduce the amount of destruction that takes place.

Our Impact on Land

How much space do you need? You will need to think about more than just your home. Think about where your food comes from, your school, and other space you use. If you start adding it all up, the amount of space you use is much larger than you may think. A simple peanut butter-and-jelly sandwich requires land to grow the wheat needed to make bread, land to grow the peanuts for the peanut butter, and land to grow the sugarcane and fruit for the jelly. A hamburger? Land is needed to raise cattle and to grow the grain that the cattle eat.

Using Land Wisely

All of the things we use in our everyday lives take some amount of land, or space, to produce. That means that every time we build a house, a mall, or a factory in a city as illustrated in **Figure 7-9,** we use a little more land. All you have to do is look at a globe, however, to see that the amount of land available for us to use is limited.

People need food, clothing, jobs, and a place to live, and each of these things takes space. But preserving natural habitats is also important. Remember, a habitat is the place where an organism lives. Once a wetland is filled in to build an apartment building, the wetland is lost.

FIGURE 7-9

Land is used for many different things other than growing food. *What takes up space in the city of Pittsburgh pictured below?*

More and more, there are laws to help protect against habitat loss and to help us use land wisely. Before major construction can take place in a new area, the land must be studied to determine what impact the construction will have on the natural habitat, the living things, the soil, and water in the area. If there are endangered organisms living there, or if the impact will be too great, construction may not be allowed. These are important studies. At stake are jobs, homes, and habitats.

Landfills

Each day, every person in the United States produces about 2.1 kg of garbage. Where does it go? About 57 percent of our garbage goes to landfills. A **landfill,** shown in **Figure 7-10**, is an area where garbage is deposited.

Most of the things we throw into a landfill are not dangerous to the environment. However, sometimes items such as batteries, paints, and household cleaners end up in landfills. These things contain potentially harmful chemicals that can leak into the soil, and eventually into rivers and oceans. Any material that can harm living things by interfering with life processes is called a **pollutant** (puh LEW tunt). Modern landfills are lined with plastic or clay to keep these chemical pollutants from leaking. However, some chemicals still find their way into the environment. If these pollutants get into the food that we eat or the water that we drink, they can cause health problems.

FIGURE 7-10

Each day, trash is put in a sanitary landfill. This trash is later covered with a thin layer of dirt and then watered down to keep the trash from blowing away.

Design Your Own Experiment
Using Land

Imagine planning a small town. Your job in this activity is to draw up a master plan to decide how 100 square units of land can be turned into a town.

PREPARE

What You'll Investigate
How should land resources be used?

Think Critically

People need homes in which to live, places to work, and stores from which to buy things. Children need to attend schools and have parks in which to play. How can all of these needs be met when planning a small town?

Goals

Design a plan in which 100 square units of land can be turned into a town.

PLAN

1. A 100-square-unit piece of land can be represented as a square divided into 100 blocks. One way to represent this is to make a square graph 10 blocks across and 10 blocks down.
2. The table on page 189 shows the different parts of a town that need to be included in your plan. The office buildings and industrial plant are places where the people of the town will work. They are each 6 blocks in size. These blocks cannot be divided but must be treated as one group. The landfill is 4 blocks in size and cannot be broken up.
3. All the other town parts can be broken up as needed. Stores and businesses are areas in which shops are located as well as medical offices, restaurants, churches, and cemeteries.

Possible Materials
- grid paper (10 squares × 10 squares)
- colored pencils

4. As a group, **discuss** how the different parts of the town might be put together. Should the park be in the center of town or near the edge of the town? Should the school be near the offices or near the houses? Where should the landfill go?

5. How will you show the different town parts on your grid paper?

Parts of Your Town	Number of Blocks Needed
Office buildings	6 blocks in one group
Industrial plant	6 blocks in one group
School	1 block
Landfill for garbage	4 blocks in one group
Houses and apartments	44 blocks—can be broken up
Stores and businesses	19 blocks—can be broken up
Park	20 blocks—can be broken up

DO

1. As a group, **plan** your town. Check over your plan to make sure that all of the town parts are accounted for.

CONCLUDE AND APPLY

1. Where did you place the office buildings and the industrial plant? Why were they placed there? Where did you place the houses, school, and the stores and businesses? Explain why you placed each one as you did.

2. Did you make one park or many parks with the land designated for park use? What are the advantages of your park(s) plan?

3. Where did you place your landfill? Will any of the townspeople be upset by its location? What direction does the wind usually blow from in your town?

4. **APPLY** Where would you put an airport in this town? Keep in mind safety issues, noise levels, and transportation needs.

189

Our Impact on Water

Did you know that you cannot live long without water? You need clean water for drinking, as well as dozens of other things, as you found out in Activity 7-1. The average person in the United States uses about 397 L of water each day. Though water is a renewable resource, in some places it is being used up faster than natural processes can replace it.

Only a small amount of Earth's water, as shown in **Figure 7-11**, is freshwater that people can drink or use for other needs. Many places around the world are running out of usable water. How do you think your life would change if your area were running out of water?

Water Pollution

People have always used water, but everyday activities can pollute water. How? When you scrub a floor with a mixture of water and a household cleaner, what do you do with the mixture afterwards? You pour it down the drain. The polluted water usually goes to a water-treatment plant, where it is cleaned before being used again.

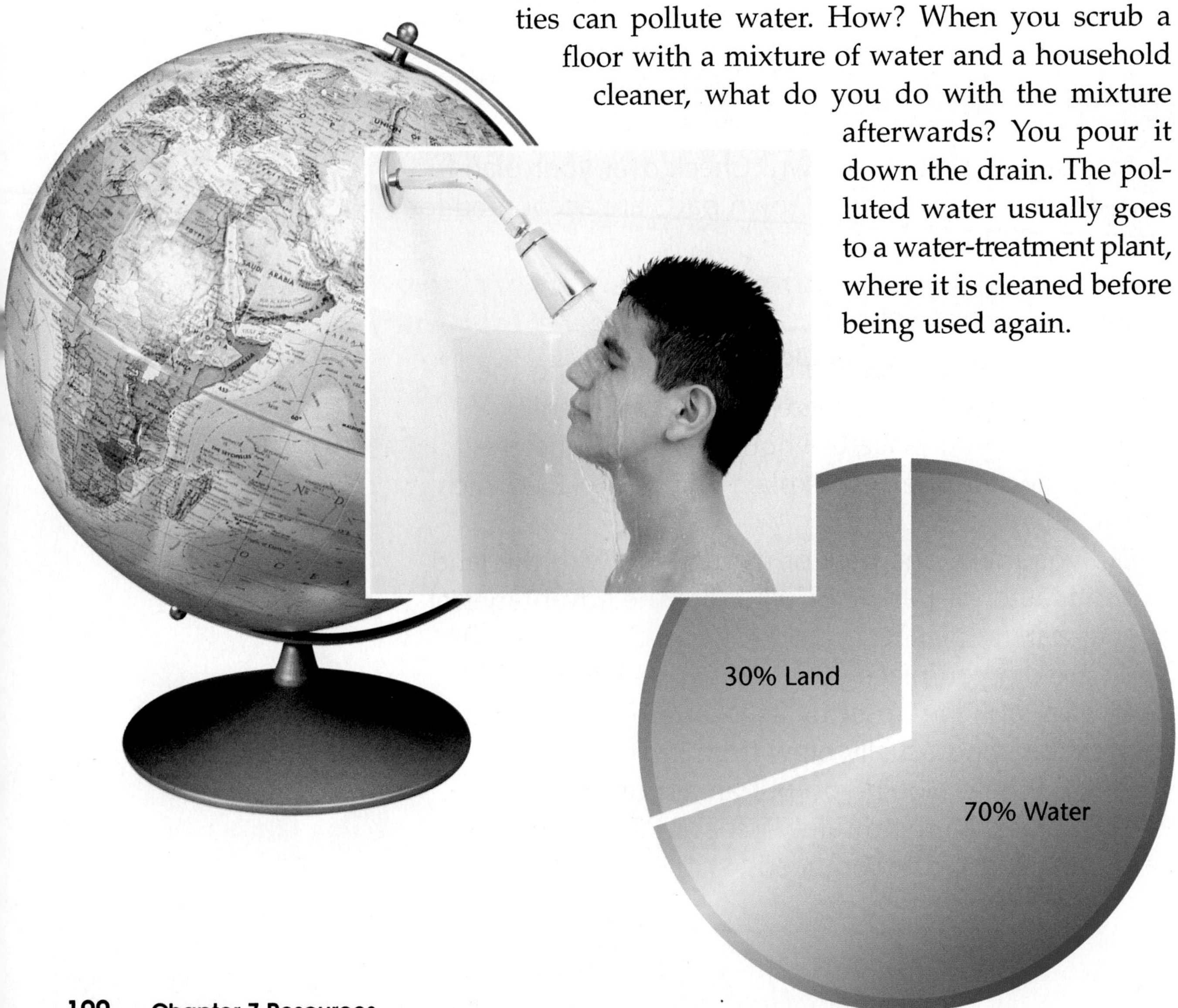

FIGURE 7-11

Although 70 percent of Earth's surface is covered by water, less than one percent is freshwater that can be used for drinking.

30% Land

70% Water

What if you poured the mixture of water and household cleaners outside on the grass or in the street? This polluted water would soak into the ground where large amounts of water are located, or it might be washed away by rain and carried into rivers and lakes. If too many people do this, it could pollute our drinking water. **Figure 7-12** shows where U.S. drinking water comes from.

Cleaning Up the Water

Countries are working together to reduce the amount of water pollution. For example, the United States and Canada have made agreements to clean up the pollution in Lake Erie, a lake that borders both countries. The U.S. government has also passed several laws to keep water supplies clean. The Safe Drinking Water Act is a set of government standards that makes sure that our drinking water is safe. The Clean Water Act gives money to the states for building water-treatment plants. Wastewater is cleaned at such plants.

Remember, Earth has a lot of water, but only a small amount of Earth's water is freshwater that people can drink or use. The best way people can protect Earth's water is by being aware of how they use it.

FIGURE 7-12

Much of our drinking water comes from rivers, lakes, and underground sources of water. This water is treated to remove impurities before it is used by people in towns and cities.

Our Impact on Air

If you live in a city, you may have noticed that on some days, the air looks hazy. Pollutants such as dust and gases in the air cause this haziness. Air pollution can be caused by natural events such as a volcano eruption that releases smoke and ash into the air. But people cause most air pollution.

Figure 7-13 shows some sources of air pollution. The two biggest sources of air pollution are cars and factories, including power plants that produce electricity. One source of pollution is the fumes that come from cars. Cars need gasoline to run. When gasoline is burned, pollutants are released into the air. The polluting fumes of cars and other vehicles cause more than 30 percent of all air pollution.

Many factories and power plants burn coal or oil for the energy they need. This activity also releases pollutants into the air. Pollutants can interfere with life processes. Air pollution can make your throat feel dry or your eyes sting. Many people have trouble breathing when air pollution levels are high. For people with lung or heart problems, air pollution can be deadly. In the United States, it is estimated that 50 000 to 120 000 deaths each year are linked to air pollution.

FIGURE 7-13

Air pollution is caused by many different activities. *What activities illustrated in this picture cause air pollution?*

People aren't the only living things that are harmed by air pollution. Acid rain causes a lot of damage to other organisms. **Acid rain** happens when the gases released by burning oil and coal mix with water in the air to form acidic rain or snow. Some scientists hypothesize that when acid rain falls on the ground, plants and trees die. When acid rain falls into rivers and lakes, it can kill the fish.

Spare the Air

The best solution for all types of pollution, including air pollution, is prevention. Reducing the number of pollutants is easier to do than cleaning up the pollution that results. Automobiles being produced today release fewer harmful gases and use less fuel than did vehicles in years past. Governments around the world are also meeting to find ways to reduce the amount of air pollutants being released into the atmosphere by factories.

It may seem at first that you have no control over sources of pollution, but think again. Think about what power plants produce. They produce electricity. When power plants burn oil or coal to make electricity, harmful pollutants enter the atmosphere and cause smog, acid rain, and other problems.

You can help protect the atmosphere by limiting the amount of energy you use at home. Conserve electricity by turning off lamps, radios, fans, and other appliances that you aren't using. Keep doors and windows closed to save heat energy in the winter or to keep a cool environment in the summer. And encourage your families to buy energy-efficient lightbulbs, like those illustrated in **Figure 7-14.**

FIGURE 7-14
Turning down your home's thermostat is one easy way to reduce the amount of energy used. Energy-efficient lightbulbs use one quarter of the energy of standard incandescent lightbulbs, and they last up to ten times longer.

FIGURE 7-15

The people who live in Anchorage, Alaska, are surrounded by many natural resources—mountains, water, and abundant wildlife.

Land, water, and air are all parts of our natural environment, as illustrated in **Figure 7-15.** To maintain the health of the environment, people need to think how their actions will affect the land, water, and air that make up Earth. Pollution is easier to prevent than to clean up.

Section Wrap-up

1. What is a pollutant? Give examples.

2. Describe some ways that humans affect land, water, and air.

3. **Think Critically:** You know that gasoline fumes cause air pollution. Describe one way that gasoline itself can cause water pollution.

4. **Skill Builder**
   Developing Multimedia Presentations
   Based on what you have learned in this section, choose one type of pollution—land, water, or air. Develop a multimedia presentation showing how human activities have affected the specific part of the environment you have chosen and what is being done to correct it. If you need help, refer to Developing Multimedia Presentations on page 564 in the **Technology Skill Handbook.**

Science Journal

In your Science Journal, list three things you did today. Write a paragraph describing how these actions might affect the environment.

Protecting the Environment

Cutting Down on Waste

The United States faces a huge waste problem. Litter gathers along highways. Landfills leak and overflow with garbage. Five billion tons is the estimated amount of solid waste thrown away each year in the United States. **Solid waste** is whatever people throw away that is in a solid or near-solid form. Look at **Figure 7-16** for examples.

Most waste is produced when coal, oil, and other natural resources are taken from the ground. Households and businesses produce only about four percent of this country's waste. However, household and business waste is still a lot of waste—nearly 200 million tons each year.

Most of the waste from our homes, schools, and businesses is paper and cardboard products. In the cafeteria at school, it's easy to see why this is so. School lunch programs all over the country depend upon paper plates, straw wrappers, milk cartons, paper bags, drink boxes, and napkins. What if individuals just tried to reduce the amount of trash they throw away each day?

Solid-waste management for individuals can be summed up by the 3 *Rs—reduce, reuse,* and *recycle* waste.

What YOU'LL LEARN

* About the problems of solid waste
* How to reduce, reuse, and recycle resources
  Science Words:
  solid waste
  recycling

Why IT'S IMPORTANT

You can do many simple things to help protect the environment.

FIGURE 7-16

Solid waste includes everything from old newspapers and pickle jars to old plastic toys and scrap metal from manufacturing processes.

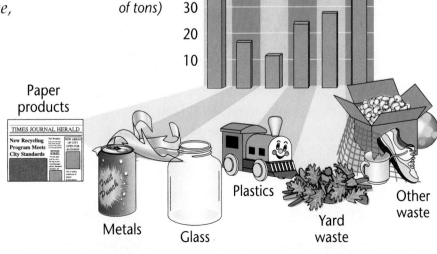

Amount of waste (millions of tons)

Paper products

Metals

Glass

Plastics

Yard waste

Other waste

Mini LAB

Making Models

Use things that are generally thrown out to make a piece of artwork.

1. Collect different items that would normally be thrown out. Such items could include newspapers, clean cans or glass, packaging, etc. Do not collect any food items or items that could be harmful. Do not take any items out of the garbage.
2. Using glue, string, or tape, create an item of artwork.
3. Give your piece of art a name.

Analysis

1. What items did you use to make your piece of artwork?
2. Is this activity an example of reducing, reusing, or recycling? Explain.

Reducing Waste

Most people would agree that there are no simple solutions to the solid-waste problem. But trying to *reduce* the total amount of waste produced is the simplest and most effective way that an individual can help. In **Figure 7-17**, you can see how you could reduce waste.

Reducing includes cutting down on the amount of trash you throw out. Reducing waste could mean buying a model car in smaller packaging. Such an activity helps protect the environment. If you buy a product that has no packaging, it means that no paper from trees or plastic from oil has been used to pack it. No energy has been used to make this packaging and no landfill will be needed for its disposal. But it may also mean that the product could break during shipping.

FIGURE 7-17

School cafeterias are busy places at lunchtime. *How could these students help reduce waste?*

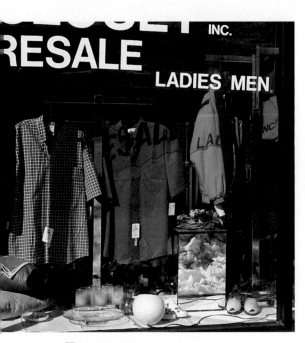

FIGURE 7-18

Secondhand stores are great places to find bargains. Like-new clothing and other items are being reused—a good way to help protect the environment.

Reusing Things

Reusing items is another way of reducing trash. Old clothes can be used as cleaning rags. Old newspapers can be used to line pet cages, to wrap gifts, or to cover the floor when painting. Some stores, as shown in **Figure 7-18,** sell used items. Books, magazines, clothes, computer disks, video games, glass jars, and cardboard boxes can also be reused. When you no longer need such items, you can give them away to somebody else who may want or need them.

Problem Solving

Uses for Plastic Things

Have you ever seen aluminum cans of beverages sold in packs of six in grocery stores? The six cans are usually held together with a plastic collar. Once the cans are released from this plastic collar, what do you do with it? Thrown away in one piece, it is dangerous to wildlife. Fish and ducks can get caught in the plastic collars, while other animals have choked on them.

Solve the Problem:

Are there other ways to deal with these plastic collars? How many ways can you think of to reuse the plastic collars in a useful way?

Think Critically:

Can you think of some other ways to package six aluminum cans of beverage together that won't hurt the environment?

FIGURE 7-19

Recycling products reduces the amount of energy used to make products.

 It takes 95 percent less energy to produce aluminum from recycled aluminum than from ore.

Recycling Things

Did you know that plastic soda bottles might have been used to make the carpeting in your home? It's true. Look at **Table 7-2.** Many items that people normally throw out—glass jars, newspapers, magazines, soda cans, and plastic milk containers—are now being recycled into other useful products. **Recycling** (ree SIKE ling) means reusing materials after they have been changed into another form. Recycling is done by bringing bottles, cans, and newspapers to a recycling center or leaving them on the curb in special recycling containers.

Buying products made from recycled materials helps protect the environment. Recycling usually saves energy, water, and other resources. **Figure 7-19** shows some examples of resources.

Table 7-2

Items to Be Recycled	Resulting Products
Newspapers, telephone books, magazines, and catalogs	Newsprint, cardboard, egg cartons, and building materials
Aluminum beverage cans	Beverage cans, lawn chairs, siding, cookware
Glass bottles and jars	Glass bottles and jars
Plastic beverage containers	Insulation, carpet yarn, textiles

B It takes 75 percent less energy to make steel from scrap than from iron ore.

C Making new glass from old glass cuts energy usage by 80 percent.

Habits for a Healthier Environment

Practicing the 3 *Rs* makes you part of the solution to the solid-waste problem. Can you see how changing your everyday habits—the way you pack your lunch, the types of items you buy, the way your throw away your trash—can have a good effect on the environment? The best way to protect the environment is to develop habits that promote a healthy environment.

Section Wrap-up

1. What does the term *reducing waste* mean? Give an example.

2. Describe how recycling a glass bottle helps protect the environment.

3. **Think Critically:** How might a person getting food from a fast-food restaurant practice reducing waste?

4. *Skill Builder*
   Recognizing Cause and Effect
   Explain how keeping a television on has an impact on the environment. If you need help, refer to Recognizing Cause and Effect on page 551 in the **Skill Handbook.**

Science Journal

In your Science Journal, make a list of specific ways you and your family can help protect the environment.

7●4 A Tool for the Environment

What YOU'LL LEARN

- About life-cycle analysis
- How to improve your note-taking skills
  Science Words:
  life-cycle analysis

Why IT'S IMPORTANT

Good note-taking skills will help you remember important things.

Are you an environmentally friendly shopper? When you buy things, do you think about how they affect the environment? Scientists wonder, too, about how the things we make and use impact the environment. That's why they've developed a tool to help them figure out the environmental impacts of products. The tool is called life-cycle analysis. **Life-cycle analysis** is a way of figuring out the environmental impact of a product through its entire life. Most scientists break down the life cycle of a product into six stages.

Life Stages of a Product

To do a life-cycle analysis, scientists start at the very beginning when natural resources are obtained to make a product. This is the first stage in the product's life. The six stages are:

1. getting the natural resources to make the product;
2. manufacturing in a factory or plant;
3. transportation to a home, store, or business;
4. use and reuse;
5. recycling;
6. disposal in a landfill or by burning.

Natural resources and energy are used during each stage of a product's life. In addition, each stage has an impact on the environment. The environmental impact might be air pollution, human health problems, use of a nonrenewable resource, or habitat loss, among others.

FIGURE 7-20

What can you do with old clothes you no longer want or fit into?

When scientists do a life-cycle analysis, they add up all the natural resources and energy that are used to get the product from stage one to stage six. Then they figure out the environmental impact of each stage in the product's life. They analyze these factors to come up with a complete picture—a life-cycle analysis—of a product.

FIGURE 7-21

Turning fabric scraps into a colorful quilt is a good way to reuse different materials.

Environmentally Friendly Shopping

Once a life-cycle analysis has been completed, scientists can compare the product to similar products to see which one is better for the environment. Companies use life-cycle analyses, too, to help them figure out how to reduce the environmental impact of the products they make. Shoppers can also use life-cycle analyses. Can you figure out how? A life-cycle analysis can help you choose which products to buy. In this way, you can be an environmentally friendly shopper.

Skill Builder: Taking Notes

LEARNING the SKILL

1. To take good notes, first read the material to identify the main ideas. The heads and subheads in the material are clues to main ideas.

2. Look for italicized or boldfaced words in the material. These are also clues to important ideas.

3. Identify details or sentences in the material that support the main ideas.

4. Using the main ideas and the details or sentences that support them, take notes about the material.

PRACTICING the SKILL

1. List three main ideas in this passage.

2. What is a life-cycle analysis? What are the six stages in a product's life?

3. What are some ways that life-cycle analyses are used?

APPLYING the SKILL

Think of a product that you would like to buy. Research the life cycle of the product. Take careful notes, and share what you've learned about the product with the class.

Xeriscaping

Shoosh, shoosh, shoosh—it's summertime and the sound of lawn sprinklers fills the air. Green grass is pretty, but it guzzles water. Half of the water used in American homes goes to keep thirsty lawns and gardens green.

Fresh, clean, drinkable water is a limited resource. As the world's population grows, the demand for water increases. Communities everywhere face shrinking water supplies and are looking for creative ways to save water.

Xeriscaping (ZEER uh skay ping) is one solution. The word *xeriscape* comes from the Greek word *xeros*, meaning "dry." But a xeriscape is not a dusty brown patch of gravel with a few cactus plants placed in the middle. A xeriscape is a cool, colorful landscape made up of many different types of water-saving plants that are native to the particular area.

Creating a xeriscape takes the right plants and careful planning. Native plants—those that grow naturally where you live—are especially good choices. Native grasses, flowers, and shrubs can often survive on rainfall alone, even during a drought.

Using specific native plants certainly limits which plants you can choose, but there are still many beautiful plants that can be used. Many of these plants that have survived dry weather conditions over the years have produced tough, colorful flowers. Such flowers are often fragrant and attract birds and bees that help in pollination.

Xeriscaping helps homeowners to reduce their water use by up to 75 percent over traditional landscaping. In fact, a successful xeriscape hardly needs to be watered at all!

SCIENCE Online

Visit Glencoe Science Online, *science.glencoe.com*, to find out how one city with an average rainfall of nine inches per year helps its citizens plan their xeriscaping.

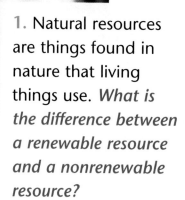

R ead the statements below that review the most important ideas in the chapter. Using what you have learned, answer each question in your Science Journal.

1. Natural resources are things found in nature that living things use. *What is the difference between a renewable resource and a nonrenewable resource?*

3. Reducing, reusing, and recycling are three things that people can do to help the environment. *When you buy a product that has no packaging, how have you helped the environment?*

4. A life-cycle analysis is a way of figuring out the environmental impact of a product through its entire life. *How does knowing this information help you to make better choices about the environment?*

2. Human activities affect the environment. *List some ways that human activities impact land, air, and water.*

Using Key Science Words

acid rain
landfill
life-cycle analysis
natural resource
nonrenewable
  resource

pollutant
recycling
renewable resource
solid waste

Match each phrase with the correct term from the list of Key Science Words.

1. a material that harms living things by interfering with life processes
2. reusing materials after they have been changed into another form
3. an area where garbage is deposited
4. things people throw away that are in solid or near-solid forms
5. when gases released by burning oil and coal mix with water in the air

Checking Concepts

Choose the word or phrase that completes the sentence.

6. An example of a nonrenewable resource is _____.
   a. sunlight   c. oil
   b. water   d. a tree

7. Using an old newspaper to line a pet cage is an example of _____ the newspaper.
   a. reusing   c. reducing
   b. recycling   d. buying

8. Collecting used paper and sending it to a factory to be made into new paper is an example of _____.
   a. reusing   c. reducing
   b. recycling   d. buying

9. Breathing polluted air can cause _____.
   a. acid rain
   b. health problems
   c. solid waste
   d. water pollution

10. A life-cycle analysis of a product will indicate its environmental impact during its _____.
    a. daily use
    b. production time
    c. entire life
    d. decay time

Thinking Critically

Answer the following questions in your Science Journal using complete sentences.

11. If people use so many paper products, why don't we run out of trees?

12. Almost 70 percent of Earth's surface is covered by water, but less than one percent can be used by people. Why?

13. When a landfill can't hold any more solid waste, it is closed down. How can a landfill be an environmental problem even though people are no longer depositing trash there?

14. Three thousand new plants grow naturally in a field every year. How must these plants be treated to make sure this type of plant is a renewable resource?

15. Some people take their own bags with them when shopping. How might this affect natural resources?

Developing Skills

If you need help, refer to the description of each skill in the Skill Handbook.

16. Concept Mapping: Use the following phrases to make a concept map showing the complete life-cycle analysis of an aluminum can: *refine the aluminum; mine aluminum ore; use the aluminum can; shape the aluminum into cans; recycle the can; melt the ore in a factory; transport the aluminum can to where it will be used.*

17. Recognizing Cause and Effect: Acid rain is caused by air pollution. But acid rain often falls on places that are far from sources of air pollution—such as factories and cars. Research acid rain in the library or on the Internet and then write a short paragraph in your Science Journal explaining why acid rain can be a problem for places that are far from sources of pollution.

18. Classifying: Classify the following resources as renewable or non-renewable: sunlight, water, oil, trees, air, coal, soil.

19. Making and Using Tables: Record the things that your family throws away for one week. Make a table listing which of these items could be recycled. In another column, list the resulting products that can be made from the recycled items. Refer to **Table 7-2** for help.

Items for Recycling	Products made from recycled material

20. Using Numbers: If everyone in the United States recycled their newspapers, 500 000 trees would be saved each week. How many trees would be saved in one year?

Performance Assessment

1. Design an Experiment: Lemon juice is an acidic liquid. Design an experiment showing the effects of lemon juice on a plant. In your Science Journal, relate the results of your experiment to what you've learned about acid rain.

2. Newspaper Article: Write a newspaper article describing an environmental problem in your own community and possible solutions.

UNIT 1

Internet Project

Life Science

Can you imagine what it would be like to see an alligator crawl out of a pond in upstate New York? How about a polar bear swimming along the southern California coast? Is there a relationship between an area's climate—its average temperature, precipitation, and other weather factors—and the types of species that can live there?

Goals

- Understand how to use the Internet as a source of information and learning.
- Understand the relationship between an area's climate and the types of species that can live there.
- Identify the type of biome you live in and describe the distribution of biomes across the United States.

Procedure

1. Identify the type of biome you live in. A biome is a large geographical area with its own distinctive climate and species. Go to Glencoe Science Online at *science.glencoe. com* to find links about biomes on the Internet.
2. Research information about your biome. What kinds of animals and plants live here? What is the climate like? Print a data table from the *Glencoe Science* page. Record this information in your data table. *If you don't have Internet access,* use the library and copy the data table on the next page.
3. Find information about the other biomes. Record the data in the data table.
4. How many different biomes are there in the United States? Research the information using the Internet. Find this Internet project at *science.glencoe.com* and print out a map of the United States. Use colored pencils, crayons, or markers to label the map to show where the different biomes are located.

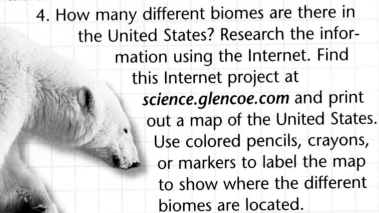

5. Identify a major city in your state. Use the Internet, newspapers, or almanacs to find annual climate data (daily high temperatures and amount of precipitation) for this city. Find average monthly temperature and precipitation for this city over a 12-month time period.

6. Use the Internet, newspapers, or the local library to find out the names of common plants and animals in your state. Post the list of species and your climate data to the Glencoe Homepage to share with students from all over the world!

7. Go to the Glencoe Homepage. Find species lists and climate data from cities within each of the remaining biomes on your map. Complete your biomes map by mounting species lists on the different biomes.

Go Further

Look up data for another country. Find out annual climate data and the species that live in different areas. Determine the biome.

Conclude and Apply

1. Compare and contrast two different biomes in North America.

2. What kinds of plants and animals live in a desert environment? What characteristics do these organisms have to survive in a desert climate?

3. Explain the relationship between an area's climate and the types of species that can live there.

Data Table					
Biome	**Common Animals**	**Common Plants**	**Climate**	**Temperature**	**Rainfall**
Tundra					
Taiga					
Tropical Rain Forest					
Grasslands					
Desert					
Temperate Forest					

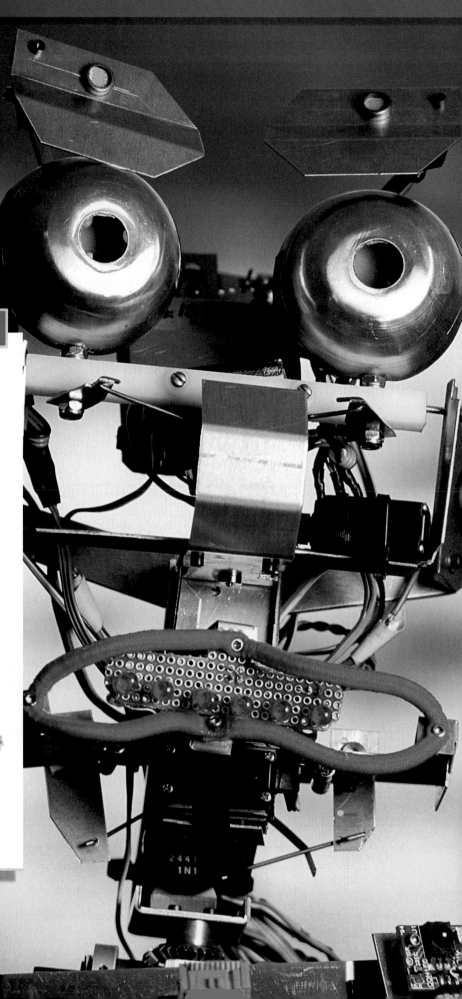

UNIT
2

Unit Contents

Unit 2 Project

Physical Science

What's happening here?

Robots are machines that may be powered by electricity, magnetism, or even by the sun. Researchers have found that people are more likely to work with robots if they have "faces." Scientists plan to use robots with faces in hospitals and nursing homes, where they will interact with patients. The robot in the larger photograph is named IT and was designed to respond to human emotions. The robot in the smaller photograph helps people pick up heavy loads. When a person puts the robot on, it measures the force of the wearer's arms, then multiplies it to control a pair of stronger robotic arms. Robots often are programmed to do work that humans find too messy, dangerous, or boring.

SCIENCE *Online*

You can visit Glencoe Science Online at *science. glencoe.com* for links to information about robots. Find out what kinds of jobs robots do and why. In your Science Journal, write about what you found out.

Chapter Preview

Skills Preview

Skill Builders
- make models
- use a computer-
ized card catalog
- take notes

MiniLABs
- infer
- observe

Activities
- hypothesize
- design
your own
experiment
- observe
- compare
- conclude

What is matter?

Perched on an enormous rock overhanging a river, you feel the gritty texture of granite under your feet and smell the pine trees along the shore. Suddenly, you realize how hot you are. Taking your empty squirt bottle down to the river, you hold the bottle under water. A stream of bubbles rises from the mouth of your bottle to the water's surface. That's puzzling, you think. If the bottle were empty, where did the bubbles come from? The Explore activity below will help you solve the puzzle.

EXPLORE ACTIVITY

Observe That Matter Occupies Space

1. Take a square of tissue paper, wad it into a ball, and tape it to the bottom of the inside of a plastic or paper cup. Turn the cup upside down. The tissue should stay inside the cup.
2. Hold the cup upside down over a bowl of water. Slowly push the cup into the water as far as you can.
3. Raise the cup out of the water, and remove the tissue.

Science Journal

In your Science Journal, describe the tissue paper before and after you put the cup in the water. Explain what you think happened. What do you think was in the cup besides the tissue paper?

8•1 Matter and Atoms

What YOU'LL LEARN

- What matter is and what makes up matter
- The parts of an atom and their charges

Science Words:
matter
atom
proton
neutron
electron

Why IT'S IMPORTANT

Everything in and around you is made up of matter.

FIGURE 8-1

Democritus was a Greek philosopher who lived from about 460 B.C. to about 370 B.C.

What is matter?

When you pulled the tissue paper out of the cup in the Explore activity, it wasn't wet. You had placed the cup in the bowl of water, so why didn't water rush into the cup and get the tissue wet? More than tissue paper was in the cup. Air was trapped inside, too.

The air took up space and kept the water from touching the tissue paper. If you could put it on a balance, you would find that air also has mass. **Matter** is the term used to describe anything that has mass and takes up space. Therefore, air is matter. Anything you can see, touch, taste, or smell is matter. Doesn't that include just about everything you can imagine? What isn't matter? Light has no mass and takes up no space. Heat has no mass and takes up no space. Emotions, thoughts, and ideas aren't matter, either. Can you think of anything else?

What makes up matter?

People have always been interested in matter. What is it made of? Is all matter alike? As the tools of science developed, so did people's ideas about matter.

An Early Idea

An early idea about matter came from Democritus (dih MAW kruh tus), shown in **Figure 8-1.** He thought that the universe was made of empty space and tiny bits of stuff (atoms). Democritus didn't really know what the tiny bits of stuff were, but he thought that they must be incredibly small—so small that they couldn't be divided into smaller pieces. In fact, the word *atom* comes from a Greek word that means "cannot be divided." In science today, an **atom** is defined as a small particle that makes up most types of matter. Democritus's idea that matter is made up of small particles and empty space was a start, but many questions still needed to be answered.

Oxygen

+

FIGURE 8-2

When a fire burns, the total mass of the fuel (the wood) and the oxygen it reacts with equals the total mass of what is produced by the fire (water vapor, carbon dioxide, and ash). No matter is gained or lost.

Water vapor and carbon dioxide

+

Later Developments

Almost 2000 years passed without much change in Democritus's idea of what makes up matter. By that time, people thought that matter could appear from nothing, or it could completely disappear. Another common belief was that the "stuff" part of matter could change its form. The French chemist Antoine Lavoisier (luh vwah zee AY) challenged these thoughts in the late 1700s.

Lavoisier showed that there was no change in mass during common chemical reactions such as burning and rusting. Have you ever sat by a fire, such as the one shown in **Figure 8-2,** watching the wood blaze brightly, then turn to embers, and finally to ashes? What starts out as a mountain of logs is just a little heap of ashes by the end of the evening. Surely, some mass has been lost in the process—some matter has disappeared. But Lavoisier showed that the total mass of the wood and the oxygen used when it burns equals the total mass of the water, carbon dioxide, and ash produced—the amount of matter stayed the same.

Rusting is another example of a chemical reaction. Rusting takes place between iron and the oxygen that is in the air. The iron gradually turns to rust. But, there is no change in the amount of matter. The total mass of the iron plus the oxygen equals the total mass of the rust when the reaction is complete. Observations like these changed the way people thought about matter. They realized that mass is never created or destroyed when things react with one another. This principle is called the law of conservation of matter.

Proust's Work

Another French chemist, Joseph Proust (PREWST), explored the "stuff" part of Democritus's space-and-stuff idea. He stated that the basic building blocks of any substance are always the same and are always put together in the same way. For example, a single unit, or molecule, of water contains two atoms of hydrogen and one atom of oxygen. If you counted all the atoms in a cup of water, there would always be a ratio (RAY shee oh) of two hydrogen atoms to every one oxygen atom. Look at **Figure 8-3.** Whether you have a drop of water or a whole ocean, the ratio of hydrogen atoms to oxygen atoms is always two to one.

The ideas of these and other scientists suggested that matter is made of small particles that behave in ways that can be predicted and measured. They began searching for a model to explain what an atom is made of and how it behaves.

Models of the Atom

To find out more about how an atom might look and act, scientists started to make models. There are many different ways to make models. Making a mental picture of something that you can't easily see is one way to make a model.

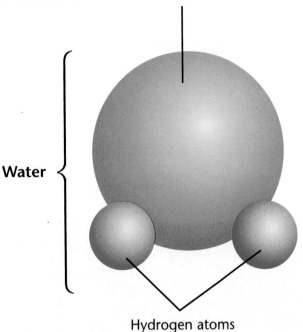

FIGURE 8-3
A ratio is a simple fraction that compares the amounts of two items.

Oxygen atom

Water

Hydrogen atoms

Ratio: $\dfrac{2 \text{ Hydrogen atoms}}{1 \text{ Oxygen atom}}$

What is a model?

Another way to make a model is to make a small working version of something larger. For example, if you wanted to design a new kind of sailboat, would you just come up with a design, build a full-sized boat, and hope it would float? It would be smarter—and safer—to first build and test a small model of your design, as in **Figure 8-4.** Then, if it doesn't float, you can change your design, build another model, and keep trying until the model works. As you could tell from the model sailboat, models can be changed as new information is gained.

A model can be a paragraph, a math equation, a drawing, a computer program, or an actual structure, like the small sailboat. A model lets you create an image in your mind of something that is too big, too small, or too complicated to understand otherwise. For example, a model of the solar system can help you see the sizes of planets in relation to each other.

A model is also used when an actual event would occur too quickly or too slowly or is too dangerous to watch directly. For example, scientists use models to study what earthquakes and fires might do to a building. Weather maps are models that help predict weather and climate changes. Scientists sometimes use models to compare the results of experiments with their ideas of what they predicted would happen.

Mini LAB

Using a Model

Use a model to infer how atoms behave.

1. Fill a pie pan with warm, whole milk.
2. After the milk has stopped moving around, add one drop each of red, blue, and yellow food coloring. Make a triangle pattern with the dots, putting them about 2 cm apart.
3. In the middle of the triangle pattern, add one drop of liquid soap.

Analysis

1. What happened to the colors in the milk?
2. Use this model to describe how freely the atoms that make up the milk move.

FIGURE 8-4

When this model is placed in water, it will show whether or not the design is a good one. If the boat doesn't sail, changes must be made.

Problem Solving

Comparing Gravel and Iron

The atoms that make up different materials determine what those materials are like and how they behave. Suppose that an artist who makes sculptures from pieces of metal lives on your block. You like to ride your bike to her house to watch her work. Because small pieces of iron may be hot as they fly off the metal she is grinding, she takes her grinder outside and uses it in the gravel driveway. But, there's a problem. You park your bike in the driveway and the small slivers of metal might give you a flat tire—or two. How can she get the small iron pieces out of the gravel?

Solve the Problem:

Think about what iron is like and how it behaves. What are some common features of iron?

Think Critically:

1. Why might a paper towel or old rag be helpful in separating the gravel from the iron? Where would you put it?
2. Explain why stirring the iron and gravel into water would not separate them.

Dalton's Model of the Atom

In 1803, English chemist John Dalton introduced a model to describe matter and atoms. He called his model "the atomic theory of matter" and defined an atom much as scientists do today—as the smallest particle that makes up most types of matter. Dalton's model said that all of the atoms in something such as gold are identical. In other words, a chunk of gold is made up of gold atoms only, and each gold atom is like every other gold atom. Dalton's model also said that atoms of different elements are different from one another. For example, gold atoms are not the same as iron atoms.

Behavior Patterns

In his studies, Dalton also noticed that reactions between different chemicals always occur in patterns that can be predicted. For example, if 2 L of hydrogen gas react with 1 L of oxygen, it can be predicted that 20 L of hydrogen gas will react with 10 L of oxygen. This verified, or supported, what Joseph Proust had predicted. Dalton's model allowed scientists to understand the findings of both Lavoisier and Proust.

What makes up an atom?

Dalton's ideas are close to the model scientists use today to describe matter and atoms. However, as scientists continued to study atoms, using better and better tools, they changed Dalton's model to make it more like what they observed. One of their conclusions is that atoms are mostly empty space. If you bump into a brick wall, you probably don't think that the atoms making up those bricks are mostly empty space. Can a model help you understand this?

Sizes of Atoms

Atoms are extremely small. It wasn't until the 1980s that improvements in electron microscopes let scientists see an atom, as in **Figure 8-5.** To give you a better idea of how small atoms are, look at **Figure 8-6.** Imagine you are holding an orange in your hand. If you wanted to use only your eyes to see the individual atoms on the surface of the orange, you would need to increase the size of the orange to the size of Earth. Then, cover it with billions and billions of marbles. Each marble would represent one of the atoms that make up the skin of the orange.

FIGURE 8-5

This photo shows six uranium atoms arranged around a central uranium atom. The atoms are enlarged about 20 million times.

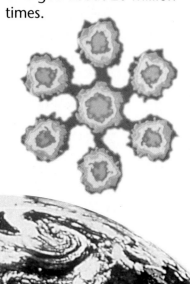

FIGURE 8-6

An atom is too small to visualize without using a model, such as this one.

What makes up each of these individual atoms? To be able to peek inside, you would have to expand your model even more. Pick one of the marbles on the surface of your Earth-sized orange, and make it as big as a football stadium. This model of an atom, shown in **Figure 8-7,** is now large enough that you can imagine the center of the atom and what surrounds this center.

The Center of the Atom

Pretend you are standing on the edge of the football field. As you look toward the center of the football field, you see what looks like a tiny cluster of grapes. This cluster represents the center of the atom, which is called the nucleus (NEW klee us). Between you and the nucleus is nothing but empty space. The same relationship of empty space to stuff exists inside an atom.

The nucleus of an atom is made of two kinds of particles—protons (PROH tahns) and neutrons (NEW trahns). A **proton** is a particle with a positive charge. A **neutron** is a particle with no charge at all. Almost all of the mass of an atom is in its nucleus. It probably looks like a bunch of grapes, with the grapes representing the protons and neutrons snuggled up close to one another.

FIGURE **8-7**

Different parts of this football stadium represent different parts of an atom.

FIGURE 8-8

This oxygen atom has eight protons and eight neutrons in its nucleus. Eight electrons, traveling at great speeds, form a cloud around the nucleus.

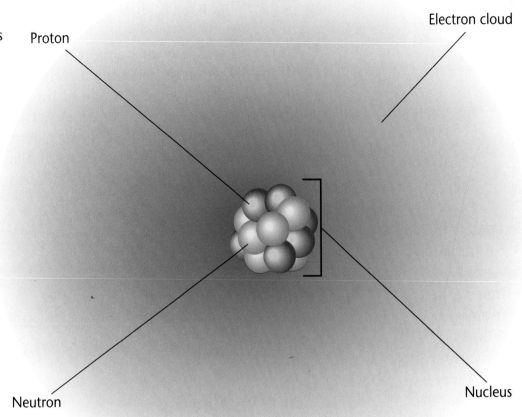

Proton

Electron cloud

Neutron

Nucleus

Around the Nucleus

You are still standing on the edge of the football field, just inside the atom. If you look up toward the top of the stadium, you'll see insects flying around the lights. They stay near the lights, but they don't fly in a set pattern. These insects represent particles around the nucleus of the atom that are called electrons (ee LEK trahns).

An **electron** is a particle in an atom with a negative electric charge. An atom has equal numbers of electrons and protons, so the negative charge of the electrons equals the positive charge of the protons. Electrons, neutrons, and protons are similar in size, but a proton or neutron is about 1800 times heavier than an electron.

Examine a model of an atom, shown in **Figure 8-8.** Because they are located toward the outside of the atom, electrons sometimes are easily removed from the atom. You'll see signs that electrons are removed from atoms in the next activity.

Design Your Own Experiment
The Electron Magnet

You can use a rubber balloon to remove electrons from the atoms of an object. After you have gathered enough electrons on the surface of the balloon, you can see their effects on other objects. An object is said to be charged when it has more or fewer electrons than usual. In this activity, you will find sources of electrons to charge the balloon and design a way to detect the presence of the electrons.

Possible Materials

- rubber balloon
- wool sweater
- hair
- metal
- wood
- pencil shavings
- small piece of aluminum foil
- foam packing peanuts
- thin stream of water
- lint

PREPARE

What You'll Investigate
What common materials are good sources of electrons? Which materials clearly show the effect of a large electron charge?

Form a Hypothesis

Make a hypothesis about which materials you think will give electrons easily to the balloon and which types will respond to the presence of a charge.

Goals

Design an experiment that allows you to charge a rubber balloon using different surfaces as a source of electrons.

Observe the effect of a charged rubber balloon on different kinds of objects.

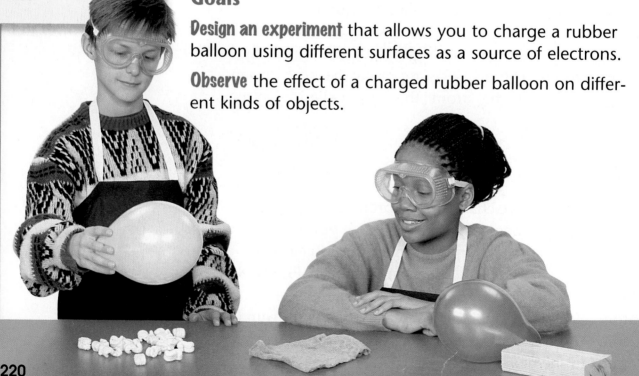

Safety Precautions

Wear goggles when using the balloons. Balloons occasionally burst.

PLAN

1. As a group, **list** the steps that you need to take to test your hypothesis. **Describe** exactly which surfaces and materials you will use to charge the balloon and which materials you will **test** once the balloon is charged.
2. Before you begin, make in your Science Journal a data table that allows room to list each material you are going to test and the results of the test.
3. **Read** over your entire experiment to make sure that all the steps are in logical order.
4. Will you **repeat** any portion of the experiment?

DO

1. Make sure that your teacher approves your plan and data table and that you have included any changes.
2. Carry out the experiment as planned and approved.
3. Be sure to **record** your observations in the data table as you complete each test in your plan.

CONCLUDE AND APPLY

1. **Compare** your findings with those of other groups.
2. **List** the materials that are good sources of electrons and those that are not. Write a general rule that will let you predict whether a material would be a good source of electrons.
3. Using your data table, **conclude** which materials are attracted to a supply of electrons.
4. **APPLY** Explain why you could be shocked when you touch a doorknob after you walk across a carpet on a cold day.

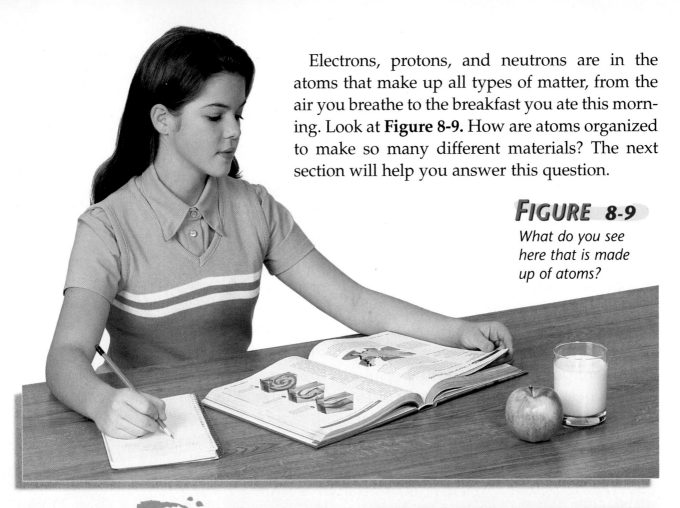

Electrons, protons, and neutrons are in the atoms that make up all types of matter, from the air you breathe to the breakfast you ate this morning. Look at **Figure 8-9.** How are atoms organized to make so many different materials? The next section will help you answer this question.

FIGURE 8-9

What do you see here that is made up of atoms?

Section Wrap-up

1. List ten things in your classroom that are made of matter.

2. An atom of the element helium contains two protons, two neutrons, and two electrons. Diagram a picture of a helium atom, using a circle to show each particle. Identify the protons with a plus sign (+), include the two neutrons with no markings, and identify the electrons with a minus sign (–).

3. **Think Critically:** Explain why a globe of Earth is a model.

4. **Skill Builder**
   Making Models Using the materials of your choice, create a model of a carbon atom. Include carbon's six protons, six neutrons, and six electrons. If you need help, refer to Making Models on page 557 in the **Skill Handbook.**

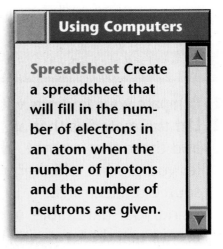

Using Computers

Spreadsheet Create a spreadsheet that will fill in the number of electrons in an atom when the number of protons and the number of neutrons are given.

Types of Matter

Elements

Most matter is made up of combinations of a basic group of building blocks. These building blocks are called elements. An **element** is matter made up of only one type of atom. For example, gold is an element made of only gold atoms, and iron is an element made of only iron atoms.

Natural and Synthetic Elements

Ninety elements are commonly found on Earth. These elements make up gases in the air, ores in rocks, and liquids such as water. Examples of elements that you are probably familiar with include nitrogen and oxygen in the air you breathe and the metals gold, copper, aluminum, iron, and silver.

Scientists also have discovered elements that are known as *synthetic* (sihn THET ihk) elements. Synthetic elements are not commonly found on Earth. They are made in machines called particle accelerators (ak SEL uh ray turs), shown in **Figure 8-10.** You'll learn more about these elements and their uses in Section 8-3.

What YOU'LL LEARN

- Definitions and examples of elements, compounds, and mixtures
- Names, symbols, and formulas for elements and compounds

Science Words:
element
periodic table
compound
chemical formula
mixture

Why IT'S IMPORTANT

If you know what types of materials make up things around you, you will be able to use these materials better.

FIGURE 8-10

This particle accelerator is at Fermilab, which is near Chicago, Illinois.

The Periodic Table

With more than 100 natural and synthetic elements to keep track of, scientists need a system of organization. Scientists use a chart based on the work of Russian chemist Dimitri Mendeleev (men duh LAY uv). This chart is called the periodic (peer ee AW dihk) table of the elements. The **periodic table** is a chart that organizes elements by the number of protons in each element's nucleus. A periodic table is found inside the back cover of your textbook. Refer to it as you examine the information included on the table. In addition to listing the elements by their numbers of protons and their symbols, the periodic table provides other useful information. If you know how to use the table, you can find out the mass of an atom, the pattern of electrons, and whether an element is a solid, a liquid, or a gas at room temperature.

Chemical Symbols

Each element is identified on the chart with a one-, two-, or three-letter abbreviation. These symbols are a chemical shorthand that allows chemists to represent elements without writing out their full names. Some of the abbreviations are initials, but others are more complicated.

Sometimes the abbreviation is a letter or two from the name of the element. For example, the element hydrogen is abbreviated H, helium is He, lithium is Li, and sulfur is S. Other times, the abbreviations come from Latin words. For example, the symbol for iron is Fe, which comes from the Latin word for iron, *ferrum*. **Table 8-1** gives more examples of elements and their symbols.

Temporary Names

Why do some elements have three-letter symbols? The three-letter symbols are based on the number of protons in the element. These elements have been discovered so recently that they do not yet have permanent names. When scientists around the world agree on a name, the three-letter symbol is replaced by a one- or two-letter symbol representing the name of the element.

Table 8-1

Some Elements and Their Symbols	
Name	**Symbol**
Aluminum	Al
Arsenic	As
Calcium	Ca
Chlorine	Cl
Gold	Au
Helium	He
Manganese	Mn
Neon	Ne
Nitrogen	N
Oxygen	O
Potassium	K
Silver	Ag
Sodium	Na
Tungsten	W
Uranium	U

FIGURE 8-11
Your favorite fruit contains sugar, which is made up of the elements carbon, hydrogen, and oxygen.

Compounds

A **compound** (KAHM pownd) is a form of matter that is made when two or more elements combine chemically. Millions of compounds can be made from combinations of elements. For example, the elements oxygen and hydrogen combine to make the compound water. Another example is shown in **Figure 8-11.** Carbon and hydrogen also make up most compounds found in living things.

USING TECHNOLOGY

Goo to the Rescue

Why would you want a compound that holds water? Can you imagine any uses for such a compound? One recently developed family of compounds is called *dehydrated gels* (dee HI dray tud • jelz). They can capture and hold more than 100 times their volume in water and other liquids.

Uses of Gels

Small packages containing dehydrated gels are packed with camera and video equipment to absorb water that might get inside the packing case and ruin the equipment. Magicians use the gels to make water seem to disappear. If you ever helped care for a baby, you can probably think of another good use for dehydrated gels—disposable diapers!

In areas with limited rainfall, gels are added to soil to increase the amount of water available to the soil. The gel holds onto water and releases it gradually into the soil, providing more useful water to crops. Can you think of uses for oil-absorbing gels?

SCIENCE *Online*

Silica gel is one of the dehydrated gels being used today. Visit Glencoe Science Online at *science.glencoe.com* for a link to a site that explains a use of this gel.

The Mystery Mixture Investigation

Cornstarch, baking powder, and powdered sugar are compounds that look alike. To avoid mistaking one for another, you may need to identify each one. You can learn chemical tests that identify different compounds.

What You'll Investigate
Which compounds are present in a mixture that contains from one to three common compounds?

Procedure

1. **Copy** the data table into your Science Journal. **Record** your results for each of the following steps.
2. **Place** a small amount of cornstarch, sugar, and baking soda on the pie pan. **Add** a drop of vinegar to each.
3. **Place** another small amount of each compound on the pie pan. **Add** a drop of iodine solution to each.
4. **Place** a small amount of each compound in a test tube. Hold the test tube with the test-tube holder. Gently **heat** the bottom of each test tube with the candle.
5. Now, **test** your mystery mixture and find out which of these compounds are in the mixture.

Goals
- Test for the presence of certain compounds.
- Decide which of these compounds are present in an unknown mixture.

Materials
- test tubes (3)
- compound samples
- dropper bottles (2)
- iodine solution
- white vinegar
- candle
- test-tube holder
- small pie pan
- matches
- mystery mixture

Data and Observations

To be tested	Vinegar fizzes	Iodine turns blue	Compound melts
Cornstarch			
Sugar			
Baking Soda			
Mystery Mixture			

Conclude and Apply

1. **Conclude** which compounds are in your mystery mixture. Describe how you arrived at your conclusion.
2. How would you be able to tell if all three compounds were absent from your sample?

A compound is usually much different from the elements that make it up. For example, table salt is very different from the gray sodium solid and the yellow-green chlorine (KLOR een) gas that it is made from. You may remember from Section 8-1 that a compound is always made of the same elements in the same ratio. Table salt is always made of one atom of sodium for each atom of chlorine. The chemical name for table salt is sodium chloride (KLOR ide).

It Takes a Formula

Just as scientists use symbols to represent elements, they write chemical formulas to represent compounds. A **chemical formula** is made up of symbols and numbers that tell what elements are in a compound and what their ratios are. How would you write a chemical formula for table salt? Remember that it contains sodium (whose symbol is Na) and chlorine (whose symbol is Cl) in a one-to-one ratio. That makes the chemical formula for table salt NaCl. Now try another familiar compound, water. Water is made of two atoms of hydrogen (H) and one atom of oxygen (O). Chemists write the formula for water as H_2O. The little 2 after the H is called a subscript. It tells you that there are two hydrogen atoms for every one oxygen atom. No subscript is used when only one atom is present. **Figure 8-12** shows the chemical formulas for some common compounds.

FIGURE 8-12

These common compounds have chemical formulas. *How many of each type of atom is in each formula?*

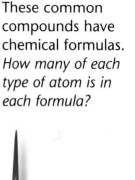

Propane
C_3H_8

Vinegar
$HC_2H_3O_2$

Battery acid
H_2SO_4

Ammonia
NH_3

8-2 Types of Matter   227

Mini LAB

Observing

Record your observations as you separate a mixture of two compounds.

1. Pour some warm water into a large drinking glass.
2. Add a mixture of salt and sand to the glass, and stir with a spoon until it appears that all the salt has dissolved. Let the sand settle.
3. Pour about 5 mL of the salt water onto a clean, dry plate. Set the plate in a warm place, and let the water evaporate.
4. Pour the sand onto a paper towel to dry.

Analysis

1. What was it about these three compounds that allowed you to separate them?
2. Explain why this method would not work for a mixture of water, salt, and sugar.

FIGURE 8-13
This brass instrument is made of a mixture that is the same throughout.

Mixtures

If two or more substances are put together but do not combine to make a compound, the result is called a **mixture.** Air is a mixture of nitrogen, oxygen, and other gases. A rock is a mixture of different minerals. What other mixtures do you see around you?

There are two kinds of mixtures. One type of mixture is the same throughout, such as air or a soft drink. You can't see the different parts in this type of mixture. An example of this type of mixture is shown in **Figure 8-13.** No matter how closely you look, you can't see the individual parts of copper and zinc that make up brass.

In the other kind of mixture, the material is made of several different items that you can see. The granite rock by the side of the river on the first page of the chapter probably contained pieces of the minerals mica, quartz, and feldspar. If you look closely at the rock, you can see the different parts. Examine another example by looking at the salad in **Figure 8-14.** Other mixtures in this category would include things like pizza, stew, a toy box full of toys, or your laundry basket at the end of the week.

FIGURE 8-14

It's easy to see that this mixture is made of different parts. A salad taken from one side of the bowl will differ from a salad taken from a different part of the bowl.

Section Wrap-up

1. Using the periodic table of elements on the inside back cover of this book for reference, write the names of each of these elements: Be, Al, P, Ar, K, Ca, V, Cu, Pb, and Au.

2. The chemical formula for table sugar is $C_{12}H_{22}O_{11}$. What does the formula tell about the makeup of this compound?

3. **Think Critically:** Explain two ways that you could separate a mixture of iron filings and table salt.

4. *Skill Builder*
   Using a Computerized Card Catalog
   Use a computerized card catalog to investigate the names of the following elements: tungsten, curium, oxygen, and another element of your choice. Find out why each element was given its name. If the symbol is not from this name, find out the origin of the symbol. If you need help, refer to Using a Computerized Card Catalog on page 562 of the **Technology Skill Handbook.**

USING MATH

Using the periodic table inside the back cover for reference, make a line graph of the number of protons in an atom versus the mass of an atom for the first 30 elements. Write down two inferences that can be supported with your graph.

8•3 Synthetic Elements

What YOU'LL LEARN

- That some elements do not occur naturally on Earth
- That taking notes provides a summary of important information

Why IT'S IMPORTANT

Taking notes on material that is unfamiliar will help you better understand what you read.

A Smashing Success

In this chapter, you've learned that an element is matter that is made of the same kind of atoms. Copper, hydrogen, sodium, iodine, and technetium (tek NEESH um) are elements. Copper, hydrogen, sodium, and iodine are commonly found on Earth. However, technetium and other elements are synthetic elements. A synthetic element is an element that is made in a machine called a particle accelerator, which was shown in **Figure 8-10.** Except for technetium and promethium, all of the synthetic elements have more than 92 protons.

The particle accelerators used to create synthetic elements speed up atoms and cause them to smash into, or collide with, other particles of matter at extremely high speeds. The tremendous force of these collisions causes atoms to join, making new elements.

Falling Apart at the Seams

Synthetic elements are unstable and break down, or decay, into other elements. Some decay very fast—in millionths of a second. Others stick around for as little as a few days or as long as millions of years. Most synthetic elements can be made only in small amounts. It may seem like the great expense of making synthetic elements is not worth the effort, but some of them have important uses, as in **Figure 8-15.**

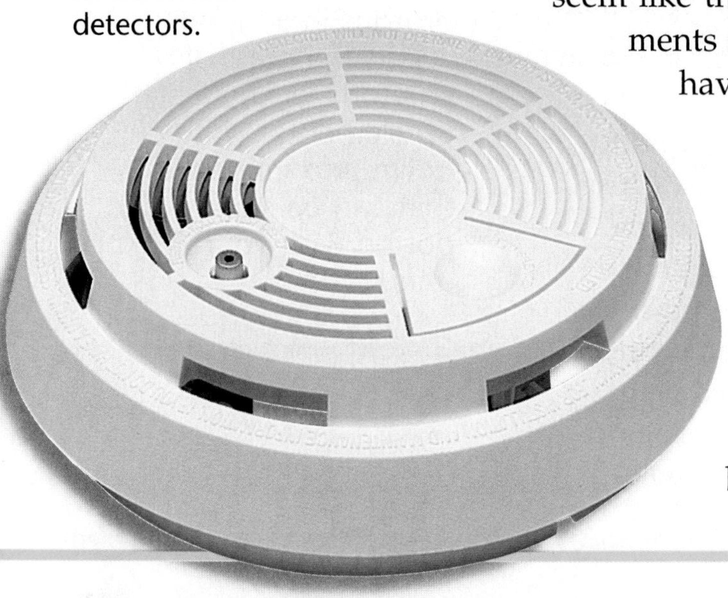

Technetium, for example, is a synthetic element that decays about every six and a half hours. This time period is short enough to cause no damage to human tissue but long enough to be tracked and detected in the human body. This element can be used to diagnose disorders of many organs of the human body—the thyroid, brain, lungs, kidneys,

FIGURE 8-15

Americium (am uh REE see um) is a synthetic element used in some home smoke detectors.

FIGURE 8-16

The person in this picture has been given a small amount of technetium, which gathered in the kidneys. Certain medical machines detect this element. A doctor can look at this picture and tell whether or not this person has a kidney problem.

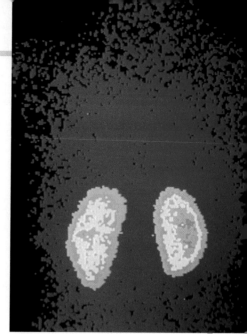

heart, and liver, to name just a few, as shown in **Figure 8-16.** When the synthetic element neptunium decays, it forms another synthetic element named plutonium (plew TOH nee um). Some forms of plutonium are used in nuclear reactors and bombs. Small amounts of another form of this element power some pacemaker batteries.

When reading about topics such as synthetic elements, it is helpful to take notes on what you read. These notes will help you organize your thoughts on the topic.

 Skill Builder: *Taking Notes*

LEARNING the SKILL

1. Read the entire text you wish to take notes on.

2. Identify the main topic in each paragraph. Write down each main idea in your own words. Leave room after each one.

3. After each main idea, jot down words and phrases that support the main idea. You may use letters or numbers to help organize your notes.

4. Read your notes aloud, using the words and phrases to make complete sentences. This reading checks your understanding of the topic.

5. If your notes don't make sense or seem incomplete, reread the passage and make any necessary changes in your notes.

PRACTICING the SKILL

1. In your own words, state the main topic of each paragraph on these two pages. List supporting details for each main idea.

2. List at least five words or phrases that describe synthetic elements.

3. Use words and phrases to state some of the advantages and disadvantages of making elements.

APPLYING the SKILL

Recall what you've learned in this chapter about elements. Use note-taking skills to briefly compare and contrast naturally occurring elements and synthetic elements.

Maya Lin's Civil Rights Memorial

Chinese-American architect Maya Lin became famous when she designed the Vietnam Veterans' Memorial in Washington, DC. Since then, she has created other monuments. One of those is the Civil Rights Memorial in Montgomery, Alabama.

This memorial honors 40 people—including Dr. Martin Luther King, Jr.—who died during the Civil Rights Movement, which started during the mid 1950s. Maya Lin was inspired by Dr. King's famous 1963 speech called "I Have a Dream." In it, King quoted: ". . . until justice rolls down like water and righteousness like a mighty stream." Lin used this quotation in the memorial.

The Right Materials

Lin used flowing water and dark granite in the memorial. The black granite wall stands in back, inscribed with the quotation, with water flowing over it. The granite, which is a mixture of the minerals feldspar, quartz, and mica, is a rock that formed when melted rock cooled. Granite is hard because the molecules that form it are held tightly together. It won't be worn down by the water that rushes over it.

In front is a round granite table on which are carved the names of those who died in the movement for racial equality. Water, which is a compound, gently flows off the edges all around the tabletop. "The water remains very still until you touch it," Lin has said. "Your hand causes ripples, which transform and alter the piece, just as reading the words completes the piece."

SCIENCE *Online*

Find out more of the story of the Civil Rights Memorial at Glencoe Science Online, *science.glencoe.com*

...UNTIL JUSTICE ROLLS DOWN LIKE WATERS AND RIGHTEOUSNESS LIKE A MIGHTY STREAM

MARTIN LUTHER KING JR

Read the statements below that review the most important ideas in the chapter. Using what you have learned, answer each question in your Science Journal.

1. Matter is anything that has mass and takes up space. The basic unit of matter is the atom. The modern atomic theory was developed by many different scientists over hundreds of years. *What was the main idea of Dalton's atomic theory?*

2. Atoms are made up of smaller particles. The two heavier particles are found in the nucleus. The lighter particles are found in the space surrounding the nucleus. *What are the three particles that make up an atom? What is the charge of each?*

3. There are several types of matter. Most elements occur naturally, but some are synthetic. A compound is made up of more than one type of element. *Describe each of the two types of mixtures.*

Using Key Science Words

atom
chemical formula
compound
electron
element

matter
mixture
neutron
periodic table
proton

Match each phrase with the correct term from the list of Key Science Words.

1. particle in the nucleus of the atom that does not carry a charge
2. made by chemically combining two or more different kinds of elements
3. an abbreviation used by chemists to identify compounds quickly
4. anything that has mass and occupies space
5. the lightest of the three main particles found in an atom

Checking Concepts

Choose the word or phrase that completes the sentence.

6. A mixture can contain which of the following kinds of matter?
   a. elements   c. compounds
   b. atoms   d. all of the above

7. All known elements are organized on the periodic table _____.
   a. by the date they were discovered
   b. in alphabetical order
   c. according to the number of protons in an atom of the element
   d. as solids, liquids, or gases

8. An atom is composed of _____.
   a. electrons, nuclei, and protons
   b. compounds, mixtures, and nuclei
   c. protons, compounds, and neutrons
   d. protons, neutrons, and electrons

9. A chemical formula is composed of _____.
   a. a variety of atoms and their weights
   b. chemical symbols and numbers of atoms
   c. various compounds and mixtures
   d. protons, electrons, and neutrons

10. The particle that is negatively charged in an atom is the _____.
    a. electron   c. neutron
    b. nucleus   d. proton

Thinking Critically

Answer the following questions in your Science Journal using complete sentences.

11. As an element, oxygen is a gas. Explain how oxygen can be the most common element in Earth's crust, yet Earth's crust is not a gas.

12. When apple juice is advertised as being *pure*, does that mean that apple juice is an element? Explain.

13. Is the art shown here a molecule of propane or a model of a propane molecule? Explain.

14. Parentheses show that a group of atoms is a unit. How many atoms are in calcium nitrate, $Ca(NO_3)_2$?

15. Oxygen from the air comes in contact with the metal in an aluminum chair. The result is a white powder coating the aluminum. Is this powder an element, a compound, or a mixture? How do you know?

Developing Skills

If you need help, refer to the description of each skill in the Skill Handbook.

16. **Interpreting Data:** Fritz had a sample of sand from the beach. He added some of it to water and stirred, then dried it. It weighed the same as it did originally. He then heated the sample and weighed it again. It was still the same. Finally, he ran a magnet around in the sand and again weighed the sand. This time, its mass was less than the original mass. What was removed from the sand? Explain.

17. **Making and Using Tables:** The name of the element gold starts with the letter *G*, but the symbol is Au. Using the periodic table of the elements, make a table listing five other elements whose symbols don't start with their first letter and give their symbols.

18. **Comparing and Contrasting:** Compare and contrast the compound sugar with the mixture chicken soup.

19. **Observing and Inferring:** Infer the effect of a negatively charged balloon, loaded with electrons, on a pile of hair that you brushed off your dog. Explain.

20. **Interpreting Scientific Illustrations:** Using the pictures of the compounds below, write the chemical formula for each one.

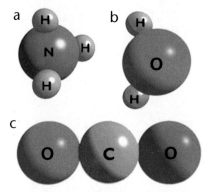

Performance Assessment

1. **Display:** Create a display of increasingly more complex kinds of matter. You may use models. Show the parts of an atom, a complete atom, a compound, and a mixture. Label the appropriate parts of each.

2. **Designing an Experiment:** You are able to separate a mixture of sand and sugar and then a mixture of iron filings and gravel. Design an experiment that allows you to separate a mixture of iron filings, gravel, and sugar.

Chapter Preview

Skills Preview

▶ **Skill Builders**
- classify
- develop multimedia presentations

▶ **MiniLABs**
- observe
- compare

▶ **Activities**
- design an experiment
- classify
- hypothesize
- observe
- compare
- conclude

Chapter 9

Properties and Changes

You've just received your favorite magazine. You open it up and see a picture of a place where volcanic ash once thundered down the sides of exploding volcanoes. Sounds as if you're looking at a Hollywood movie set. No, it's the Cascade Mountains of Oregon and Washington. What else could you see there? Cinder cones formed by bubbly lava. Lava tubes once filled with oozing melted rock, or magma. Before you look any further, try the activity below to compare some of the kinds of rocks you'll see.

EXPLORE ACTIVITY

Compare Characteristics

1. Obtain samples of the volcanic rocks obsidian (ahb SIH dee un) and pumice (PUH mus). The samples should be about the same size.
2. Which sample is heavier?
3. Compare the colors of the two rocks.
4. Look at the surfaces of the two rocks. How are the surfaces different?
5. Place each rock in water and observe whether or not it floats.

Obsidian

Pumice

Science Journal

In your Science Journal, make a table that compares the observations you made about the rocks.

9•1 Physical and Chemical Properties

What YOU'LL LEARN

- Physical properties and chemical properties of matter
- The properties of acids and bases and the differences between them

Science Words:
physical property
density
state of matter
chemical property

Why IT'S IMPORTANT

When you learn about physical and chemical properties, you can better describe the world around you.

Physical Properties

Would it surprise you to know that both of the rocks that you examined in the Explore activity started out as the exact same kind of lava? So why are they so different? As you will find out later, they were produced by two different kinds of volcanic activity. Characteristics that you observe about matter are often related to how that matter was formed.

In the Explore activity, you used your senses to classify different types of matter. This classification helps you better understand what the types of matter are and how they were formed.

Common Physical Properties

Do you have a favorite souvenir—an unusual rock or seashell, a funny hat, or a special cup? Take a minute to describe that souvenir in as much detail as you can. What features did you use in your description—color, shape, and hardness? These features are all properties, or characteristics, of the souvenir. Scientists use the term **physical properties** to describe characteristics you can detect with your senses. All matter, such as the hat in **Figure 9-1**, has physical properties. Practice identifying physical properties by listing some physical properties of your science book.

For most of your life, you have been aware of some physical properties, such as color, shape, smell, and taste. You may not be as familiar with others, such as texture and density. Texture is how rough or smooth something is.

FIGURE 9-1

What are some of the physical properties of this hat?

Density relates the mass of something to how much space it takes up, which is its volume.

Density

In the Explore activity, you compared the densities of the rock samples when you decided which equal-sized rock was heavier. Which rock was more dense—pumice or obsidian? If someone asked you which is heavier, the grocery bag in **Figure 9-2** or a grocery bag full of cans of soda, you'd probably say that the bag of canned soda is heavier. It is heavier because the density of the cans of soda is greater than the density of the snacks. If you found the mass of each full bag and divided each mass by its volume, you could find the density of what was in each bag.

Whether you realize it or not, you see examples of density every day. Ice floats in a soft drink because the density of ice is less than that of the soft drink. You shake a bottle of oil-and-vinegar salad dressing before using it. Oil is less dense than vinegar, so the oil rises to the top.

Density also can be used to identify compounds and elements because each has its own density. For example, suppose you have 20 mL (volume) of wood alcohol. It has a mass of 15.8 g. You can find the density of this alcohol by dividing its mass by its volume.

$$\text{Density} = \text{mass}/\text{volume}$$

$$D = m/v = 15.8 \text{ g}/20 \text{ mL} = 0.79 \text{ g/mL}$$

Mini LAB

Compare Densities

Compare the densities of some common liquids.

1. Lightly color about 20 mL of water with red food coloring and about 20 mL of white corn syrup with blue food coloring.
2. Pour the water into a clear, colorless, plastic cup.
3. Slowly add the corn syrup to the cup by pouring it down the side.
4. Next, in the same way, add about 10 mL of rubbing alcohol to the cup.

Analysis

1. What did you observe?
2. Explain what happened in terms of the densities of each liquid that you added.

FIGURE 9-2

This bag is easy to carry because its contents have a low density.

Calculating Density

Example Problem:

An astronomer found a small meteorite. The astronomer knew that nickel and iron are often found in meteorites. Nickel has a density of 8.9 g/mL, and iron has a density of 7.9 g/mL. The meteorite had a mass of 176 g and a volume of 20 mL. What is the density of the meteorite? Is the meteorite mostly nickel or mostly iron?

Problem-Solving Steps

1. Find the density of the meteorite by dividing its mass by its volume.

$$176 \text{ g} \div 20 \text{ mL} = 8.8 \text{ g/mL}$$

2. Compare this density to the densities of iron and nickel. The density is closer to the density of nickel.

Solution: The density of the meteorite is 8.8 g/mL, which is close to the density of nickel. Therefore, the meteorite is mostly nickel.

Practice Problem

On a trip to the Southwest, you bought a bracelet that was labeled as pure silver. Is it really pure silver? The density of silver is 10.5 g/mL. The bracelet has a mass of 42 g and a volume of 5 mL. What is the density of the metal in the bracelet? Is the metal pure silver? How do you know?

Formation

The physical properties of many materials are related to the way they form. Remember the two rocks you described in the Explore activity. The two rocks are made of the same materials, but their physical properties are different. One is dark; one is light in color. One sinks; one floats. One is smooth; the other is rough and jagged. Their physical properties are different because they formed in different ways. Take a closer look at what happened.

Pumice formed when melted, or molten, rock came into contact with water while it was still under the ground. This extremely hot mixture of water and molten rock exploded out the top of the volcano. The water then boiled, forming lots of little holes in the rock. The lava cooled quickly, preserving the little holes as it hardened. The result is a brown-colored rock that is so light it floats on water. Obsidian, on the other hand, is a dark rock that sinks. It came from the same kind of lava, and it also cooled quickly. But it didn't contact water as it formed, so no bubbles resulted.

FIGURE 9-3

All three states of water are present here—solid, liquid, and gas.

State of Matter

You know that solid pumice is different from liquid magma. State of matter is another physical property. The **state of matter** tells you whether a sample of matter is a solid, a liquid, or a gas. For example, if you hold an ice cube in your hand, you have water in its solid state. Water exists in the liquid state in oceans, in rivers, and in your bathtub. It also exists as a gas in the air. In each case, each molecule of water is the same—two hydrogen atoms and one oxygen atom. But water appears different because it exists in different states, as shown in **Figure 9-3**.

Physical Properties of Acids and Bases

What do you think of when you hear the word *acid?* Do you picture a dangerous chemical that can burn your skin, make holes in your clothes, and even destroy metal? Some acids, such as hydrochloric acid, are like that. But not all acids are possibly harmful. One example is shown in **Figure 9-4**. Every time you eat a citrus fruit such as an orange or a grapefruit, you eat citric and

FIGURE 9-4

When you sip a carbonated soft drink, you drink carbonic and phosphoric (faws FOR ihk) acids.

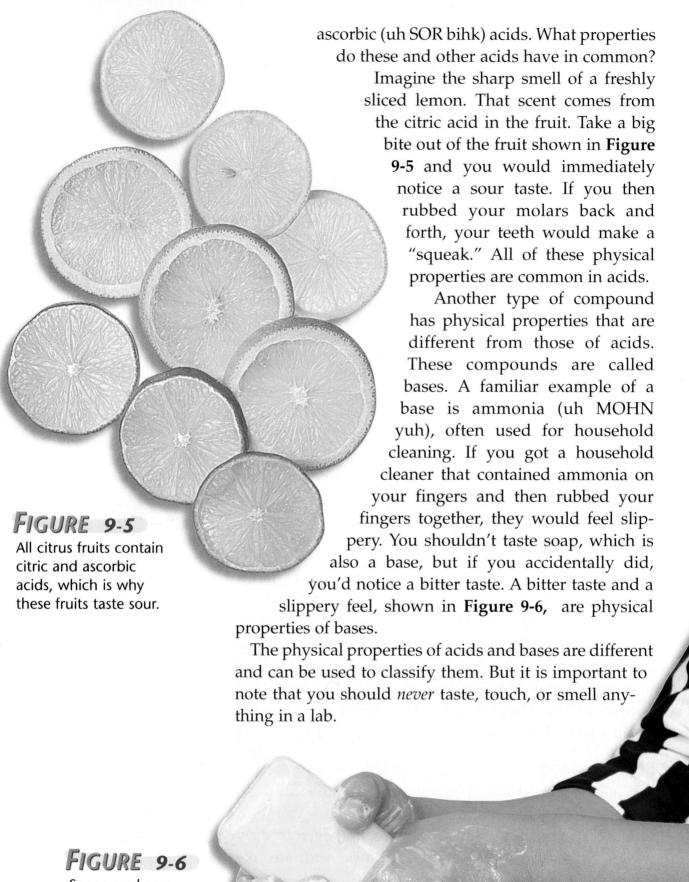

ascorbic (uh SOR bihk) acids. What properties do these and other acids have in common? Imagine the sharp smell of a freshly sliced lemon. That scent comes from the citric acid in the fruit. Take a big bite out of the fruit shown in **Figure 9-5** and you would immediately notice a sour taste. If you then rubbed your molars back and forth, your teeth would make a "squeak." All of these physical properties are common in acids.

Another type of compound has physical properties that are different from those of acids. These compounds are called bases. A familiar example of a base is ammonia (uh MOHN yuh), often used for household cleaning. If you got a household cleaner that contained ammonia on your fingers and then rubbed your fingers together, they would feel slippery. You shouldn't taste soap, which is also a base, but if you accidentally did, you'd notice a bitter taste. A bitter taste and a slippery feel, shown in **Figure 9-6,** are physical properties of bases.

The physical properties of acids and bases are different and can be used to classify them. But it is important to note that you should *never* taste, touch, or smell anything in a lab.

FIGURE 9-5

All citrus fruits contain citric and ascorbic acids, which is why these fruits taste sour.

FIGURE 9-6

Soaps are bases, which is why they are slippery.

Chemical Properties

You've noticed the density and the state of an ice cube. You've described the color and texture of rocks. You've noticed the taste of acid in a lemon and the slippery feel of a base such as ammonia. However, a description of something using only physical properties is not complete. What type of property describes how matter behaves?

Common Chemical Properties

If you strike a match on a hard, rough surface, the match will probably start to burn. The element phosphorus (FAWS for us) and the wood in the match combine with oxygen in the air to form new materials. Why does that happen? The phosphorus and the wood both have the ability to burn. The ability to burn is a chemical property. A **chemical property** is a characteristic of a substance that allows it to change to a new substance.

You see an example of a chemical property when you leave a half-eaten apple on your desk, and it turns brown. The property you observe is the ability to react with oxygen. Two other chemical properties are shown in **Figure 9-7**.

FIGURE 9-7

The chemical properties of a material often require a warning about its careful use.

A Gasoline is flammable. Gas pumps warn customers not to get near them with anything that might start the gasoline burning.

B Toxicity (tahk SIS uh tee), another common chemical property, indicates how poisonous something is. Workers who use toxic chemicals have to be careful.

FIGURE 9-8

Gold and iron have different chemical properties that make them suitable for different uses.

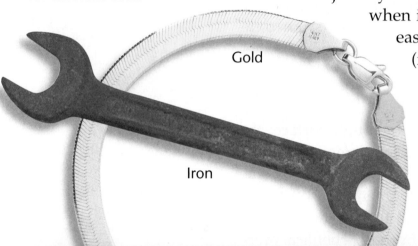

Gold

Iron

Look at **Figure 9-8.** Would you rather wear a bracelet made of gold or one made of iron? Why? Iron is less attractive and less valuable than gold. It also has an important chemical property that makes it unsuitable for jewelry. Think about what happens to iron when it is left out in the air. Iron rusts easily because of its high reactivity (ree ak TIV uh tee) with oxygen in the air. Reactivity is how easily one thing reacts with something else. The low reactivity of silver and gold, in addition to their desirable physical properties, makes those metals good choices for jewelry.

USING TECHNOLOGY

Properties and Pools

Did you ever wonder why chlorine is added to swimming pools? Chlorine compounds change the properties of the pool water. Chlorine reacts with cellular substances, killing bacteria, insects, algae, and plants.

Any time you have standing water, mosquitoes and other insects can lay eggs in it. Various plants and algae can also turn a sparkling blue pool into a slimy green mess. Bacteria are another problem. When you go swimming, you bring along millions of uninvited guests—the normal bacteria that live on your skin. The chlorine compounds kill the bacteria—as well as insects, algae, and plants that may be in the pool.

Chlorine makes the water more acidic, but it can cause other problems as well. Have your eyes ever burned after swimming in a pool? The chlorinated water can irritate the skin and eyes of swimmers.

SCIENCE
Online

Pool owners need to keep track of several different chemicals as they care for their swimming pools. Visit Glencoe Science Online at *science.glencoe.com* for a link about pool maintenance.

Chemical Properties of Acids and Bases

You have learned that acids and bases have physical properties that make acids taste sour and bases taste bitter and feel slippery. But the chemical properties of acids and bases are what make them both useful and sometimes harmful.

Many acids react with, or *corrode*, certain metals. Have you ever used aluminum foil to cover leftover spaghetti and tomato sauce? The next day, you might find small holes in the foil where it has come into contact with the tomatoes in the sauce, as shown in **Figure 9-9.**

The acids in tomato sauce, oranges, carbonated soft drinks, and other foods won't hurt you. However, many acids can damage plant and animal tissue. Nitric (NITE rihk) acid and sulfuric (sulf YER ihk) acid are found in small amounts in rain in areas that have a lot of pollution in the air. This rain, called acid rain, harms plant and animal life in areas such as the one shown in **Figure 9-10,** where acid rain falls. Sulfuric acid that has no water mixed with it is useful in many industries because it removes water from certain materials. However, that same property causes burns on skin that touches sulfuric acid.

A strong base also can damage living tissue. It is not uncommon for someone who smells strong ammonia to get a bloody nose from the fumes. And the reason that ammonia feels slippery to the touch is that the base actually reacts with fat that lies under the top layer of skin cells in your fingertips.

FIGURE 9-9

Aluminum reacts easily with acids, which is why acidic food, such as tomatoes, should not be cooked or stored in aluminum.

FIGURE 9-10

These trees show damage caused by acid rain.

Activity 9-1

Design Your Own Experiment
Homemade pH Scale

The stronger an acid or base, the more likely it is to be harmful. A scale, called the pH scale, is used to measure how strong acids and bases are. Solutions with pH of 1 to 6 are acids; pH 7 is neutral; and a pH of 8 to 14 is a base. The strongest acids have a pH of 1, and the strongest bases have a pH of 14. In this activity, you will measure the pH of some things you commonly use. Use treated paper that, when dipped into a solution, turns a color. Check the color against the chart below to find the pH of the solution.

Possible Materials
- vial of pH paper, 1–14
- pH color chart
- distilled water
- fruit juices
- vinegar
- salt
- sugar
- soft drinks
- household cleaners
- soaps and detergents
- antacids

PREPARE

What You'll Investigate
How acidic or basic are some common household items?

Form a Hypothesis

Think about the properties of acids and bases. In your group, **make a hypothesis** about which kinds of solutions you are testing are acids and which kinds are bases.

Goals

Design an experiment that allows you to test solutions to find the pH of each.

Classify a solution as an acid or base according to its pH.

Safety Precautions

Never eat, taste, smell, or touch any chemical during a lab.

Data and Observations

Solution to be tested	pH	Acid, base, or neutral

pH	Color	pH	Color
1		8	
2		9	
3		10	
4		11	
5		12	
6		13	
7		14	

PLAN

1. As a group, **decide** what materials you will test. If a material is not a liquid, **dissolve** it in water so you can test the solution.
2. **List** the steps and materials that you need to **test** your hypothesis. Be specific. Will you **repeat** any part of the experiment?
3. Before you begin, **copy** a data table like the one shown into your Science Journal. Be sure to leave room to **record** results for each solution tested. If there is to be more than one trial for each solution, include room for the additional trials, too.
4. **Reread** the entire experiment to make sure that all the steps are in logical order.

DO

1. Make sure that your teacher approves your plan and data table. Be sure that you have included any suggested changes.
2. Carry out the experiment as planned and approved. Wash your hands when done.
3. Be sure to **record** your observations in the data table as you **complete** each test.

CONCLUDE AND APPLY

1. **Compare** your findings with those of other student groups.
2. Were any materials neither acids nor bases? How do you know?
3. Using your data table, **conclude** which types of materials are usually acidic and which are usually basic.
4. **APPLY** Perhaps you have been told that you can use vinegar to dissolve hard-water deposits because vinegar is an acid. If you run out of vinegar, which of the following—lemon juice, ammonia, or water—could you most likely use instead of vinegar for this purpose?

What happens in reactions between acids and bases? Acids and bases are often studied together because they react with each other to form water and other useful compounds called *salts*. Look at **Figure 9-11.** That familiar stuff in your salt shaker—table salt—is the most common salt. Table salt, sodium chloride, is formed by the reaction between the base sodium hydroxide and hydrochloric acid. Other useful salts are calcium carbonate, which is chalk, and ammonium chloride, which is used in certain batteries.

FIGURE 9-11

These everyday items contain salts.

Section Wrap-up

1. Make a chart and list ten things in your home. On the chart, include the following physical properties for each item: color, state of matter, texture, and hardness.

2. Design three separate symbols that could be put on containers to warn chemists that a certain chemical could burn easily and be toxic and corrosive.

3. **Think Critically:** Explain why the temperature at which wax melts is a physical property and not a chemical property of wax.

4. **Skill Builder**

   Classifying Classify each of the following properties as being either physical or chemical.
   a. Iron will rust when left out in the air.
   b. Lye feels slippery.
   c. Iodine is poisonous.
   d. Solid sulfur shatters when struck.

   If you need help, refer to Classifying on page 542 of the **Skill Handbook.**

Science Journal

Think about safety factors you should check around your home. Are they based upon physical properties or upon chemical properties? Explain.

✔ Check battery in smoke alarm

✔ Plan an escape route if there is a fire

The *Hindenburg* Disaster

On May 6, 1937, the *Hindenburg* drifted down to land at Lakehurst, New Jersey. The huge German airship had just flown across the Atlantic Ocean with 97 people on board. The *Hindenburg* was one of the largest airships ever built. It was 245 m long and 41 m wide.

Airships were the first flying machines that had a crew and could be steered. Most had a wooden or metal framework inside the balloon covering.

Airships were filled with lighter-than-air gas. Hydrogen, which has the physical property of being the least dense of all gases and is therefore the lightest element, provided the greatest amount of lift. But hydrogen gas has chemical properties that can make it dangerous to use. Hydrogen is flammable, and it burns explosively.

Up In Flames

As the *Hindenburg* prepared to land, onlookers saw a small burst of flame near the ship's tail. Suddenly, the entire tail section was on fire. The fire roared forward along the ship. In seconds, the *Hindenburg* was completely engulfed in flames. When it crashed to the ground, 36 people were killed.

To this day, no one is exactly sure what caused the *Hindenburg* to catch on fire. One hypothesis is that a spark of electricity in the air near a leak in the ship's hull ignited the gas. Another hypothesis is that the skin of the airship first caught on fire, which then set the hydrogen burning. Whatever its cause, the fiery *Hindenburg* disaster marked the end of airships as a way for people to travel.

SCIENCE *Online*

In the Netherlands, university students have been studying the environmental benefits of bringing back the use of airships. Visit Glencoe Science Online at *science.glencoe.com* for a link about their findings.

Science & Society

9•2 Road Salt

What YOU'LL LEARN

- How road salt can both make travel safer and damage the environment
- Why main ideas must be identified to understand scientific writings

Why IT'S IMPORTANT

Knowing the properties of materials helps you decide how to use them.

Salt and Ice

In many climates, weather changes with the seasons. Summers are hot and winters are cold. Areas that have such climates can receive a lot of snow in the winter. Changes in temperature can cause the snow to melt, then refreeze, turning it to ice. Ice is the solid form of water. It can be dangerous to cars and trucks traveling on icy roads and to people walking on ice-covered sidewalks.

In some areas, such as the one shown in **Figure 9-12,** people spread rock salt on slick sidewalks and roads. Why? Salt changes one physical property of water—the temperature at which it freezes. When salt is added to water, the salt causes the water to freeze at a temperature lower than the freezing temperature of pure water, which is 0°C. Applying salt to slick areas melts any ice present, unless the ground temperature is colder than the normal freezing temperature of water.

Pass (on) the Salt, Please

Salt can make winter travel safer. But it also can cause problems. If they have not been protected from salt, cars driven on salted roads can begin to rust after only a few winters. Another rust problem is shown in **Figure 9-13.**

In addition to damage that you can easily see, salt causes other problems in the environment. As snow melts, salty water runs off the roads and sidewalks onto nearby land. Too much salt in the soil makes it difficult for many types of plants to grow. Salty water also flows into waterways, polluting nearby rivers,

FIGURE 9-12

This salt will help melt the snow and ice on the highway, making it safe for travelers.

16908

lakes, and streams that provide much of the water needed by towns, cities, and farmlands.

Salt Substitutes

Because of the damage done to the environment, many places now use sand, cinders, or crushed gravel to provide the traction needed to prevent slipping on snowy roads and walkways. These materials do not harm the environment as salt can.

Chemicals called deicers also can be used rather than salt. A deicer keeps snow or ice from attaching to the pavement. A chemical called CMA is a deicer that causes little or no damage to steel. CMA also breaks down into harmless materials and does not hurt plants or drinking-water supplies.

FIGURE 9-13

Salt also speeds the rusting of steel structures, such as bridges, guardrails, and lampposts.

 ## Skill Builder: Identifying the Main Idea

LEARNING the SKILL

1. Read carefully. Use the title and headings to get an idea of what the whole article is about.

2. Reread each paragraph, one at a time. Find the main idea of each paragraph by choosing a single sentence that best states the topic of the paragraph.

3. If a single sentence does not clearly state the main idea of a paragraph, reword what is in the paragraph to state the main idea.

4. Pick out details that support the main idea. These details can be phrases or words from the paragraph, pictures, or captions.

PRACTICING the SKILL

1. What is the main idea of the first paragraph in the section called "Salt and Ice"?

2. **a.** What are deicers? How do they compare with salt?

   b. Which question in part **a** refers to the main idea? Which question refers to supporting details?

3. Reread "Road Salt." Study your answers to the questions. In your own words, what is the main idea?

APPLYING the SKILL

Cloud seeding is another process that can help humans but could harm the environment. Find the main ideas in readings about cloud seeding. From these main ideas, find out how cloud seeding relates to weather and how it affects the environment.

Physical and Chemical Changes

What YOU'LL LEARN

- How to identify physical and chemical changes
- Examples of how physical and chemical changes affect the world we live in

Science Words:
physical change
chemical change

Why IT'S IMPORTANT

If you understand the changes in the things around you, you will better be able to use these changes.

Physical Change

It's picnic time. You offer to help carry the food to the picnic table—grilled chicken, potato salad, baked beans, and a big watermelon. On the last trip—with the watermelon—you suddenly lose your hold on the large, slippery fruit. It goes crashing to the ground and splits open, becoming an example of a principle that you will learn in this section—physical change.

What is physical change?

Most matter can undergo **physical change,** which is any change in size, shape, or form. The identity of the matter stays the same. Only the physical properties change. Look at **Figure 9-14.** The watermelon underwent a physical change. It went from being one large, round melon to being many smaller pieces splattered all over the floor. It is still watermelon; it just looks different.

Examples of Physical Changes

How can you recognize a physical change? Just look to see whether or not the matter has changed size, shape, or form. If you cut a watermelon into chunks, the watermelon has changed both size and shape. That's definitely a physical change. If you pop one of those chunks into your mouth and bite it, you have changed the watermelon's size and shape again.

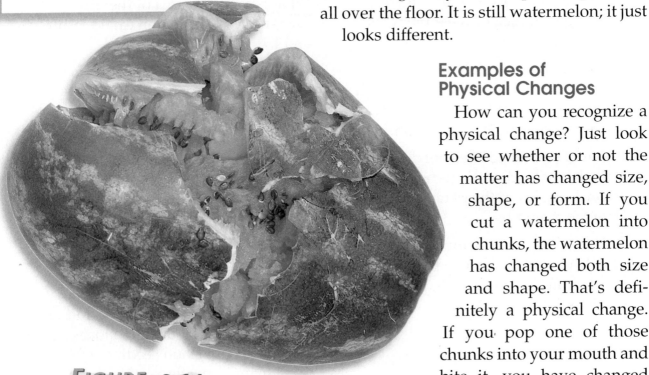

FIGURE 9-14

The physical change this watermelon underwent changed its form.

There's another way that matter can undergo a physical change. It can change from one state to another. After dropping the watermelon, suppose you save the picnic by bringing out Popsicles for dessert. It's a hot day, and the Popsicles start to melt and drip onto the patio as you and your friends eat them. What kind of physical change is happening? A Popsicle is a frozen, solid mixture of water, sugar, food coloring, and flavoring. In the hot sun, the water in the Popsicle changes state and melts, turning into a drippy liquid. As the drops of melted Popsicle on the patio sit in the sun, the water changes state again, evaporating to become a gas. In each case, the Popsicle is composed of the same ingredients. But the water it contains is in different states as it changes from a solid to a liquid to a gas. Other examples of change of state are shown in **Figure 9-15.**

FIGURE 9-15
The four most common changes of state are shown here.

C A solid will melt, becoming a liquid.

B As it cools, this liquid metal will become solid steel.

A Water vapor in the air changes to liquid water when dew forms.

D Liquid water in perspiration changes to a gas when it evaporates from your skin.

FIGURE 9-16

You can see dramatic examples of physical weathering caused by water and wind on rocky coastlines and in deep canyons.

Some physical changes, such as the melting Popsicle, happen quickly. Others take place over a long time. Physical weathering is a physical change that is responsible for much of the shape of Earth's surface. Examples are shown in **Figure 9-16**. Examples also can be found in your own schoolyard. All of the soil that you see comes from physical weathering. Wind and water erode rocks, breaking them into small bits. Water fills cracks in rocks. When it freezes, the ice splits the rock into smaller pieces. No matter how small the pieces of rock are, they are still made up of the same things that made up the bigger pieces of rock. The rock has simply undergone a physical change. Gravity, plants, animals, and the movement of land during earthquakes also help cause physical changes on Earth.

Chemical Changes

Your bicycle gets a chip in the paint. Soon, the bike has a spot of rust. A shiny copper penny becomes dull and dark. An apple left out too long begins to rot. What do all these changes have in common? In each of these examples, the original materials that make up the bike, the penny, and the apple change into other materials that have different properties; such a change is called a **chemical change.**

Examples of Chemical Change

Chemical changes are going on around you—and inside you—every day. When you eat, food undergoes chemical changes so that your body can use it. When plants use water and carbon dioxide to make sugar and oxygen, a chemical change occurs. Many industries make use of chemical changes to manufacture useful products from raw materials. Most products—from the toothpaste you use every morning to the clothes you wear—are produced by chemical changes.

The surface of Earth is a product of chemical changes. Remember that physical weathering breaks down rocks. Chemical changes occur as chemical weathering. Caves, as shown in **Figure 9-17,** are formed by chemical weathering. Acid rain also brings about chemical weathering. Look at a statue that has been outside for a long time. How do you think it has changed over the years? The acid in acid rain is responsible for damaging marble statues and building materials, as well as for making some lakes toxic to wildlife.

FIGURE 9-17

Over many years, rainwater slowly reacts with layers of limestone rock. It forms caves and collects minerals that it later deposits as cave formations.

Mini LAB

Compare Chemical Changes

Show how salt speeds up the chemical change between oxygen and iron.

1. Separate a piece of fine steel wool into two halves.
2. Dip one half in tap water and the other half in the same amount of salt water.
3. Place both pieces of steel wool on a paper plate and let them sit overnight. Observe any changes.

Analysis

1. What happened to the steel wool that was dipped in the salt water?
2. What might be a common problem with machinery that is operated near an ocean?

Signs of Change

Ice melts, paper is cut, metal is hammered into sheets, and clay is molded into a vase. Seeing signs of these physical changes is easy—something changes shape, size, form, or state.

Sometimes, it's just as easy to tell that a chemical change has occurred. When wood burns, you see it change to ash and other products. When you combine soap and some types of tap water, soap scum forms a ring in your sink. Another example is shown in **Figure 9-18.** When this cake is baked, changes occur that make the cake become solid. The chemical change that occurs when baking powder mixes with water forms bubbles that make the cake rise. Raw egg in the batter undergoes changes that make the egg solid.

In all of these examples, you know that a chemical change occurs because you can see that a new substance forms. It's not always so easy to tell when new substances are formed. What are other signs of chemical change? Do Activity 9-2 to find out.

FIGURE 9-18

Chemical changes are common when food, such as cake, is cooked.

Sunset in a Bag

How do you know when a chemical change occurs? You'll see some evidence of chemical change in this activity.

What You'll Investigate
What is evidence of a chemical change?

Procedure

1. **Add** 20 mL of warm water and 5 mL of phenol red solution to the plastic bag. **Seal** the bag, and gently **slosh** the solution around to mix it.
2. Now, **add** a teaspoon of calcium chloride to the solution in the bag. Again, **seal** the bag and **slosh** the contents to mix the solution. **Record** any change in temperature.
3. **Open** the bag and quickly **add** a teaspoon of baking soda. Again, **seal** the bag and **slosh** the contents to mix the ingredients together. **Observe** what happens.

Conclude and Apply

1. What evidences of chemical change did you **observe**?
2. Is it always easy to tell whether or not energy is released? **Give an example** of a chemical change that does not show an obvious energy change.

Goals
- Observe a chemical change.
- Identify some signs of chemical change.

Materials
- baking soda
- calcium chloride
- phenol red solution
- warm water
- teaspoons, 2
- resealable plastic bag
- graduated cylinder

257

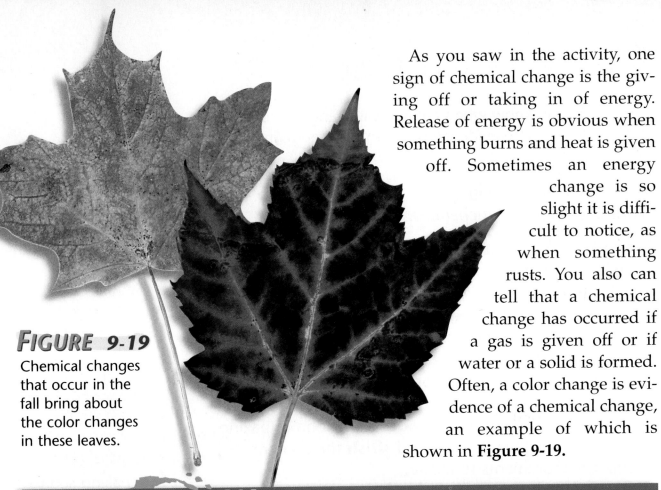

As you saw in the activity, one sign of chemical change is the giving off or taking in of energy. Release of energy is obvious when something burns and heat is given off. Sometimes an energy change is so slight it is difficult to notice, as when something rusts. You also can tell that a chemical change has occurred if a gas is given off or if water or a solid is formed. Often, a color change is evidence of a chemical change, an example of which is shown in **Figure 9-19.**

FIGURE 9-19
Chemical changes that occur in the fall bring about the color changes in these leaves.

Section ◯ Wrap-up

1. List five physical changes you can observe in your home. Explain how you decided that each change is physical.

2. When you cook an egg, what kind of change occurs—physical or chemical? Explain.

3. **Think Critically:** Which of the following involves a change of state: grinding beef into hamburger, pouring milk into a glass, making ice cream, or allowing soup to cool in a bowl? Explain.

4. *Skill Builder*
   Developing Multimedia Presentations
   Prepare a multimedia presentation that shows the steps in preparing, lighting, and extinguishing a wood fire. Identify each step as a physical or a chemical change. If you need help, refer to Developing Multimedia Presentations on page 564 of the **Technology Skill Handbook.**

USING MATH

No mass is lost during a chemical change, even if it seems like there might be. If 1000 g of wood burn in 16 g of oxygen, 98 g of ash are produced. How many grams of gases are also produced?

Read the statements below that review the most important ideas in the chapter. Using what you have learned, answer each question in your Science Journal.

1. A physical property can be observed without changing the makeup of the material. A chemical property indicates whether or not a material can undergo a chemical change. *Metals are flexible and shiny. Are these properties physical properties or are they chemical properties of metals?*

2. Acids and bases have properties that can be used to identify them. They react with each other to produce water and a salt. *Name one physical property and one chemical property of an acid.*

3. A physical change is a change in the size, shape, form, or state of matter. The makeup of the matter stays the same. In chemical changes, new materials are formed. *What are some signs that a chemical change has occurred?*

Using Key Science Words

chemical change   physical change
chemical property   _ physical property
density   state of matter

Match each phrase with the correct term from the list of Key Science Words.

1. mass divided by volume
2. describing an object using color, shape, size, texture, odor, or form
3. snowballs melting in the sun
4. acid rain damaging marble statues
5. matter as a solid, a liquid, or a gas

Checking Concepts

Choose the word or phrase that completes the sentence.

6. The temperature at which something boils is a _____.
   a. chemical change
   b. chemical property
   c. physical change
   d. physical property

7. A chemical change might be identified by release of a gas or _____.
   a. change of state
   b. energy
   c. liquids
   d. physical properties

8. New compounds are formed during a _____ change.
   a. chemical   c. seasonal
   b. physical   d. state

9. Solid, liquid, and gas are _____.
   a. physical changes
   b. physical properties of soil
   c. chemical changes
   d. three states of matter

10. A broken window would be considered a _____ change.
    a. chemical   c. neutral
    b. weathering   d. physical

Thinking Critically

Answer the following questions in your Science Journal using complete sentences.

11. Think about what you know about density. Could a bag of feathers be heavier than a bag of rocks? Explain.

12. Sugar dissolves in water. Is this a physical property or a chemical property of sugar?

13. When butane burns, it combines with oxygen in the air to form carbon dioxide and water. What two elements must be present in butane?

14. Identify each of the following as either a physical property or a chemical property.
    a. Sulfur shatters when hit.
    b. Gasoline burns explosively.
    c. Baking soda is a white powder.

15. Identify each of the following as either a physical change or a chemical change.
    a. Metal is drawn out into a wire.
    b. Sulfur in eggs tarnishes silver.
    c. Baking powder bubbles when water is added to it.

Developing Skills

If you need help, refer to the description of each skill in the Skill Handbook.

16. **Comparing and Contrasting:** Compare and contrast physical and chemical changes.

17. **Interpreting Scientific Illustrations:** Review the pictures below and determine whether each is a chemical change or a physical change.

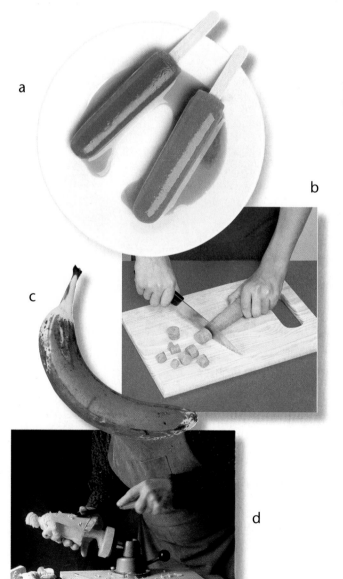

a

b

c

d

18. **Measuring in SI:** Suppose you have a rock that has a mass of 92 g. You pour water into a graduated cylinder to the 37-mL mark. After adding the rock, the water level reads 83 mL. What is the density of the rock?

19. **Observing and Inferring:** When you mixed two substances together, you observed that heat, gas, and some light were produced. Is this change chemical or physical? Explain.

20. **Concept Mapping:** Make a concept map that uses the terms *matter*, *physical properties*, *chemical properties*, *physical changes*, and *chemical changes*. Include an example for each term. Use the connecting terms *has* and *undergoes*.

Performance Assessment

1. **Display:** Create a display that demonstrates the characteristics of a chemical change. Be sure your display shows release of energy, change of color, and the formation of a solid.

2. **Designing an Experiment:** You know that some acids react with metal to produce a chemical change. Design an experiment that allows you to decide whether acids react with all metals or whether only a few combinations produce chemical changes.

Chapter Preview

Skills Preview

▶ **Skill Builders**
- infer
- recognize cause and effect
- use E-mail

▶ **MiniLABs**
- estimate
- analyze data

▶ **Activities**
- observe
- collect data
- calculate
- predict
- hypothesize

Chapter 10

Forces and Motion

Skateboarders fly up into the air, turn, and come down onto the ramp as if skateboarding were simple. Studying such motion helps you understand forces at work. After you've studied this chapter, you'll understand a lot more about the forces behind the motions of skateboarding. You can start learning about forces in the activity below.

EXPLORE ACTIVITY

Marble Skateboard Model

1. Using the picture as a guide, staple two pieces of construction paper together at one end to make a ramp.
2. Use the smallest ring binder you have to make a slope. Tape one edge of the ramp to the binder. Set a book on the other side to hold the ramp in place.
3. Roll a marble down the slope so that it travels up the ramp. Use the pen to mark the marble's highest point on the ramp.
4. Put a book under the ring binder, and roll the marble again.

Science Journal

In your Science Journal, tell which slope made the marble roll higher up the ramp. What property of the slopes made the marble roll to different heights?

10•1 What is gravity?

What YOU'LL LEARN

- How gravity affects everything
- The difference between weight and mass
- How weight and mass are measured

Science Words:
gravity
weight
mass

Why IT'S IMPORTANT

On Earth, gravity affects you every moment of every day.

Why Things Fall

To move an object, you usually have to do something to it. You have to push or pump a skateboard, open doors, or turn the pages of this book. But, sometimes all you have to do is let go of the object. It moves without being pushed, turned, or pumped. In the Explore activity on the previous page, the marble rolled down the ramp by itself. When a skateboarder jumps into the air, he or she always comes back to the ground. There is a reason why things fall or roll without being thrown or pushed.

Throughout history, people have tried to explain why things fall. The modern explanation for why things fall was first proposed in the 1600s. A scientist named Sir Isaac Newton said that everything—you, a marble, your desk, the moon—pulls on everything else. He called this pull gravity (GRA vuh tee). **Gravity** is a pull that every object exerts on every other object. **Figure 10-1** shows the pull of gravity in action.

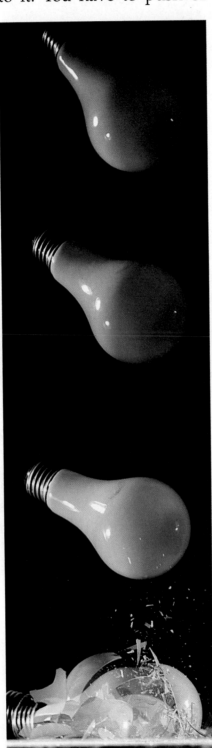

FIGURE 10-1
This lightbulb falls because of gravity between the lightbulb and Earth.

Gravity Treats Everything the Same

Newton said that gravity acts on everything, large or small, in the same way. For example, if a tennis ball and a bowling ball are dropped at the same time, which will hit the ground first? The answer is that they will both hit the ground at the same time. When the astronauts walked on the moon, they demonstrated this fact with a feather and a lead weight. On Earth, air slows a feather's fall, just as it slows the fall of the skydiver in **Figure 10-2.** On the moon, there isn't any air to get in the way. Puffs of dust on the moon let everyone know exactly when the feather and the lead weight hit the ground. Newton was right! The two objects hit the moon at exactly the same time.

You Have Gravity!

Remember, Newton's explanation said that everyone and everything has its own gravity. If that's true, why aren't you pulled toward a desk at the other side of the room? The answer has to do with matter. The more stuff, or matter, something has, the more pull it has. Earth is huge compared to a desk. The pull of Earth's gravity is so great that you must notice it. It keeps your feet on the ground. That desk across the room is pulling, but the pull is so small that you never notice it.

The More Matter, the More Gravity

Our moon and Jupiter are good examples of how matter affects gravity. The moon's gravity is about one-sixth as strong as Earth's. Jupiter, on the other hand, is much larger than Earth. Jupiter's gravity is about two and one-half times as strong as Earth's. Imagine how different skateboarding would be on the moon or on Jupiter.

Have you seen movies of the astronauts walking on the moon? Lower gravity and bulky space suits make it hard to walk like they do on Earth, so they hop. It seems dangerous because they jump so high. How high do you think astronauts could jump if they were on solid ground on Jupiter?

FIGURE 10-2

Air slows the fall of this skydiver. Without the parachute, he or she would fall much faster.

Problem Solving

Weighty Matters

You've done an activity with slopes and ramps. Now it's time to design two ramps of your own. One will be on the moon and one will be on Jupiter.

You'll need to design a jump ramp for the moon, where there is less gravity than on Earth, so things weigh less than on Earth. It won't take much force to shoot you high up into the dark sky.

Next, design one for Jupiter, where gravity is two and one-half times what it is here. You'll be heavy, but gravity will provide great acceleration. On Jupiter, it will be hard to launch the skateboard high.

Solve the Problem:

1. **How much steeper will the jump ramps need to be for the moon? Remember, the moon has only one-sixth of the gravity of Earth.**
2. **How much less steep could the ramps be for Jupiter?**
3. **Will you have to use different materials to support the different weights?**

Think Critically:

In what ways could both ramps be just like one on Earth? In what ways will they need to be different?

How do you measure gravity?

How do you suppose scientists know that the moon's gravity is one-sixth of Earth's gravity? How can you measure Earth's gravity? You measure the pull of gravity every time you use a bathroom scale.

Weight is a measure of how much Earth's gravity pulls down on an object. Every time you weigh yourself on a scale, you are measuring the pull of Earth's gravity. If you weigh 100 pounds on Earth, your weight on the moon would be 17 pounds. How much would you weigh on Jupiter? The answer is 250 pounds. Take your own weight and multiply it by 2.5. Imagine what it would be like to jump, run, or even walk on Jupiter.

Astronomers learned that Jupiter is larger than the moon by observing their sizes using telescopes. But you know that gravity doesn't depend on size—it depends on the amount of matter present. **Mass** is the amount of matter in an object. Mass doesn't change from place to place. Your weight would be less on the moon, but your mass wouldn't change. You have to take away or add something to change an object's mass. For example, you could shrink the mass of a candy bar by taking a bite from it.

How do you measure mass?

Mass is measured with a tool called a balance. See **Figure 10-3.** A balance is like a small, precise seesaw. To measure your mass, you could sit on one end of a seesaw while someone else stacked objects on the other end until the seesaw balanced on its pivot point. On Jupiter or the moon, it would still take the same mass of objects to balance your mass. Mass doesn't change from place to place.

It's important to remember that a balance measures mass, but a scale measures the effect of gravity on an object by measuring weight. A balance and a scale are not the same thing.

FIGURE 10-3

Both a balance and a seesaw can be used to find the mass of an object. When they balance, masses are equal on both sides.

Pivot points

Mini LAB

Estimating Mass
Your body is going to be a balance.

1. Get two plastic grocery bags—one with a kilogram of mass in it and one that is empty.
2. Hold one bag in each hand.
3. Have someone put cans of food into the empty bag until you think the two bags have the same mass and balance each other.
4. Add the masses of the cans together.

Analysis
1. A kilogram equals 1000 grams. How close did you come to balancing the kilogram of mass?
2. Would this bag balance be accurate enough for a science lab? Explain.

You may have heard someone say he or she "weighs" 50 kilograms. This isn't really correct. Mass and weight are two different things, as explained in **Figure 10-4.** It's important to use different units for them. Mass is measured in grams (g) or kilograms (kg). Weight is measured in newtons (N), named in honor of Sir Isaac Newton. On Earth, a medium-sized apple weighs about 1 N. A kilogram weighs about 10 N on Earth. So if your mass is 50 kg, your weight is about 500 N on Earth.

FIGURE 10-4
The mass of each of these things always stays the same. Their weights, however, can be different in different places.

Section Wrap-up

1. If an astronaut could hop 40 cm high on Jupiter, how high could he or she jump on Earth? On the moon?

2. **Think Critically:** If Earth's mass were smaller, how would Earth's gravity be affected?

3. **Skill Builder**
   Inferring Because gravity on Mars is about one-third of that on Earth, could you infer that Mars has more mass than Earth? Why? If you need help, refer to Inferring on page 550 in the **Skill Handbook.**

USING MATH

Imagine that you had a mass of 60 kg and a weight of 600 N. What would your weight and mass be on Mars if the gravity there were one-third as strong as Earth's?

How fast is "fast"?   10•2

Faster and Faster

Think of skating down Dead Man's Hill. Your heart starts to pound as you go faster and faster. You feel every bump in the pavement through your feet and into your knees as the curving path seems to move faster and faster past you. It's the pull of gravity that makes the trip down so much fun and the climb up such hard work. Keep thinking about this trip as you read on.

How fast were you going?

What was your speed near the bottom of the hill? Thirty kilometers per hour? Five kilometers per hour? Speed is a measure of how far you move in a given time. To find your speed, you need to know both time and distance. The path down Dead Man's Hill is 100 m long. Suppose a friend timed your trip down and found that it took you 20 s. Divide 100 m by 20 s and you've found your average speed: 5 meters per second, or about 18 kilometers per hour. "Meters per second" is abbreviated *m/s*. "Kilometers per hour" is abbreviated *km/h*. Speed can be read directly from a speedometer, such as the one shown in **Figure 10-5.**

What YOU'LL LEARN

- How to talk about the movement of objects
- What things affect movement

Science Words:
average speed
acceleration
friction

Why IT'S IMPORTANT

You can use speed, acceleration, and friction every time you move or cause something to move.

FIGURE 10-5
The speedometer in this car uses units of mi/h, or mph, and units of km/h.

269

Remember the Explore activity at the beginning of the chapter? At the top of the ramp, the marble wasn't moving at all. It gradually moved faster as it rolled down the ramp. Your trip down the hill was similar. You weren't moving at 5 m/s at the beginning, and you were moving faster than 5 m/s at the end. But your average speed was 5 m/s.

Average speed describes the movement for an entire trip. If you traveled at exactly 5 m/s from top to bottom of the hill, the 100-m trip would take 20 s. Look at **Figure 10-6** for another example of average speed.

FIGURE 10-6

Look at the graph. If the path of the skateboarder from start to finish is 52 m, what is his average speed?

Time Trials

Before a big car race, all the contestants must pass the time trials. Time trials are races against the clock instead of against other cars.

What You'll Investigate
Can time trials be used to predict the winner of a car race?

Procedure
1. **Construct** a racetrack using the two metersticks. Use the picture to help you.
2. **Test** your track with your car. If the car runs into the metersticks a lot, move the metersticks farther apart.
3. **Wind up** a car. Start it at the zero marks on the metersticks, and time its trip to the end of the metersticks. **Repeat** this three times and **record** your data.
4. **Calculate** the average time.
5. **Calculate** the car's average speed using the average time and the length of the track.

Conclude and Apply
1. **Compare** the average speed of your car to those of the other cars in your class.
2. **Predict** which car would win a race based on the results of your time trials. Test your prediction.
3. **Explain** whether time trials accurately predict which car will win a race.

Goals
- Conduct time trials with wind-up toys.
- Test the speed predictions from the time trials.

Materials
- metersticks (2)
- timer
- wind-up cars or other wind-up toys
- tape

Car # _____	Time (s)	Distance (m)
Trial 1		
Trial 2		
Trial 3		
Average		

FINISH

START

How do you measure "speeding up"?

Think back to your skateboarding trip down Dead Man's Hill. Instead of timing your trip like you timed the wind-up cars in the activity, you could use a speedometer or radar gun to measure your speed on the hill. These tools would give your speed at one particular moment in time. At the top of the hill, your speed was 0 m/s. Near the bottom of the hill, it was close to 10 m/s, the fastest speed you went.

As you were going downhill, you were going faster and faster. As you went down the hill, you followed the curving path. At the bottom, you slowed down and stopped. All these changes in speed and direction are examples of acceleration (ak sel uh RAY shun). **Acceleration** is a change in speed or direction. An acceleration can speed you up, slow you down, or change

USING TECHNOLOGY

Making Skateboards

As you might guess, a skateboard's wheels play an important part in its design. What the board is best used for is determined by the size and hardness of the wheels.

When Speed Is King

Skateboard wheels are made out of urethane, a kind of plastic. Some urethanes are softer than others. Softer wheels give a better grip. They are used for "skating" hills where the skater likes to carve his or her way down a hill—where the skater likes to pick up as much speed as possible. Softer wheels are also wider and taller. The increased height makes them faster because they travel farther than a short wheel each time they go around.

When Style's the Thing

Harder, smaller wheels are used by skaters who want to do tricks, such as tail slides, sliding the back of the board on curbs, and cutbacks, which are quick turns. The smaller size helps keep the board under control.

SCIENCE *Online*

In both in-line skating and skateboarding, forces and motion come into play. Visit Glencoe Science Online at *science.glencoe.com* for a link to a site discussing the physics of in-line skating.

your direction of travel. From top to bottom of the hill, you accelerated from 0 m/s to about 10 m/s. When you stopped, you accelerated from 10 m/s to 0 m/s.

Use your trip down Dead Man's Hill as an example to help explain acceleration. As you were going down the hill, you went a little faster each second—first 0 m/s, then 1 m/s, then 2 m/s, and so on. Your speed was increasing 1 m/s each second. Your acceleration was 1 m/s per second.

Stopping Is Important

At the bottom of the hill, it would be wise to stop before you run into a tree! Stopping is a change in speed, so it is also a kind of acceleration. Traveling down the hill, you kept moving faster. That was adding a bit of speed each second. Braking is moving slower and slower. That is like subtracting speed each second.

Gravity speeds you up as you come down the hill, but what makes you stop? When you are on a skateboard, you drag a foot or slide the board sideways to stop. If you are using in-line skates, you might use a heel stop. Each of these ways involves rubbing something against the ground to slow down and stop. This is using friction (FRIHK shun) to help you stop. **Friction** is the push or pull that opposes motion between two surfaces that are touching each other. Think of rubbing together two pieces of coarse sandpaper. Friction makes it hard to slide the two pieces back and forth.

Mini LAB

Analyzing Friction Data
Skateboards, bicycles, and in-line skates are no fun when there's too much friction.

1. Use your ring binder for a slope as you did in the Explore activity.
2. Make the slope so steep that a flat metal washer will slide all the way to the bottom.
3. Put different types of surfaces on your slope such as plain paper, sandpaper, and waxed paper. Find the angle of the slope for each surface that is just steep enough to make the metal washer slide all the way to the bottom. For each material, measure the angle of the slope with a protractor, as shown below, and record your data in a data table.

Analysis
1. Of the surfaces, which one required the smallest angle of slope for the ring to slide to the bottom of the ramp? What makes that surface different from the others?
2. How could you change the surfaces to make the angle of incline needed to move the washer smaller?

FIGURE 10-7
The rosin increases friction, and the oil reduces friction on the movable parts of these musical instruments.

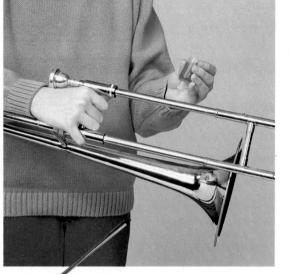

Uses of Friction

In the MiniLAB, you explored friction with several materials. When you stop a skateboard, you notice the friction between your foot and the ground, but there is also friction between your other foot and the skateboard. Think of rubbing two ice cubes together. The surfaces of the cubes are smooth and wet, and there is little friction. If the skateboard were as smooth as ice, think how hard it would be to stop without slipping off the board and falling. Friction can be useful, or it can be a problem, as shown in **Figure 10-7.** It's useful when you are trying to stop or stand up, but it slows down the wheels. You want the wheels to turn as easily as possible. Otherwise, skating will be more work and you won't be able to go as fast.

Section Wrap-up

1. If you rode with your cousin for a two-hour trip to a town 60 km away, what was your average speed in km/h? In m/s?

2. **Think Critically:** Why is friction an important part of stopping a car?

3. **Skill Builder**
   Recognizing Cause and Effect On a skateboard, where do you think friction is a problem? What will it do to the skateboard after a lot of use? What could you do to reduce the amount of friction? If you need help, refer to Recognizing Cause and Effect on page 551 in the **Skill Handbook.**

USING MATH

A race car moves from standing still to 50 m/s in 7 s. Is it accelerating faster or slower than a sports car that can accelerate from 0 m/s to 70 m/s in 8 s? Explain.

How do things move?

It's a Law

Remember Sir Isaac Newton's explanation of gravity from earlier in this chapter? During the last 400 years, Newton's theories have been tested again and again here on Earth and in space. When a theory has been tested many times and the results are the same each time, it may be called a **scientific law.** A scientific law isn't like the laws enforced by a government. A scientific law is an accurate description of some important part of the natural world. Three important laws about how things move are named after Sir Isaac Newton. These three laws apply to anything that moves—spaceships, bikes, skates, or people.

Before we begin talking in detail about Newton's laws, you need to learn an important word: *force.* A **force** is a push or a pull. Gravity is a force. Friction is a force. Every time you walk, open a door, or turn a page, you are using force to move something. Look at **Figure 10-8.** Now, let's explore how Newton's laws relate to forces and to movement.

What YOU'LL LEARN

- How Newton's three laws of motion apply to everyday life

Science Words:
scientific law
force
inertia

Why IT'S IMPORTANT

Newton's laws of motion explain how most sports happen, including skateboarding, biking, in-line skating, and wheelchair racing.

FIGURE 10-8

The people who hold this hide provide an upward force greater than the downward force of gravity on the girl.

275

Newton's First Law

Have you ever noticed that once you get your bike stopped, it's difficult to get it moving again? You may have to pedal hard to accelerate it back up to speed. Newton's first law of motion explains why this happens: An object at rest tends to stay at rest, and a moving object tends to keep moving in a straight line until it is affected by a force. **Inertia** (ih NUR shuh) is the name for this resistance to a change in motion. Another way to think of Newton's first law is: Things don't stop or start by themselves. If a bike is not moving, it will stay that way until some force makes it move.

If Newton's laws are true, why doesn't the bike shown in **Figure 10-10** coast forever in a straight line? The reason is friction. On Earth, everything eventually slows down and stops because of friction and gravity. Gravity pulls things down to the ground. On the ground, they eventually roll or drag to a stop because of friction between them and the ground.

FIGURE 10-9

In space, Earth's gravity has little effect on these astronauts.

FIGURE 10-10

Inertia keeps this bicycle rider going until another force—usually friction—causes her to stop.

In space, it's easy to see Newton's first law at work, as in **Figure 10-9.** You may have seen astronauts moving equipment outside a spaceship. Even a gentle push starts an object moving, and it doesn't stop until it hits something or the astronaut retrieves it.

Balanced and Unbalanced Forces

It's possible for many forces to act on the same thing at once. If all the forces cancel out, or balance, nothing will change. Think about stopping a bike on a hillside, **Figure 10-11.** You use the brakes. After you've stopped, you put your feet down and let up the brakes a little. Gravity is still pulling the bike down the hill. Both your feet and the bike's brakes are using friction to balance that pull down the hill. If you release the brakes completely or lift your feet, the forces become unbalanced and the bike begins to move again unless you hold it back.

SCIENCE *Online*

Isaac Newton helped explain many natural laws. Visit Glencoe Science Online at *science.glencoe.com* to find out more about Newton.

Newton's Second Law

Newton's second law involves force, mass, and acceleration. You know that it's harder to move a large, massive bike than to move a smaller, less massive one. Newton's second law says: Acceleration depends on the mass of an object and the force pushing or pulling the object. Here are two mathematical ways to write Newton's second law:

$$\text{Force} = \text{Mass} \times \text{Acceleration}$$
$$F = m \times a$$

or

$$\text{Acceleration} = \text{Force}/\text{Mass}$$
$$a = F/m$$

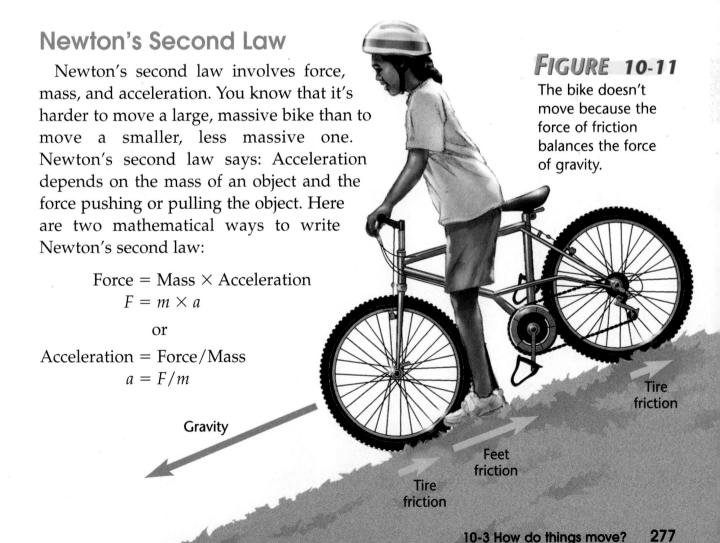

FIGURE 10-11

The bike doesn't move because the force of friction balances the force of gravity.

Gravity

Tire friction

Feet friction

Tire friction

Here are two simpler ways to think of Newton's second law. Apply them to **Figure 10-12.**

1. It takes less force to accelerate something with a small mass as fast as something with a large mass. A large, massive truck would need a powerful motor to start moving from a complete stop and keep up with a small sports car.

2. If you want to go faster, you have to use more force. You have to pedal harder to accelerate a bike to top speed in 5 s than to do the same in 2 minutes.

FIGURE **10-12**

If both riders pedal with the same force, the rider moving less mass accelerates faster.

Newton's Third Law

If you are walking down the hall and give your friend a playful shove, your friend may react by pushing you back. What happens if you walk over to a wall and give it a push? Does the wall push you back? Believe it or not, it does. Remember, if forces aren't balanced, then something has to change speed or direction. If the force of your push were not balanced by the force of the wall pushing back, you would make the wall move. Newton's third law explains such balanced forces. The third law says: For every action force, there is an equal and opposite reaction force. Another way to think of this law is: Forces always come in pairs. When you push against a wall, as in **Figure 10-13,** it pushes back just as hard.

Does the floor push up on you?

Sure it does, just like the wall! You know your weight is pushing down on the floor, so the floor must be pushing up on you. Think about quicksand. It looks like it will hold you up, but when you step on it, you sink. The sand does push up with an opposite force, but that force is smaller than your weight, and you sink. Any time there is movement, when you sink or fall, the forces acting on you are out of balance. When there is no change in movement, the forces are equal and balanced, such as your force on the floor and the floor's force on you. When you step into quicksand, the forces are not equal and there is definite movement: you sink. In the activity on the next pages, you can find out how action-reaction forces affect rocket movement.

FIGURE 10-13

As this girl pushes on the wall, the wall pushes back with equal force.

Activity 10-2

Design Your Own Experiment
Rocket Races

Going into space for the first time was an exciting moment for humanity. Scientists used rockets to test new technology outside our atmosphere for the first time. Rocketry uses Newton's third law. In this experiment, you will have a chance to design your own rocket and race it against others.

Possible Materials

- balloons of various shapes and sizes
- drinking straws
- tape
- construction paper
- waxed paper
- foil
- rubber bands
- scissors
- string or fishing line

PREPARE

What You'll Investigate
Design a balloon rocket that travels fast enough and has enough fuel to cross a finish line first.

Form a Hypothesis

Develop a hypothesis about the design and size of your rocket.

Goals

Design and **build** an efficient rocket that uses Newton's third law for its propulsion.

Compare results.

Safety Precautions

Be sure you review and understand all safety precautions before beginning the activity.

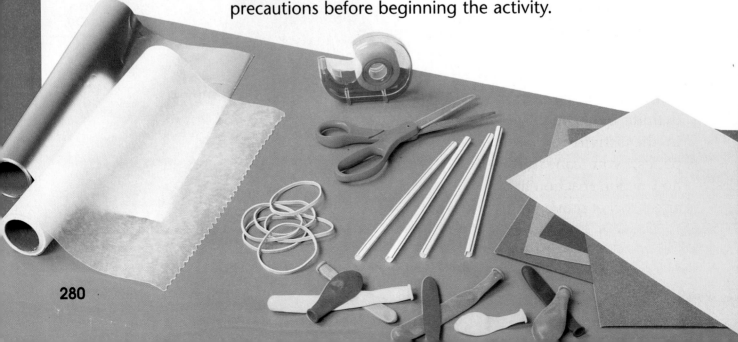

PLAN

1. Make sure your group has agreed on a **hypothesis** statement.
2. **Sketch** a design of your rocket. See the photo and the list of possible materials for ideas.
3. **List** detailed, orderly steps that you will follow to build your rocket.
4. **Choose** your balloon carefully. Which balloon has the best mass? Remember, $F = m \times a$. Will it have enough fuel to finish the race? Will the neck size be important? Which shape do you think will fly the fastest?
5. How long a straw will you attach to your balloon, and how many pieces will you need to guide it along the string?
6. What shape will the rocket have? How will its shape affect its flight?
7. Make sure your teacher approves your plan and that you have included any suggested changes in the plan.

DO

1. **Gather** materials and **build** the rocket as planned.
2. **Race** it against your classmates' rockets.

CONCLUDE AND APPLY

1. **Describe** the winning rocket. What gave it an advantage over the other rockets?
2. What parts of your design were successful? Why?
3. What would you change on your design next time? Why?
4. **APPLY** Which of Newton's laws played an important part in this experiment? **Explain.**

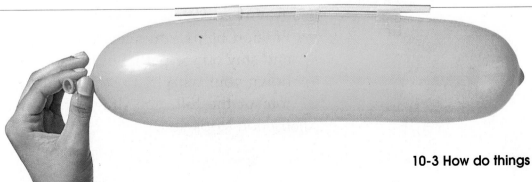

FIGURE 10-14

Force pairs are part of the movement of this rocket and this student.

A The thrust of the rocket is a greater force than gravity, so the rocket moves up.

B The force the girl exerts on the backpack is balanced by the force of gravity acting on the backpack. The forces are balanced, so the backpack stays in place.

What do these laws mean to you?

As you found with the rockets in the activity and can see in **Figure 10-14,** force pairs are part of all movement. When you ride a skateboard or bicycle, you must control the forces that affect your movement. When you push off at the top of a hill, you are overcoming the inertia of the skateboard. The unbalancing force of gravity accelerates you downhill. Friction helps you keep your footing and stay on the path, but it also slows down your wheels. To stop at the bottom of the hill, you drag a foot and friction stops you.

What do Newton's laws mean to someone in a wheelchair?

To some people, dealing with gravity is a challenge every day. Imagine being in a wheelchair, such as the boy in **Figure 10-15,** and having to go up and down ramps all day. The friction of the wheels plus gravity make it hard to get up a ramp.

Later, keeping control as the wheelchair accelerates back down the same ramp takes effort. Inertia makes the chair difficult to start and stop. People in wheelchairs understand Newton's laws and how to use each one to their advantage.

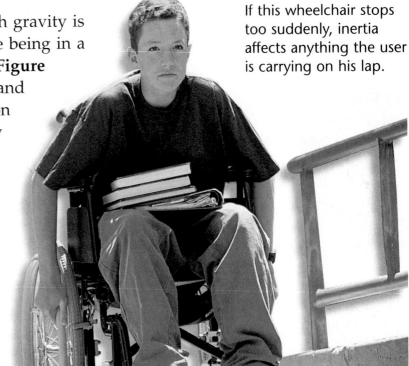

FIGURE 10-15
If this wheelchair stops too suddenly, inertia affects anything the user is carrying on his lap.

Section Wrap-up

1. When a person in a wheelchair stops suddenly, why might the books in his or her lap sometimes fly onto the floor?

2. Tell which of Newton's three laws of motion was the most difficult for you to understand and why. Then, tell how you'd explain one of the laws to a younger student.

3. **Think Critically:** Use a diagram to explain the balanced and unbalanced forces that allow you to sit at your desk and write.

4. *Technology Skill Builder*
   Using E-mail Use E-mail to contact your state or local government. Find out the requirements for building ramps for public buildings. If you need help, refer to Using E-mail on page 565 in the **Technology Skill Handbook.**

Science Journal

In this chapter, a bicycle was used to help explain $F = m \times a$. In your Science Journal, use a skateboard to explain this same law.

10•4 Air Bag Safety

What YOU'LL LEARN

- Air bag design and safety issues
- How to analyze news media

Why IT'S IMPORTANT

You'll be able to think critically about the news you read or hear.

Why the concern about air bags?

To help reduce traffic deaths, the U.S. government required air bags in all new cars by 1998. Air bags are bags that inflate inside a car immediately on impact. They prevent the driver and the front-seat passenger from smashing against the steering wheel, front panel, or windshield of the car because of their inertia.

While air bags have reduced traffic deaths, they have also caused some deaths, particularly among children. Read the following fictional newspaper article to learn more about the safety issues surrounding air bags. After you've finished reading, you'll learn how to analyze news media for fairness and accuracy. News media—radio, television, news magazines, and newspapers—are the sources from which we get our news. You need to stay informed to be a good citizen, but you also need to think critically about the news you read or hear.

Air Bag Alarm

WASHINGTON, DC — Since the law was passed requiring air bags in all new cars, parents have expressed alarm over their children's safety. In one six-month period alone, the force of inflating air bags caused the deaths of 15 children.

But Fred Jones, government spokesperson, claimed that air bags have saved more lives than they've taken. He admitted that air bags can hurt children, but countered that children should not be riding in the front seat.

"Children under 12 should always ride in the back seat, securely buckled in a seat belt," he said.

Jones added that in cars that don't have enough rear space for children, the government will allow car companies to install cut-off switches that can turn off air bags on the passenger sides of cars.

One problem is that many people forget to turn off the switch when they enter their cars. Because of this, car companies are developing "sensors" that can detect the presence of a child's car seat. These sensors would deactivate the air bags auto-

maticlly. The problem with this solution is that all car seat manufacturers would have to agree to install the sensors in their car seats. In the meantime, some car companies are "depowering" air bags so that they inflate with less force. It remains to be seen whether this will reduce injuries to children.

 ## Skill Builder: Analyzing News Media

LEARNING the SKILL

1. To gather news, reporters talk to people who are involved in an issue. These people are called sources. Reporters should identify their sources. If you know the sources, you can look up the facts to see whether they are true.

2. News reporters are supposed to be unbiased. That is, they should not include their own views in a news report unless the report is meant to be an analysis of an issue. Still, a reporter's biases sometimes come through. Look for words like *admits* and *claims,* or look for instances where the reporter stated something without backing it up with a source.

3. Ask yourself whether the report is balanced. Does it fairly represent both sides of the issue? Are some sources quoted while others are not?

PRACTICING the SKILL

1. Why are parents concerned about air bags? What are some of the things being done to address these concerns?

2. Did the reporter identify all sources by name?

3. Did you see any instances of bias in the article? If so, list them.

4. Were all sides of the issue fairly represented? Why or why not?

APPLYING the SKILL

If you follow the news, you know that people disagree on many issues. Choose an issue that interests you and bring news articles about the topic to class. Examine the different views presented by sources in the articles. Analyze the news articles for examples of biases.

People & Science

Josh Pankratz, Student of Aikido

Q **What is Aikido, and what does it emphasize?**

A Aikido is a Japanese martial art that focuses on self-defense, on fending off attackers without injuring them. Aikido uses only self-defensive moves and skills.

Q **Could you describe some of those moves and how they use forces?**

A In Aikido, you learn to redirect, rather than oppose, the force of an attacker's motions. Imagine that someone is coming at you with an arm raised to hit you. In Aikido, you could step forward quickly and deflect the person's arm as it comes down. You also could step backwards at an angle and simply avoid the blow. Or you could step in from the side and grab the person's hand. Then, by redirecting the force he or she is exerting, you can turn, unbalance, and throw your attacker to the ground.

Q **Does a person need to be really strong to practice Aikido?**

A No, not at all. The whole idea is not to use force—to use as little muscle power as possible. It's mostly ways of moving, really, that allow you to turn and throw someone or to use his or her motion to deflect an attack. Physical size and strength don't matter very much.

Q **Is Aikido considered a competitive sport?**

A No, Aikido isn't competitive at all. There are no tournaments or contests. In Aikido classes, partners take turns throwing and being thrown by each other. It's really a partnership between two people. The idea of "trying to win" simply isn't there.

Career Connection

Think about careers that deal with the use of motion. Interview someone in your community who has such a career. Make a list of all the careers that were considered by other students in the class.

Read the statements below that review the most important ideas in the chapter. Using what you have learned, answer each question in your Science Journal.

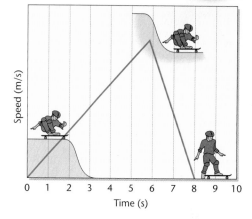

1. Gravity pulls all things toward Earth in the same way. *What are two examples that show this?*

2. Average speed during a trip is the total distance traveled divided by the total time of the trip. *How would stopping at a gas station and playing an arcade game affect your average speed during a car trip?*

4. Air bags have saved lives but can be dangerous under special conditions. *What are two ways to make air bags safer?*

3. According to Newton's first law, objects at rest tend to stay at rest, and objects that are moving tend to keep moving in a straight line. *How does this law relate to inertia?*

Using Key Science Words

acceleration
average speed
force
friction
gravity

inertia
mass
scientific law
weight

Use one of the Key Science Words above to answer each of the questions below.

1. What is the word for a push or a pull?
2. What word means that something is speeding up or slowing down?
3. What is the force that pulls all objects toward each other?
4. What is the force that resists motion when two surfaces are sliding on each other?
5. What is the result of dividing measured distance by measured time?

Checking Concepts

Choose the word or phrase that answers the question.

6. When you are sinking in quicksand, which of the following is true?
   a. The forces on you are balanced.
   b. The forces on you are unbalanced.
   c. The sand is giving an equal and opposite reaction force.
   d. Inertia is causing you to sink.

7. When you are accelerating from the top of a hill, what helps you stop?
   a. friction   c. gravity
   b. inertia   d. mass

8. On Jupiter, what would you expect to experience?
   a. Gravity would decrease.
   b. Your weight would stay the same.
   c. Your mass would change.
   d. Gravity would be stronger.

9. On the moon, which would be true?
   a. Your weight would be less.
   b. Your weight would stay the same.
   c. Your mass would change.
   d. Gravity would be stronger.

10. To go faster on a skateboard ramp, what would you do?
    a. Put a rough surface on the ramp for a better grip.
    b. Leave your safety equipment behind to lighten the load.
    c. Add more weight to the board so you will have more acceleration at the bottom of the slope.
    d. Make the slope steeper to increase the acceleration.

Thinking Critically

Answer the following questions in your Science Journal using complete sentences.

11. What would you need to measure to test how good brakes are? Explain.

12. If someone told you he or she had discovered a new scientific law, what would that person need to show you to prove it?

13. For a long drive, a car's average speed was 60 km/h, but its speed halfway there was measured at 90 km/h. How is this possible?

14. Two cars are identical, but one carries an extra 100 kg of equipment. Which car will probably win a race? Explain.

15. You're pulling a cart. The forces between your hand and the handle are equal and opposite. The cart is accelerating. What must be true?

Developing Skills

If you need help, refer to the description of each skill in the Skill Handbook.

16. **Concept Mapping:** Use the following words to make a network tree: *Newton's laws, action and reaction forces, gravity, Newton's first law, Newton's second law, Newton's third law, inertia, F = m × a.*

17. **Making and Using Graphs:** Make a bar graph showing your weight on Earth, the moon, Mars, and Jupiter.

18. **Comparing and Contrasting:** You did one experiment with an accelerating marble. Then you did another lab with a sliding, flat, metal washer. What are the similarities and differences between the marble's movement and the washer's?

19. **Recognizing Cause and Effect:** In terms of forces, list causes for each of the following effects: (a) a baseball flies out of the ballpark; (b) a car brakes and slides to a stop; (c) a person in a wheelchair stops suddenly and her books fly off her lap; (d) a marble falls and rolls downhill; (e) you lean against a wall and the wall doesn't fall down; and (f) you sink into quicksand.

20. **Interpreting Data:** Use the following data to answer the questions.

Marble Motion		
Time	**Speed**	**Distance**
Start	0	0
1 second	1 m/s	0.5 m
2 seconds	2 m/s	2 m
3 seconds	3 m/s	4.5 m

Was this marble speeding up? When was it going its average speed? What was its speed at 2 s?

Performance Assessment

1. **Predicting:** Use the data presented in question 20 above to predict the speed and distance of the marble after 6 s.

2. **Designing an Experiment:** Design an experiment that determines what shoe has the correct tread for an activity such as running.

Chapter Preview

Skills Preview

▶ **Skill Builders**
  • recognize cause and effect
  • compare and contrast
  • sequence

▶ **MiniLABs**
  • calculate
  • compare

▶ **Activities**
  • hypothesize
  • design an experiment
  • draw scientifically
  • make a model

Chapter 11

Work and Machines

An outdoor vacation can be a lot of fun, but getting ready for it isn't always easy. There is so much to do—meals to plan and supplies, tents, and equipment to organize. Don't forget the fun stuff like canoes, paddles, and fishing poles. Everything needs to be lifted, moved, and packed. This involves a great deal of work. That's what this chapter is about—work and how to make it easier. You can start learning about work in the Explore activity.

EXPLORE ACTIVITY

Inferring

1. On this vacation, you will carry a canoe. Practice by using two or three books to represent the weight of the canoe. Stand up and hold the books with both hands.
2. Lift the books up over your head three times.
3. Walk around the room twice with the books over your head. Don't let the books rest on the top of your head.
4. Which do you infer was more work, lifting the canoe or carrying it? Why do you think so?

In your Journal, discuss the easiest way to load everything into a van for a class camping and canoeing trip.

11•1 What is work?

What YOU'LL LEARN

- What *work* really means
- How to figure out how much work you've done

Science Words:
work
joule

Why IT'S IMPORTANT

If you know what work is, you can learn how to make it easier.

Doing Work

What do you think of when you hear the word *work*? You may think that reading this chapter is work. On a vacation, carrying a canoe probably seems like work, too. Most adults talk about work every day. Would you be surprised to learn that none of these examples is work in the scientific sense?

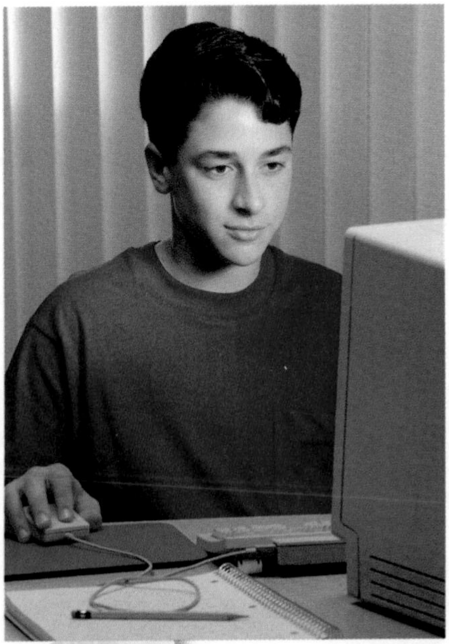

FIGURE 11-1
The student is "working" on a school project. *Is he doing work in the scientific sense? Explain.*

When are you working?

Work. The very word makes some people tired. That's because their idea of work is anything that takes effort. That can mean homework, as shown in **Figure 11-1,** or running a marathon. But effort does not always equal work in the scientific sense. If you push against a wall until you are tired but the wall doesn't move, a scientist would say you haven't done any work on the wall. **Work** is done when a force moves something through a distance and the force and the motion are in

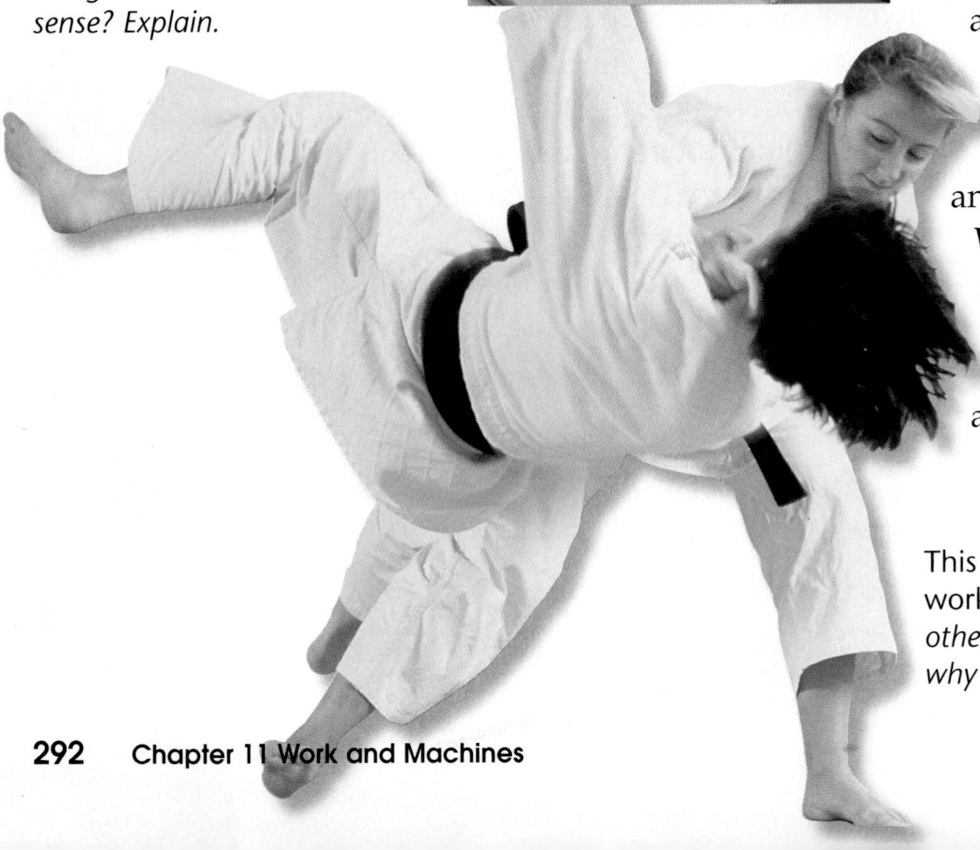

This judo throw looks like work. *Is the girl throwing the other girl doing work? Explain why or why not.*

the same direction. An example of this definition of work is lifting a canoe, as shown in **Figure 11-2.** The force is in an upward direction and the canoe moves up, so it agrees with both parts of the definition of work. If you carry the canoe along a level path, you are still holding up the canoe, but the canoe is moving forward. The force and the movement are not in the same direction. Even though carrying the canoe feels like work, it doesn't fit the scientific definition.

How much work?

Work depends on force and distance. You can find the amount of work by multiplying the force used by the distance over which the force is acting.

If the canoe weighed 300 newtons (N) and you lifted it 1.5 meters (m), then you did 450 N•m of work on the canoe. This is read as 450 newton-meters.

$$W = F \times d$$
$$W = 300 \text{ N} \times 1.5 \text{ m}$$
$$W = 450 \text{ N•m}$$

To make it easier to talk about work, there's a unit of measure with a shorter name. A **joule** (J) is one newton-meter (N•m), so you did 450 J of work on the canoe when you lifted it. The joule (JEWL) was named in honor of James Joule, a scientist who studied work, energy, and heat.

FIGURE 11-2

Work is done when the canoe is lifted up because the canoe is moved, and the force and motion are in the same direction.

Mini LAB

Calculating Work

Suppose your teacher asked you to move ten books from the floor to the desk. What would be the easiest way to do the work?

1. Lift all ten books to the desk.
2. Lift two books at a time to the desk.

Analysis

1. Which way of lifting was easier?
2. Assume the books weighed about 10 N each and you lifted them 1 m. How much work did you do when you lifted the whole stack at once? How much work did you do when you lifted two books at a time?
3. Which was more work?

Is going uphill work?

Now instead of picking up and carrying a canoe, imagine carrying your own weight up a mountain. When you hike up the mountain, you do work by moving your weight up the trail just as you did work lifting the canoe. This time, you won't be walking on a level path. You are going to hike up a hill moving your weight. You will be going up, so you will be doing work. How do you calculate the amount of work done? Remember, $W = F \times d$. You are lifting yourself up the hill with every step you take. The distance equals how high you went up, and the force is your weight.

FIGURE 11-3

Who does more work getting to the top, the person hiking up the trail or the person climbing up the cliff? If they start at the same place and both reach the top, they will do the same amount of work even though it is much harder to climb up the cliff.

Effort Doesn't Always Equal Work

Which seems like more work, hiking or mountain climbing as shown in **Figure 11-3?** You can compare the amount of work and effort involved in the two activities. Suppose you climb straight up the side of a mountain, climbing a rock cliff using just your fingers and toes. Meanwhile, a friend takes the easy way up to the mountaintop, strolling along a hiking trail. Which of you did more work? Before you answer, think back to the way you learned to calculate the amount of work done. It doesn't seem possible, but you and your hiking friend did exactly the same amount of work, assuming you both weigh the same. You both lifted the same amount of weight (force) and you both traveled up the same distance. Even though the trip up the hiking trail was easier than climbing, the same force and the same distance make the same work.

Section Wrap-up

1. You picked up a 10-kg (98-N) barbell and lifted it 1.5 meters. How much work did you do on the barbell? What is the force? What is the distance?

2. Which of the following is work: pushing your little brother on his tricycle, lifting a box, holding books over your head, climbing up stairs, doing homework, or carrying a box uphill?

3. **Think Critically:** When an astronaut is weightless and lifts an object over her head, is work being done? Explain.

4. *Skill Builder*
   Recognizing Cause and Effect You do the same amount of work on two identical lunch boxes by moving them. However, one box traveled farther than the other. Are the forces you used to lift the two boxes different? How do you know? If you need help, refer to Recognizing Cause and Effect on page 551 in the **Skill Handbook**.

Science Journal

You are helping tear down an old garden shed, but one wall refuses to fall. You have pushed it and pulled it until you are tired. Did you do any work? Explain.

11•2 Simple Machines

What YOU'LL LEARN

- That simple machines make work easier
- How to use different simple machines

Science Words:
simple machine
inclined plane
screw
wedge
lever
wheel and axle
pulley

Why IT'S IMPORTANT

Simple machines make work easier.

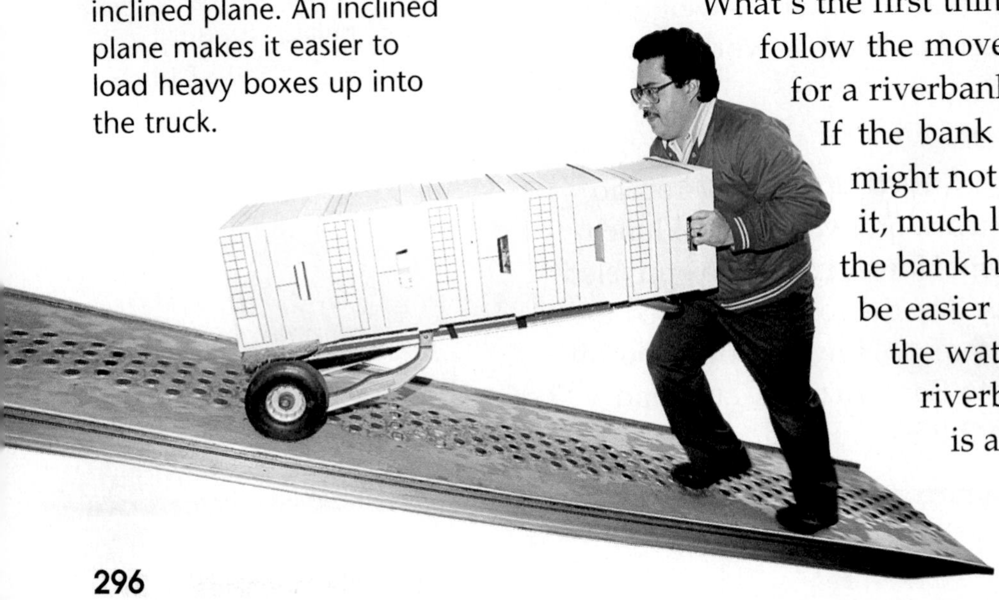

FIGURE 11-4

What simple machine is being used here to help load the boxes into the truck? The ramp is an inclined plane. An inclined plane makes it easier to load heavy boxes up into the truck.

Simpler Than You Think

Remember the hiker and climber from the last section? You probably didn't notice the machine that made work easier for the hiker. The word *machine* probably makes you think of a complicated gadget with metal, plastic, gears, and motors. But the hiking trail was a kind of machine—a simple machine. A **simple machine** is a device that does work with only one movement. Simple machines are the most basic form of useful tool. In this section, you'll learn to recognize the six types of simple machines that are all around you. The six simple machines are the inclined plane, screw, wedge, levers, the wheel and axle, and the pulley. As you read on, try to figure out which type of simple machine the hiker used.

What is an inclined plane?

On moving day, the first thing movers do is put a ramp up to the back of the truck, as shown in **Figure 11-4**. A ramp makes it easier to move things up and into the truck. This type of ramp is called an inclined plane. An **inclined plane** is a sloped surface used to make lifting things easier.

Imagine that you're canoeing and you need a place to lift the canoe out of the water and onto the riverbank. What's the first thing you'll look for? You'll follow the mover's example and search for a riverbank that looks like a ramp. If the bank has straight sides, you might not even be able to walk up it, much less drag a canoe up it. If the bank has a gentle slope, it will be easier to pull the canoe out of the water and up the bank. The riverbank with a gentle slope is an inclined plane.

296

What is a screw?

The inclined plane made it easier to unload the moving truck and get the canoe out of the water. But if you change the shape of an inclined plane, you can do even more jobs with it. One way to change its shape is to make it into a screw. You've probably seen screws in a hardware store or used them to hold things together, but have you ever looked at one closely? If you do, you'll notice that a **screw** is an inclined plane wrapped around a rod as shown in **Figure 11-5.** In the MiniLAB, you'll learn more about how screws and inclined planes are related.

Just as ramps can have different slopes and lengths, screws come in different sizes and shapes. In the MiniLAB, some screws had small, gently sloped, closely spaced threads and some had large, steep, widely spaced threads. Screws are designed for what they do—holding wood or metal, going through thick or thin pieces of material.

Screws aren't found just in toolboxes—screws are found all around you. When you unpack the tent, you may find a new type of tent stake. It looks like a big spiral. The directions say it's easier to use and holds better than the old, straight kind. You try it and sure enough, it is easy to put into the ground. Then, you give it a good, hard tug and it doesn't move at all.

Mini LAB

Comparing Ramps and Screws

1. Make an inclined plane out of a piece of paper. Cut a piece of paper in half diagonally from corner to corner.
2. Use markers or a pencil to draw a dark line along the cut edge.
3. Start with the biggest end and wrap the paper around a pencil so that you can see the dark line. Observe the dark line along the edge of the inclined plane.
4. Now, observe the screws your teacher gives you.

Analysis

1. What geometric shape does the line around the pencil form?
2. How are the screws similar to the pencil and paper?
3. What would you have to change about the paper and pencil to make them more closely match each of the screws?

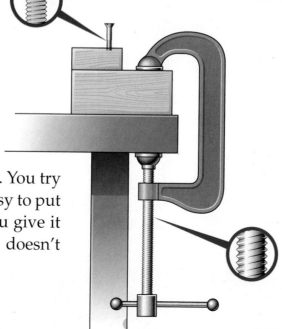

FIGURE 11-5

The threads in a screw or a clamp are an inclined plane wrapped around a rod.

What is a wedge?

After the tent is up, it's time to clear a space for the campfire. You need to cut up some dead tree branches lying on the ground for firewood, so you get out a hatchet to cut the branches into small enough pieces to carry. When you use a hatchet, you are using another version of an inclined plane called a wedge. A **wedge** is an inclined plane with one or two sloping sides. Pushing the hatchet into the branch forces the wood apart. Can you think of any wedges that you use every day? A kitchen utensil drawer and a carpenter's toolbox are good places to look. Anything with a sharp edge is a wedge—how about scissors, a knife, a can opener, a chisel, and a saw?

What is a lever?

Your campsite is almost finished. You have pitched the tent and cleared the area except for one large rock. It would make a great seat by the campfire, but it's too far away, and it's buried too deep to move by hand. Another

FIGURE 11-6

You may not realize that you use simple machines every day.

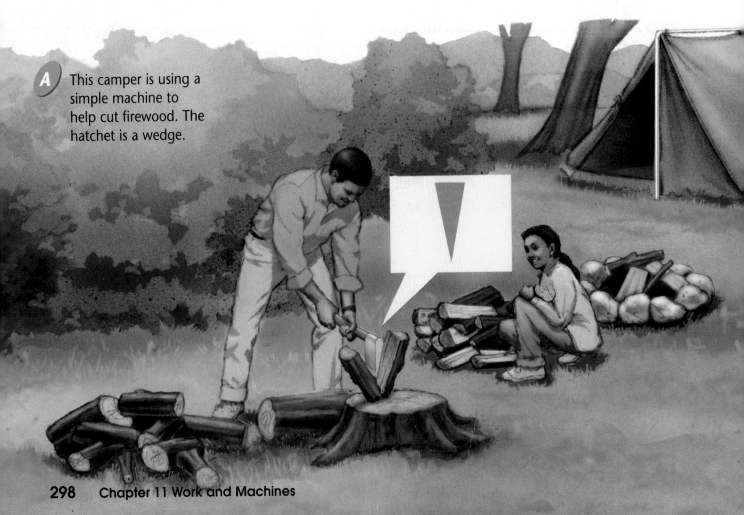

A This camper is using a simple machine to help cut firewood. The hatchet is a wedge.

type of simple machine can help you—a lever. A **lever** is a bar that's free to move around a pivot point. The pivot point is called a fulcrum (FUL krum). In a lever, the work you do on your end (force you exert times the distance your end moves) equals the work on the other end (weight of the rock times the distance that end moves). There are three kinds, or classes, of levers. Moving the rock is a job for a first-class lever, as shown in **Figure 11-6.**

First-Class Levers

Find a thick branch and put one end under the rock. Put a smaller rock between the branch and the ground to be the fulcrum. Now, push down on the branch. You can easily pry the big rock up and out of the ground. A first-class lever allows you to move things that you could not move without it.

A seesaw is another example of a first-class lever. The fulcrum is in the middle, and each end has a force pushing down. Other types of levers have the fulcrum in different places. Let's look at some other types of levers.

B This camper is using another simple machine to move the heavy rock. The branch is a lever, and the small rock is a fulcrum. By using the lever and fulcrum, the rock can be moved more easily.

Second-Class Levers

A second-class lever also increases your force. Suppose you need to move a canoe closer to camp. The canoe is still loaded with supplies and is too heavy for one person to lift and carry. Instead, you pick up one end and drag the canoe closer to camp.

It's easier to lift one end of the canoe because the canoe is a second-class lever as shown in **Figure 11-7.** The ground where the tip of the canoe touches is the fulcrum. The lever is the whole canoe plus the fulcrum. A second-class lever transfers some of the weight to the fulcrum, making it easier for you to lift and move the canoe, supplies and all. Another example of a second-class lever is a wheelbarrow. The wheel at the front is the fulcrum and carries some of the weight while you lift the handles at the back.

FIGURE 11-7

The forces in and out are in different places in second-class and third-class levers.

Force out

Force in

Fulcrum

A The person dragging the canoe up to the campsite is using a second-class lever.

Third-Class Levers

The third and last class of lever doesn't increase your force, but it can make some kinds of work easier. It also can help you have fun.

Now that the hard job of setting up camp is done, it's time to go fishing. You cast a baited hook into the river, and soon you've hooked a fish. The fishing rod you use is a third-class lever as shown in **Figure 11-7B.** This class of lever does not make your force larger, it makes things move faster. The far end of the rod makes large, fast movements. Your wrist is the fulcrum. Moving the bottom of the rod a little makes the top of the rod move a lot. That fast motion is just what you need to cast your baited hook far out into the water. To see how much this type of lever helps you, imagine fishing with only a baited hook and some line. Can you think of other examples of third-class levers? A canoe paddle, a baseball bat, and a broom are three examples.

It's smart to use a simple machine any time it will make your work faster and easier. In the activity on the next two pages, you can explore other ways to use simple machines.

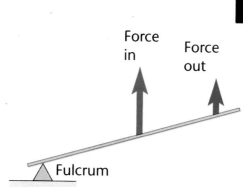

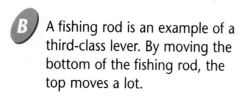

B A fishing rod is an example of a third-class lever. By moving the bottom of the fishing rod, the top moves a lot.

Activity 11-1

Design Your Own Experiment
The First Wonder of the World

The Pyramids of Egypt, the Hanging Gardens of Babylon, and the Great Wall of China are some of the amazing things constructed by early civilizations. Ancient engineers and architects knew how to use simple machines to greatly multiply the force their workers could apply.

Possible Materials
- wood block
- paper clip
- tape
- spring scale
- pencil
- ruler
- clay
- notebook
- 3-ring binder
- books

PREPARE

What You'll Investigate
How can you use levers and inclined planes to move a model block of stone?

Form a Hypothesis

The Egyptian workers who built the great pyramids cut tunnels into quarry walls to remove large blocks of stone. They put the blocks of stone onto wooden sleds and pulled them to the pyramid. Then they pulled and pushed the block up the half-finished pyramid to place it in its layer. **Make a hypothesis** about how to use simple machines to remove the block from the quarry and move it up the side of a half-built pyramid.

Goals

Design an experiment to test your hypothesis. Choose an object to represent the block. Use other objects to represent levers or inclined planes.

Measure the force exerted with and without simple machines using a spring scale.

PLAN

1. **Discuss** what materials will represent the stone block, a lever and fulcrum, an inclined plane, and a half-built pyramid.
2. **Decide** how you will **measure** the force exerted in each task with and without a simple machine.
3. **Draw** a diagram of the experiment.
4. **Plan** how you will **record** the design and results of your experiment.

DO

1. Make sure your teacher has approved your design.
2. Carry out the experiment.
3. **Record** your observations.

CONCLUDE AND APPLY

1. What was the force exerted without a simple machine to move the block onto a sled? Up the pyramid? What was the force with a simple machine?
2. **Compare and contrast** your results for each case.
3. **APPLY** Predict a way to modify your simple machines and exert even less force.

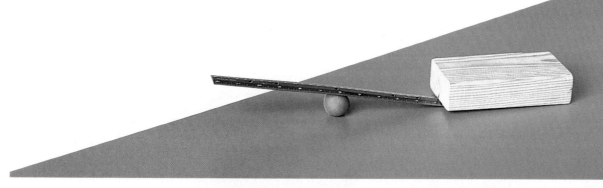

FIGURE 11-8

Doorknobs and fishing reels are two examples of a wheel and axle.

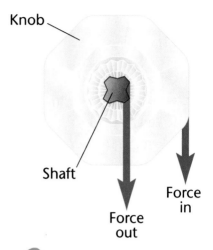

Knob

Shaft

Force out

Force in

A A small force moving a large distance (around the wheel) turns into a large force moving a small distance (around the axle).

B In a doorknob, without the knob (wheel), the shaft (axle) is hard to turn.

Working Smart

When you use levers to lift something, as you did in the activity, is the amount of work greater or less than if you didn't use a machine at all? It's the same. Simple machines usually let you use less force but cause you to go a longer distance. For example, on page 294, the shorter path to get to the top of the mountain was straight up the steep side, but the longer route on the gentle slope of the hiking trail was easier. The purpose of simple machines is to make work easier—to allow you to use less force, or move a smaller distance, or change the direction to get the same amount of work done. Some people call this working smarter, not harder.

What is a wheel and axle?

As you stand on the riverbank, casting your baited hook and reeling in fish, the lever isn't the only simple machine you're using. The reel on your fishing rod is a simple machine called a wheel and axle. A **wheel and axle** is a simple machine made with two different-sized wheels that are connected and turn together.

Usually, a wheel and axle makes your force greater. A doorknob is a wheel and axle. If you took the knob off a door, you'd see the small, square axle, as shown in **Figure 11-8.** When you turn the knob, the axle turns, too, and opens the latch or lock. Then you can pull the door open. The large knob on a smaller axle makes your force larger. Without the knob, turning the small axle with your fingers is difficult.

On the reel of a fishing rod, the handle turns a large wheel. The large wheel turns a smaller wheel, and it pulls in the fishing line. Because it's a big wheel turning a smaller wheel, it also makes your force greater. Can you imagine trying to catch a fighting fish by hand without the help of a wheel and axle?

C A fishing reel is an example of a wheel and axle. The crank handle turns the axle.

What is a pulley?

The fishing was good, and you've just finished the last bites of a fine fish dinner. Now it's time to put the supplies away so that raccoons, skunks, and bears can't get into them overnight. You throw a rope over a tree limb, tie it to the supplies, pull them up off the ground, and leave them hanging in midair. You just used another simple machine called a fixed pulley. A **pulley** is a surface with a rope or chain going around it. Pulleys are used for many purposes today, but the first pulley ever used was probably a tree limb with a rope over it, just like yours.

Fixed Pulleys

The limb is a fixed pulley because when you pull down on the rope, the limb doesn't move, but the supplies do. This type of pulley changes the direction of your force, but not its size, as shown in **Figure 11-9** on the next page. When you pull down, the supplies go up. Think how much more effort it would take if you had to climb the tree to pull up the supplies.

What could you do if the supplies were too heavy to pull up with a pulley? You could add another type of pulley to the fixed one.

Problem Solving

The Force of Flowing Water

Flowing water can have a lot of force. Imagine trying to hold back the water by covering the faucet with your hand. But you can use two simple machines to stop the flow. Look at the diagram of a faucet below.

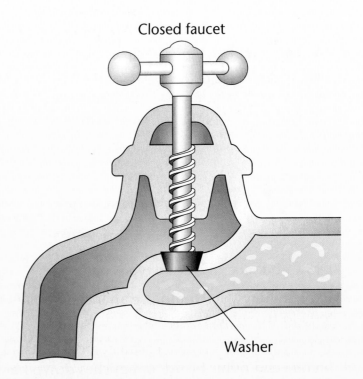

Closed faucet

Washer

Solve the Problem:
1. **What part of the faucet is a wheel and axle?**
2. **What part of the faucet is an inclined plane?**

Think Critically:
Describe what would happen if each simple machine were replaced. Would it be easier or harder to turn the faucet on and off?

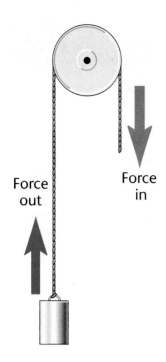

Force out

Force in

FIGURE 11-9

A pulley makes it easier to lift heavy things. *Why do most people find it easier to pull down than lift up?*

Movable Pulleys

Because the supplies are too heavy, you will need to make your force greater. One way is by using two pulleys. First, attach a pulley to the supplies. Then, tie one end of a rope to the tree limb and run the other end through the pulley attached to the supplies. Now, throw the rope up and over the limb just like last time. This time the new pulley moves with the supplies. Because it doesn't stay in one place, it is called a movable pulley.

Having two ropes supporting the supplies instead of one made pulling them up easier. You had to pull about half as hard. Each rope is supporting half of the weight of the supplies.

Helpful Machines

Are you surprised that machines made your camping trip easier? Think about how many simple machines you were able to use during this trip. It's a lot of simple machines for one trip, but each one provided a different way to make the trip easier and more fun.

Section Wrap-up

1. Draw and label the three types of levers.

2. Explain how you would use an inclined plane tool to hold two objects together.

3. **Think Critically:** When a skateboarder puts a foot on one end of his board and pushes down hard, the other end of the board pops up. Draw and label a diagram that shows which type of lever the skateboarder is using when he does this.

4. *Skill Builder*
   Comparing and Contrasting Compare a wheel and axle to a wheel *on* an axle. Tell how they are different and how they are alike. If you need help, refer to Comparing and Contrasting on page 550 in the **Skill Handbook.**

Science Journal

In your Journal, list at least five simple machines that you have used today and tell how you used each one.

What is a compound machine?

Putting It Together

On your camping trip, you used pulleys, levers, and inclined planes. You also use them every day. Now that you know what they are, you'll notice simple machines in many places. Often, however, more than one machine is used at a time. A **compound machine** is two or more simple machines used together. Look at **Figure 11-10** for some examples of compound machines.

What YOU'LL LEARN

- That simple machines can be put together to do work

Science Words:
compound machine

Why IT'S IMPORTANT

Combinations of simple machines make work easier to do.

FIGURE 11-10

A compound machine is two or more simple machines used together.

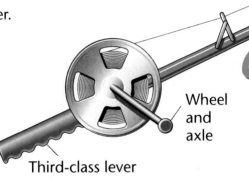

Wheel and axle

Third-class lever

B A fishing rod and reel is a compound machine because it combines two machines— a wheel and axle and a lever. The entire rod is a third-class lever.

A The illustration shows how the person is using two levers as a compound machine to row the boat.

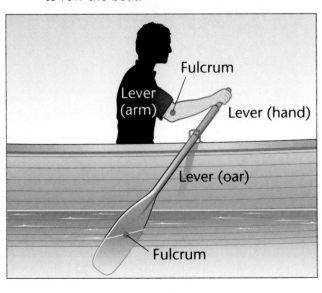

Fulcrum

Lever (arm)

Lever (hand)

Lever (oar)

Fulcrum

C Scissors are two levers, a fulcrum, and two wedges.

Fixed pulley

Movable pulley

When simple machines are combined to form a compound machine, each machine helps make work a little easier. The block and tackle shown in **Figure 11-11** combines two pulleys. Each pulley makes the force the person has to provide less. Even someone who isn't strong can lift a load weighing several thousand newtons. The two simple machines, the pulleys, which form the compound machine in **Figure 11-11,** are easy to see.

Sometimes the simple machines that combine to form compound machines are not so easy to see. Look around your classroom. How many simple machines can you find?

FIGURE 11-11

This block and tackle combines many pulleys to lift heavy things more easily.

USING TECHNOLOGY

Exercise Equipment

Many apartment complexes, hotels, recreation centers, and sports clubs have fitness centers. In these centers, you'll find every possible combination of levers, pulleys, and weights to work out almost every muscle. There are different ideas about what the best exercise plan is.

Fitness technology is changing so fast that fitness centers quickly become outdated. As fast as the fitness industry is changing, one thing stays the same: regular, moderate exercise is still necessary for good health, and many ways to get exercise involve simple machines.

SCIENCE Online

Inventors have fun with simple and compound machines. Learn more about them on the Internet when you visit the link at Glencoe Science Online. *science.glencoe.com*

Food out of Reach

A problem on any camping trip is keeping the animals out of the food. This is where a compound pulley machine can help.

What You'll Investigate

How can you make a model of a way to keep food out of the reach of animals?

Procedure

1. **Place** two chairs so that the chair backs can support the dowel rod.
2. Use a twist tie to **hang** a pulley from the dowel rod.
3. **Set up** the string as shown. It needs to pass around a pulley and then be tied to the dowel rod.
4. **Tie** one bag of marbles to the second pulley as shown.
5. Slowly let go of the bag and observe.
6. **Tie** on a second bag of marbles, slowly let go, and **observe** what happens.

Conclude and Apply

1. What happened when the marbles and pull were the same? What happened when you doubled the amount of marbles?
2. Do the pulleys allow you to pick up the most marbles with less effort? Explain your answer.
3. **Draw a diagram** of the setup. Add labels indicating the forces on the strings. Explain why it is easier to lift the marbles with two pulleys.

Goals
- Construct a pulley machine.
- Compare two uses of pulleys.

Materials
- string
- twist ties
- bags of marbles or other weights
- 2 pulleys
- dowel rod or broomstick

FIGURE 11-12

This is a water pump proposed by the Greek inventor Archimedes more than 2000 years ago. The turning screw (inclined plane) of the pump winds water up from a lower level to a higher one. The handle is a wheel and axle.

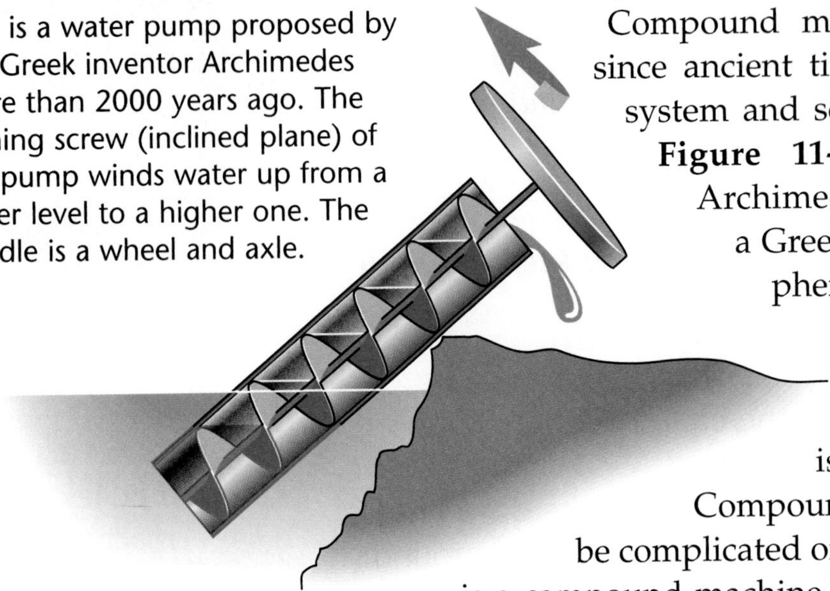

Compound Machines

Compound machines have been used since ancient times. The water-pumping system and screw conveyors shown in **Figure 11-12** were proposed by Archimedes (ar kih MEE deez), a Greek mathematician, philosopher, and inventor.

Any machine or device that is made up of two or more simple machines is a compound machine. Compound machines don't have to be complicated or fancy. A hand can opener is a compound machine. The crank is a wheel and axle, the handles are levers, and it has a circular wedge that cuts down into the can. Look at a bicycle and see how many simple machines you can find. A bicycle is another example of a compound machine.

The two-pulley system was only one of the compound machines you used on the camping trip. You may not have noticed several others. The biggest was the car that took you and your canoe to the river. The fishing rod and reel and the hatchet are two more examples.

Where else would you look for examples of compound machines, like the ones shown in **Figures 11-13** and **11-14?** Try the nearest gym, farm, kitchen, grocery store, or recreation center. Combinations of pulleys, inclined planes, and levers help people in many ways to make their lives easier.

FIGURE 11-13

This grain conveyor is another example of a compound machine. The corkscrew is a screw and a wheel and axle.

A The lever system is at rest before the key is played.

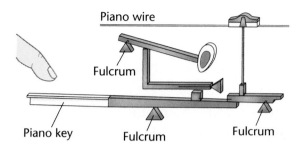

Piano wire

Fulcrum

Piano key   Fulcrum   Fulcrum

B When the key is played, the levers cause the hammer to strike the wire.

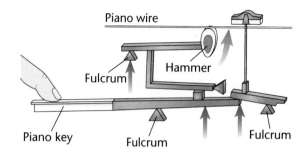

Piano wire

Fulcrum   Hammer

Piano key   Fulcrum   Fulcrum

FIGURE 11-14

A piano is a compound machine. When a key is played, a system of levers causes a felt-tipped hammer to strike the piano wire. This causes the note that you hear.

Section Wrap-up

1. What is a compound machine?

2. Why are scissors a compound machine?

3. **Think Critically:** Explain why a hatchet is a compound machine, and list all of the simple machines that are part of it.

4. ![Skill Builder icon] **Skill Builder**
   Sequencing Two pulleys form a compound machine. A movable and fixed pulley were used to lift the supplies on the camping trip. Starting at the tree, describe the path of the rope all the way from the pulleys to the hands of the camper. If you need help, refer to Sequencing on page 543 in the **Skill Handbook.**

Science Journal

Pick any compound machine you use at home or in the classroom. Draw it in your Science Journal and tell which simple machines it uses.

11•4 Access for All

What YOU'LL LEARN

- That simple machines make many sports and recreation activities accessible to all
- That an outline is an orderly, written summary of a text

Why IT'S IMPORTANT

Outlining helps you organize information in a meaningful way.

FIGURE 11-15

These athletes are playing bank-shot basketball.

The Americans with Disabilities Act

Have you ever been on crutches? If you have, you know how hard it is to go up and down stairs and get around at home. Some people have to deal with those problems every day. The Americans with Disabilities Act (ADA) is designed to stop discrimination against people with disabilities (dihs uh BIHL uh teez). A disability is a physical or mental limitation. The ADA became a law in 1990. The ADA requires that public buildings and services be easy for disabled persons to use. Constructing new equipment and buildings or changing older ones has made many places and activities available to all.

Happy Campers

The use of simple machines in day and summer camps over the past few years has helped many campers. Portable ramps make hiking nature trails possible for people who use wheelchairs. In some camps, levers replace doorknobs. Levers also make it easier to turn light switches on and off.

Be a Good Sport!

Simple and compound machines allow persons with disabilities to play along. Bank-shot basketball, for example, is a non-running sport that requires strategy, concentration, and hoop-shooting skill. Bank-shot basketball is similar to traditional basketball, as shown in **Figure 11-15.** But, bank shot features 18 stations whose degree of difficulty depends on levers, ramps, and other simple machines.

On the golf course, golf clubs, which are third-class levers, are used to increase the speed of the golf ball. Golfers usually stand while hitting the ball. Some disabled golfers must hit from a sitting position. A golf cart with a turning seat that uses a screw may be the answer for these golfers.

Skiing is another sport in which simple machines allow persons with a variety of disabilities to hit the slopes, as shown in **Figure 11-16.** Some ski boots have wedges attached to their bottoms that aid stability and assist people with poor balance. A ski slant board does the same thing. The slant board is a metal base attached to a ski binding. The board can be tilted forward, backward, or side to side. This clever device allows a disabled skier to position himself or herself in a balanced position. The slant board can also be used by skiers whose legs are different lengths.

FIGURE 11-16

This skier is competing in the Winter Paralympics in Lillehammer, Norway.

 ## Skill Builder: Outlining

LEARNING the SKILL

1. Carefully read the material to be outlined.

2. Choose at least two or three ideas from the passage that summarize each section. Depending on the length of the information to be summarized, your outline may have more than two or three main ideas.

3. Now, pick out key words and phrases that are important to the passage. These supporting details are written under the main idea headings.

4. Arrange the key words and phrases under the correct titles. Use numbers and letters to sequence the words and phrases.

5. Check the accuracy of your outline by comparing the outline to the passage. Is the order correct? Does the outline make sense? Make any necessary changes.

PRACTICING the SKILL

1. Make an outline using the bold-faced titles in the article as your main ideas.

2. Use what you have learned to make another outline of the information on pages 312 and 313. Use these main ideas: levers, pulleys, and inclined planes.

APPLYING the SKILL

Find an article in the library about how simple machines make many playgrounds accessible to all. Make an outline of the main points in the article.

Rube Goldberg's Wonderful, Wacky Machines

Do you like gadgets? Many little devices we use every day make it easier to do jobs. Think of a pencil sharpener or a can opener. But if you are like most people, you've probably used at least one gadget that was so hard to operate that it would have been simpler to do the job without the gadget.

Rube Goldberg (1883-1970) understood that problem. Goldberg was an American cartoonist who poked fun at gadgets and machines. He thought that technology often made life more complicated than it should be. From the 1920s until the early 1960s, Goldberg drew hundreds of cartoons that featured funny and complicated inventions. These enormously complex machines were designed to do simple things like lick a stamp, crack open an egg, or scratch a mosquito bite.

Each machine had many steps and combined tools, household objects, and animals in odd ways. For example, Goldberg's "Automatic Screen Door Closer" used flies, a spider, a potato bug, a mechanical soldier, and a circus monkey (who was an expert bowler) just to close a door!

Rube Goldberg entertained newspaper audiences for many years with his imaginative cartoons. They were considered humorous and weird, but they also made people think. Even today, his cartoon machines remind us that the best approach to solving a problem is often the simplest one.

SCIENCE Online

How many steps does it take to scratch your back? Visit a link through Glencoe Science Online at **science.glencoe.com** to see how many steps Rube Goldberg used in one of his cartoons.

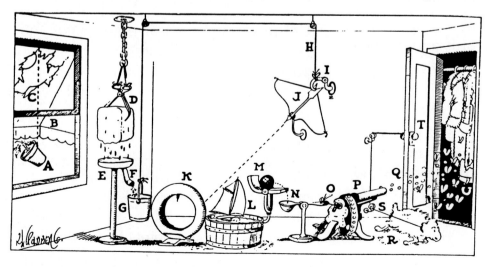

Read the statements below that review the most important ideas in the chapter. Using what you have learned, answer each question in your Science Journal.

1. *Work* is defined as a force moving an object through a distance. The direction of the force must be in the same direction that the object moved. *Is this girl working?*

2. Simple machines are devices that make work easier by changing the size or direction of a force. *What simple machine is shown in this image, and how does each change force?*

4. Compound machines use several simple machines working together to accomplish a task. *What simple machines are working together here?*

3. There are three classes of levers. Each lever has a different arrangement of the fulcrum, force, and load. First- and second-class levers make your force larger. Third-class levers increase speed. *Which class of lever is shown in this example? How do you know?*

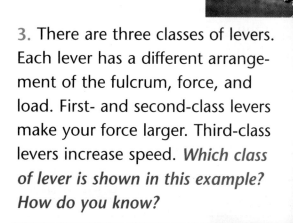

Using Key Science Words

compound machine
inclined plane
joule
lever
pulley
screw
simple machine
wedge
wheel and axle
work

Using the list above, replace the underlined words with terms that include the correct key science word.

1. <u>A bar or rod free to rotate around a single point</u> can be used to lift heavy objects.
2. Pulleys, inclined planes, and levers are all examples of <u>a device that does work with one movement</u>.
3. A hatchet is an example of <u>two inclined planes combined</u>.
4. <u>An inclined plane wrapped around a rod</u> is often used to hold objects together.
5. Lifting a barbell is <u>exerting a force over a distance in the same direction as the object's motion</u>.

Checking Concepts

Choose the word or phrase that completes the sentence.

6. The _____ is the unit used for work.
    a. joule   c. pound
    b. newton   d. meter

7. A _____ is a compound machine.
    a. screw
    b. fishing rod and reel
    c. ramp
    d. wheel and axle

8. An example of an inclined plane is a(n) _____.
    a. mountain climber
    b. elevator
    c. seesaw
    d. moving van ramp

9. A _____ uses a lever.
    a. vise
    b. stairway
    c. seesaw
    d. moving van ramp

10. _____ is not work.
    a. Carrying a canoe on level ground
    b. Lifting a canoe
    c. Carrying a canoe uphill
    d. Paddling a canoe across a lake

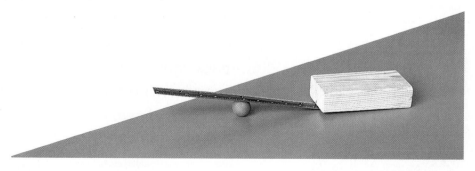

Thinking Critically

Answer the following questions in your Science Journal using complete sentences.

11. Does using a wheelbarrow to lift a load of mulch make you do more, less, or the same amount of work? Explain.

12. Which is easier to use to sweep a large area, a broom with a long handle or a broom with a short handle? Why?

13. What kind of simple machine is a wrench? How does the length of a wrench affect your force?

14. How are a needle and an ax related?

15. Explain how a screwdriver and a screw form a compound machine.

Developing Skills

If you need help, refer to the description of each skill in the Skill Handbook.

16. **Observing:** Make a list of the simple machines that make up a manual can opener. Refer to **Table 11-1** for help.

17. **Using Numbers:** Elena exerted a force of 200 N over a distance of 4 m in the direction of the force. How much work did she do?

18. **Concept Mapping:** Make a network tree concept map that includes the six types of simple machines and two examples of compound machines.

19. **Comparing and Contrasting:** Compare and contrast a first-class lever and a wheel and axle.

20. **Predicting:** Raul ties a bag weighing 400 N to one end of a rope. He tosses the rope over a tree branch and pulls down with 200 N of force. What happens?

Performance Assessment

1. **Invention:** Design a compound machine that will give a dog a treat any time the dog wants one.

2. **Scientific Drawing:** Draw each of the classes of levers, identify all the parts, and explain how they make tasks easier.

Table 11-1

Simple machine	Example
Inclined plane	the hiking trail up the mountain
	the gentle slope to pull the canoe onshore
	the screw anchors to hold the tent in place
	the wedge on a hatchet
Lever	First-class—crowbar to pry up the rock
	Second-class—canoe used to haul supplies
	Third-class—fishing rod
Wheel and axle	reel on the fishing rod
Pulley	fixed—rope over a limb
	movable—pulley on the supplies
	combination—using the tree limb and the movable pulley to lift the supplies

Chapter Preview

Skills Preview

▶ **Skill Builders**
- make a graph
- make a model
- make generalizations

▶ **MiniLABs**
- observe
- compare

▶ **Activities**
- observe
- measure
- sequence
- design an experiment
- graph
- compare
- predict

Energy

With the wind rushing past her and the snow flying up around her, this snowboarder feels the thrill of moving at high speed. If you could watch her in action, you would notice that she can change the direction and speed of her board by changing her body position. A surfboard rider feels the same sense of connection between his or her body position and the board's motion. So does a skier, a cyclist, and even a young child on a swing. All are aware of something changing—position, speed, or direction. Do you realize that energy plays a part in these changes?

EXPLORE ACTIVITY

Observe Energy in Action

1. Obtain a new wide rubber band.
2. Hold the rubber band in both hands and touch it to your lower lip.
3. Quickly stretch the rubber band several times, and touch it to your lip again.
4. What difference do you notice?

Changes such as the one you noticed in the Explore activity occur all around you. In your Science Journal, list other such changes that you have observed today.

12•1 How does energy change?

What YOU'LL LEARN

- What energy is and the forms that it takes
- The difference between potential energy and kinetic energy

Science Words:
energy
kinetic energy
potential energy

Why IT'S IMPORTANT

The more you know about energy and how it changes, the better you can use energy.

Energy

Energy (EN ur jee) is a term you probably have used many times. You may have said that eating a plate of spaghetti gives you energy or that a peppy Olympic gymnast has a lot of energy. But do you know that a burning fire, a bouncing ball, and a gallon of gasoline also have energy? Exactly what is energy?

What is energy?

Energy is the ability to change things. Energy can change the temperature, shape, speed, position, or direction of an object. Soccer players use energy to change the direction of soccer balls by hitting them with their feet or with their heads, as shown in **Figure 12-1.** Energy can change the shape of modeling clay or the temperature of a cup of water. You can use energy to make your muscles change the speed of a bicycle, and a snowboarder uses energy to change position on the path the snowboard takes.

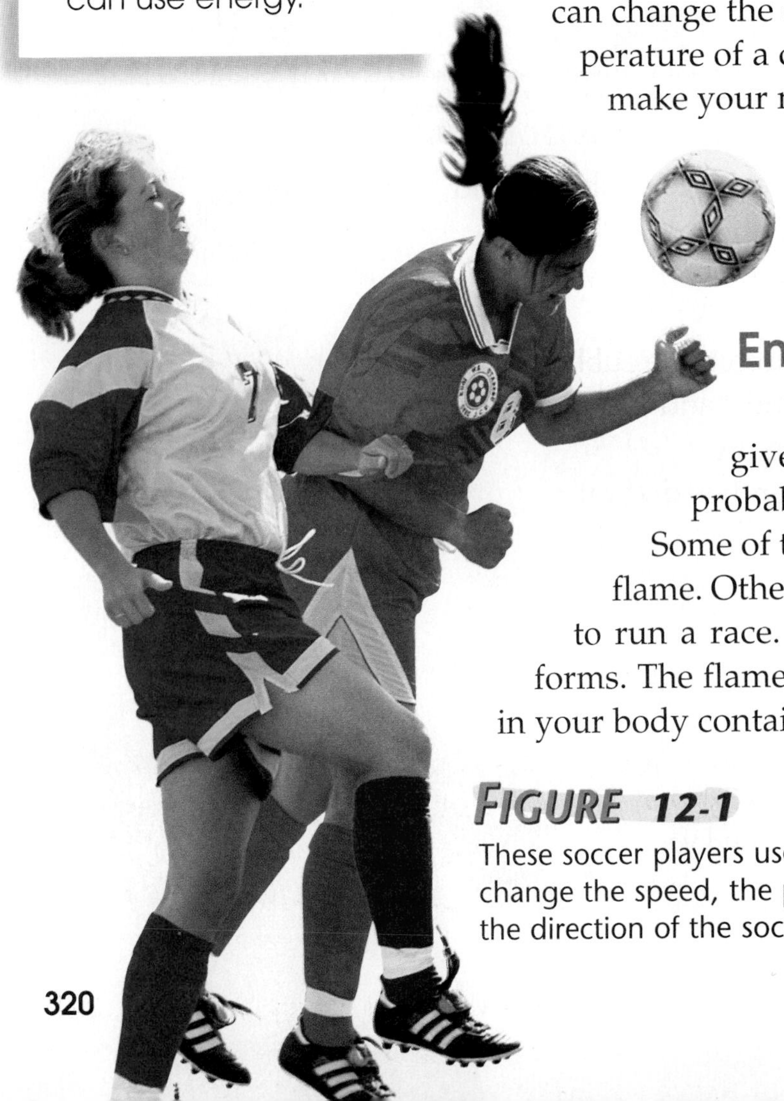

Energy Transformations

If you were to ask your friends to give examples of energy, you would probably get many different answers. Some of them might mention the energy in a flame. Others might suggest the energy needed to run a race. Energy occurs in several different forms. The flame gives off heat and light. Fat stored in your body contains chemical energy.

FIGURE 12-1

These soccer players use energy to change the speed, the position, and the direction of the soccer ball.

Change Can Cause Change

In causing other things to change, energy itself often changes from one form to another. Any change of energy from one form to another is called an energy transformation. Energy transformations take place all around you every day. When a car sits in the sun all day, the energy in light waves changes to a form of energy that warms the inside of the car. In fireworks, chemical energy in the ingredients transforms into the energy in light, sound, and heat.

During an energy transformation, such as when light causes an increase in temperature, the total amount of energy stays the same—no energy is lost or gained. Only the form of energy changes, not the amount.

Energy Changes in Your Body

You may not realize it, but you transform energy every time you eat. Like the logs, twigs, and leaves on the burning campfire shown in **Figure 12-2,** the corn on the cob, tossed salad, and baked potato you eat for dinner contain chemical energy. The chemical energy from any meat you eat originally also came from plants that stored the energy from sunlight in their leaves. What happens to the chemical energy in the food you eat? Energy contained in your food is transformed into energy that moves your muscles and keeps your body warm.

Have you ever noticed how warm you become while playing, walking, running, or working hard? How many times have you thought it was cool outside until you started moving around? You soon found yourself taking off your jacket or sweater because your body was releasing energy as heat when your muscles contracted.

Fat stores chemical energy that all of your muscles need, including your heart and those that help cause air to move into your lungs.

FIGURE 12-2

In a campfire, the chemical energy in the wood is changed, or transformed, into heat and light.

FIGURE 12-3

Low-fat items provide a small amount of fat to your diet.

Look at the low-fat items in **Figure 12-3.** All healthy bodies need a certain amount of fat—not too much, but not too little. Grocery store shelves are full of products labeled "fat free" that claim to be good for your health. But you would not be as healthy as you could be if you were on a completely fat-free diet.

Useful Changes

Since the earliest times, human beings have used energy transformations in ways that are useful to them. Early humans learned to use chemical energy to obtain heat and light when they built a fire. Today, electrical energy produces heat and light in a lightbulb. Chemical energy in gasoline runs the engine in your family car. The water heater in your home transforms chemical or electrical energy into the energy in hot water. Electrical energy is changed to energy of motion in an air conditioner. Hydroelectric and wind power plants transform the energy of moving water and wind into electrical energy.

FIGURE 12-4

A sled can have either potential energy or kinetic energy.

Kinetic and Potential Energy

You've seen that energy can take many useful forms, such as light, heat, and motion. Two types of energy that relate to motion and position are called kinetic (kih NET ihk) energy and potential (poh TEN shul) energy.

Kinetic Energy

If you were asked whether a baseball flying toward center field has energy, you might say that it does because it is moving. Moving objects have a type of energy called **kinetic energy.** As shown in **Figure 12-4,** a sled moving down a hill has kinetic energy. Molecules in air that is blown through a musical instrument also have kinetic energy.

Not all moving objects have the same amount of kinetic energy. Which would have more kinetic energy, a sled or a molecule of oxygen in the air? The amount of kinetic energy an object has depends on the mass and speed of the object. If the two objects were traveling at the same speed, the sled would have more kinetic energy than the oxygen molecule because the sled has more mass. Look at another example in **Figure 12-5.**

Potential Energy

If you were asked whether the sled sitting at the top of the hill in **Figure 12-4** has energy, what would you say? An object does not have to be moving to have energy— it may have what is called potential energy. **Potential energy** is energy that is stored. This is not energy that comes from motion; it is energy that comes from position or condition. The sled at the top of the hill has potential energy because even though it is not moving, it has potential for movement because of its position. As you can see, potential and kinetic energy are related to each other. Just as you can fill a water bottle to be used later at the basketball court, you can store potential energy to be used later when you need it.

FIGURE 12-5

If both have the same mass and each is running at its top speed, which would have more kinetic energy, a cheetah—one of the fastest land animals— or a dog?

FIGURE 12-6

This clock contains a pendulum. *When does the pendulum have the greatest potential energy?*

Pendulum

Arc

Potential to Kinetic

Potential energy becomes kinetic energy when something acts to release the stored energy. For example, it takes a lot of energy to tightly coil the wire in a coiled spring toy. Energy is stored in the coiled wire. When the toy walks down a flight of steps, the stored energy of the spring allows the movement to continue. It needs only the slightest help from gravity to keep it going.

One of the easiest ways to see the difference between potential and kinetic energy is to work with a pendulum (PEN juh lum), as shown in **Figure 12-6.** A pendulum is a weight that swings back and forth from a single point. A child's swing is an example of a pendulum. You can add potential energy to a swing by pulling back on it. Similarly, you can add potential energy to a pendulum by moving the weighted end to one side. When you let go of a swing or any other pendulum, gravity releases the potential energy, and the pendulum swings down in a curved path, called an arc. The instant the pendulum moves, potential energy is transformed into kinetic energy.

The more potential energy something has, the more kinetic energy results from the transformation of potential energy to kinetic energy. When a ski lift takes a skier to the top of a ski run, the skier's potential energy increases.

FIGURE 12-7

A bowling ball ready to be rolled down the alley has potential energy.

 A When you roll it down the alley, your muscles give the bowling ball kinetic energy.

That energy becomes kinetic energy when the skier moves down the mountain. The higher the ski run, the greater the potential energy the skier has at the top and the more kinetic energy she has skiing down the slope.

Transfer of Kinetic Energy

Kinetic energy can be transferred from one object to another when those objects interact. Look at the changes and transfer of energy shown in **Figure 12-7.** Energy can even be transferred from one bowling pin to another, knocking them all down with one roll of the ball, even though the ball does not touch all of the pins.

It is important to know the difference between potential and kinetic energy. It is this difference that allows us to store energy for later use.

B When it hits the pins, the ball transfers its energy to them.

Problem Solving

Observing Kinetic Energy

Baseball players both love and hate a fastball. A fastball is a pitch that comes straight across the plate at the fastest rate of speed the pitcher can deliver. Speeds of 90 mph are not unheard of when a good pitcher delivers a fastball. A team values a pitcher who can throw good fastballs. A good batter has mixed feelings about a fastball. The ball is harder to hit as it whizzes by, but some of the most dramatic home runs hit by batters are hit off fastballs.

Solve the Problem:
It may be difficult for a batter to connect with a fastball. But when it does happen, why is there a good chance that the ball will be hit a great distance? The answer lies in the presence of potential and kinetic energy in the ball and the bat.

Think Critically:
If you were a major league ballplayer, would you want to face a fastball pitcher? Explain.

Activity 12-1

Colliding Objects

Have you ever dashed around a corner and crashed into someone coming from the other direction? In such a case, it's obvious that energy is transferred. In this activity, you will experiment with colliding balls to examine energy transfer.

Goals

- Observe two objects colliding.
- Measure the results of the collisions.

Materials

- small, rubber, high-bouncing ball
- volleyball
- large, rubber playground ball
- tennis ball
- 3-m fabric measuring tape
- masking tape

What You'll Investigate

How does the combined mass of two balls affect their collision?

Procedure

1. **Copy** the data table into your Science Journal.
2. **Hang** the measuring tape on the wall. Be sure its starting point is 1 m off the floor.
3. **Hold** the bottom of the tennis ball 1 m off the floor. **Place** the small ball directly on top of the tennis ball.
4. **Drop** the balls at the same time. Use the measuring tape to **measure** the height the small ball rebounds into the air. **Repeat** steps 3 and 4 three times, and **average** the data. **Record** all data.
5. **Repeat** steps 2 and 3 for the volleyball and the small ball, and then the playground ball and the small ball.

Conclude and Apply

1. **Rank** the balls from lightest to heaviest. What is the relationship between the mass of the two balls dropped and the height the small ball rebounds into the air?
2. Based on your observations, is the mass of an object related to the amount of potential energy it has?

Data and Observations

Small ball and . . .	Trial 1	Trial 2	Trial 3	Average
tennis ball				
volleyball				
playground ball				

FIGURE 12-8
Both the cars and the people have kinetic energy and potential energy several times during a roller coaster ride.

Section ☀ Wrap-up

1. Think of a roller coaster like the one shown in **Figure 12-8** climbing to the top of the steepest hill on its track. When does the first car have the greatest potential energy? When does it have the greatest kinetic energy?

2. Use the relationship between weight and kinetic energy to explain why some amusement parks have weight limits for their rides.

3. **Think Critically:** You get up in the morning, get dressed, eat breakfast, walk to the bus stop, and ride to school. List three different energy transformations that have taken place.

4. **Skill Builder**
   Making and Using Graphs A pendulum was found to swing seven times per minute. If the string were half as long, the pendulum would swing ten times per minute. If the original length were twice as long, the pendulum would swing five times per minute. Make a bar graph that shows these data. Draw a conclusion from the results. If you need help, refer to Making and Using Graphs on page 547 in the **Skill Handbook.**

Science Journal
In your Science Journal, write a paragraph about what energy transformations took place when last night's dinner was prepared.

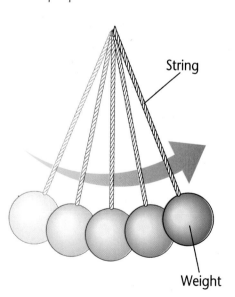

String

Weight

12•2 How do you use thermal energy?

What YOU'LL LEARN

- The difference among thermal energy, heat, and temperature
- Important uses of thermal energy and how thermal energy moves

Science Words:
thermal energy
heat
radiation
conduction
convection

Why IT'S IMPORTANT

Life would not be possible if thermal energy did not exist and did not move.

Thermal Energy

Imagine that you have just come inside from playing outside in freezing weather. What do you think about? A cup of hot cocoa? A nice, warm fire in the fireplace or a warm stove? What is it about these things that make you think they will warm you?

What is thermal energy?

Hot cocoa, as shown in **Figure 12-9,** and a fire will warm you because they have high amounts of the form of energy known as thermal (THUR mul) energy. **Thermal energy** is the total amount of energy—both potential and kinetic—of the atoms that make up a material. The amount of thermal energy that a cup of cocoa contains is determined by the amount of cocoa in the cup and the amount of energy in the atoms that make it up.

Thermal energy moves from a warmer object to a cooler one. Thermal energy in the cup of hot cocoa will transfer to your body and you will become warmer.

FIGURE 12-9

Drinking warm cocoa transfers thermal energy from the cocoa to your body.

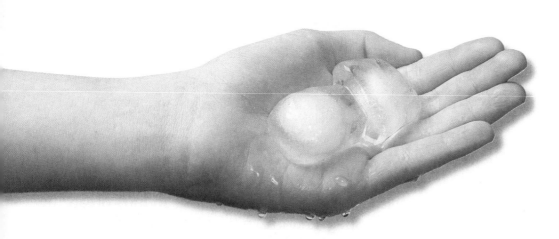

FIGURE 12-10

Thermal energy moves from a warmer object to a cooler one. *Is the air around the ice cubes colder than, warmer than, or the same temperature as the rest of the air in the room?*

How much thermal energy does it have?

An object's kinetic energy doesn't determine its thermal energy. For example, a ball at 25°C sitting in your backyard has the same amount of thermal energy as it does at 25°C heading through the air toward your friend. The energy the atoms have, not the energy of the object itself, determines the amount of thermal energy.

Different materials may have different thermal energies. If you get into a car that has been closed up on a sunny, summer day, where will you sit? If you can choose between sitting on a fabric-covered seat and sitting on a vinyl-covered seat, you'll find out that the vinyl has more thermal energy, even though the temperatures inside the cars are the same and the masses of the seats are similar. This difference is based on the way the atoms in each material are attracted to each other.

Heat and Thermal Energy

Thermal energy is what you may have been calling heat. Heat and thermal energy are related, but they are not the same. Look at **Figure 12-10.** Suppose you pick up an ice cube to place it in a drink. As you hold the ice cube, your hand becomes colder. Some of the ice melts because thermal energy from your hand transfers to the ice. The thermal energy that moves from a warmer object to a cooler one is **heat.**

Mini LAB

Observe a Strained Flame

See what happens when heat is transferred.

1. Light a candle.
2. Hold a metal strainer over the flame. Try to make the flame come through the wire mesh of the strainer.
3. Extinguish the candle.

Analysis

1. What happened to the flame when the strainer was in it?
2. In terms of heat transfer, suggest an explanation for your observations.

Design Your Own Experiment
Does thermal energy affect kinetic energy?

You are in charge of dinner, and spaghetti sounds great. You put the dry spaghetti in a pot of boiling water and cover it. What soon happens? If you have ever seen steam lift the lid off a pot of boiling water, you've witnessed thermal energy transforming to energy of motion.

PREPARE

Possible Materials
- identical balls, at least 3
- meterstick
- thermometer
- paper
- pencil
- freezer

What You'll Investigate
Will the thermal energy in a ball affect how high it will bounce?

Form a Hypothesis

Think about what you know about how thermal energy affects matter, such as the air inside a ball. Remember that a measure of temperature indicates the amount of thermal energy the ball contains. In your group, **make a hypothesis** about how the temperature of a ball affects the height of its bounce.

Goals

Design an experiment that tests several balls to see whether the temperature of each ball when it is dropped makes it bounce to different heights.

Observe how energy is transformed from thermal energy to kinetic energy.

Safety Precautions

Wear goggles. Do not use any temperature beyond the range of what would normally be found in any safe area of your classroom.

PLAN

1. **Choose** a way to test your group's hypothesis. Remember to test only one thing.
2. **Decide** how many balls you are going to test and at which temperatures. Decide how long each ball will be kept at that temperature before you test it.
3. Pick a height from which to drop each ball. The height from which each ball is dropped must always be the same. **Decide** how to measure the bounce of the ball and how many times you are going to drop each ball.
4. **Prepare a data table** in your Science Journal that will show the temperature of each ball and the height of each bounce. **Decide** how to graph the results so the data are easy to see and compare.
5. **Read** over your experiment and see whether you can picture what is happening. Do any parts seem unclear? These parts will need more planning.

DO

1. Make sure your teacher has approved your plan and your data table before you proceed.
2. Carry out your plan.
3. While doing the experiment, **record** any observations and **complete** your data table.

CONCLUDE AND APPLY

1. **Graph** temperature versus height of bounce for each ball. **Compare** your graphs to those of other students in the class.
2. How did the temperature of the ball affect its bounce? **Summarize** your findings in a few sentences.
3. **Compare** your results and your hypothesis. From your results, what is the relationship between thermal energy in the ball and kinetic energy in the bounce?
4. **APPLY** Use your results to **predict** how the air temperature when you are playing tennis might affect how hard you have to swing the racket.

Temperature and Thermal Energy

Temperature also is related to thermal energy and heat. You are aware of temperature every day. When you get up in the morning, the temperature of the air outside helps you decide what to wear. When you are not feeling well, you take your body temperature to find out whether you have a fever, as shown in **Figure 12-11**.

Temperature Is an Average

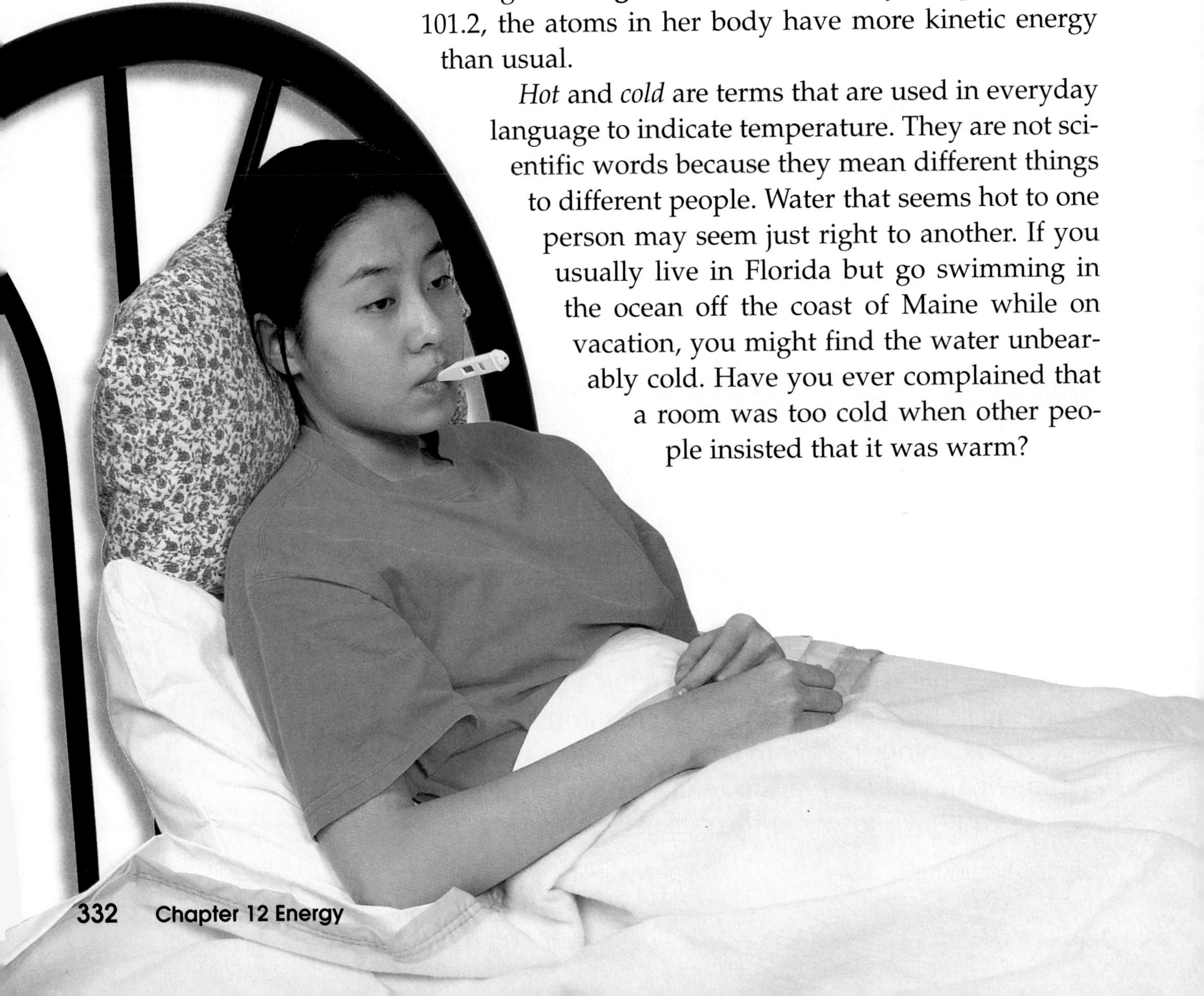

FIGURE 12-11

In the United States, thermometers that use the Fahrenheit scale are used to find body temperature.

But what is temperature? If you have a sample of a material, that sample is made up of many atoms. Each of those atoms has a certain amount of kinetic energy and is moving. If you measured the kinetic energy of each atom in the sample and averaged them, this average kinetic energy would be the temperature of the sample. For example, normal human body temperature is 98.6°F. If the girl in **Figure 12-11** has a body temperature of 101.2, the atoms in her body have more kinetic energy than usual.

Hot and *cold* are terms that are used in everyday language to indicate temperature. They are not scientific words because they mean different things to different people. Water that seems hot to one person may seem just right to another. If you usually live in Florida but go swimming in the ocean off the coast of Maine while on vacation, you might find the water unbearably cold. Have you ever complained that a room was too cold when other people insisted that it was warm?

Temperature Scales

Because everyone experiences temperature differently, you cannot accurately measure temperature by how it "feels." Recall that temperature is the average kinetic energy of all the atoms in a material. But there is no way to measure the kinetic energy of each atom and then calculate an average. Instead, scientists use various temperature scales designed to measure the average kinetic energy of the particles in a material. These scales divide changes in kinetic energy in atoms into regular intervals. Look at **Figure 12-12.** The scales most commonly used in daily life are the Celsius (SEL see us) and Fahrenheit (FAYR un hite) scales.

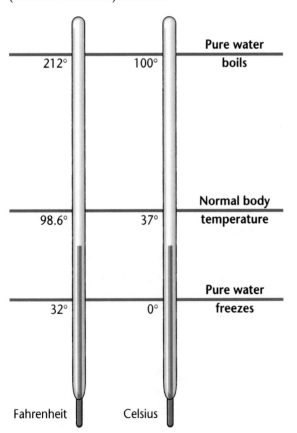

Fahrenheit | Celsius

212°	100°	Pure water boils
98.6°	37°	Normal body temperature
32°	0°	Pure water freezes

Mini LAB

Comparing Energy Content

Find out which has more kinetic energy, hot water or cold water.

1. Pour equal amounts of hot, cold, and room-temperature water into each of three transparent, labeled containers.
2. Measure and record the temperature of the water in each container.
3. Use a dropper to gently put a drop of food coloring in the center of each container.
4. Observe after 2 minutes to see how quickly the color has spread through each container.

Analysis

1. Based on the speed at which the food coloring spreads through the water, rank the containers from fastest to slowest. What can you infer about how water temperature affected the movement of the food coloring?
2. If air acts much like water, where on Earth would you expect to see the most energetic weather develop?

Go to Glencoe Science Online at *science.glencoe.com* for a link to a site that compares temperature scales and gives conversion factors.

FIGURE 12-12

These thermometers show at what temperatures water freezes and boils and normal human body temperature on the Fahrenheit and Celsius scales.

Thermal Energy on the Move

Thermal energy has many properties that affect our lives in important ways. What is the weather like today? Weather occurs because thermal energy causes air in the atmosphere to move. When you cook food, as in **Figure 12-13,** heat moves through the food, causing changes in it. Your home is heated and cooled by this same movement of thermal energy. Thermal energy moves in three distinct ways—by radiation, by conduction, and by convection.

Radiation

Radiation (ray dee AY shun) is energy that travels by waves in all directions from its source. The sun transfers energy to Earth through radiation. Have you ever been on a hike on a cool day and chosen to lean against a large boulder because it was warm from the sunlight? The boulder was heated by radiation. A microwave oven and a solar cooker cook food because of transfer of energy by radiation.

Conduction

Have you ever picked up a spoon that was in your soup and discovered that the spoon was hotter than it was when you placed it there? The spoon handle became hot because of conduction (kun DUK shun). **Conduction** is the transfer of kinetic energy from one molecule to another, much the same way kinetic energy is transferred from a bowling ball to the pins. When you put a pan on the stove to make a grilled-cheese sandwich, as shown in **Figure 12-14,** the heat from the stove's heating element transfers energy to the pan, making its molecules move

FIGURE 12-13

Popcorn can be cooked by either radiation or conduction.

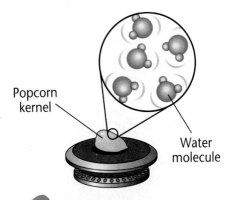

Popcorn kernel

Water molecule

A A kernel of popcorn contains water molecules.

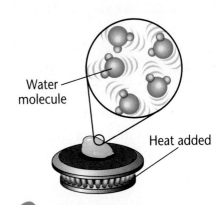

Water molecule

Heat added

B When heat is added to the kernel, the water molecules in the kernel move faster. They become water vapor and cause pressure inside the kernel.

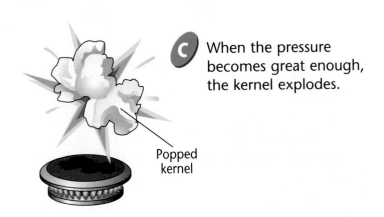

C When the pressure becomes great enough, the kernel explodes.

Popped kernel

faster. Some of these molecules bump into the molecules in the bread and butter that are on the surface of the pan. Energy is transferred to the sandwich. This transferred energy causes the sandwich to cook. Even though conduction is a transfer of kinetic energy from particle to particle, the molecules involved don't travel from one place to another. They simply move in place, bumping into each other and transferring energy from faster-moving particles to slower-moving ones.

FIGURE 12-14

Energy transferred by conduction from the pan to the sandwich causes the cheese to melt and the bread to toast.

USING TECHNOLOGY

The Microwave Oven

When people's lives became busier, they needed faster ways to prepare food. The microwave oven became the answer for many households. It is easy to use, quick, and relatively inexpensive.

The microwave oven uses microwave radiation to transfer thermal energy to the inside of a food item, cooking it from the inside out. Microwaves increase the motion of the molecules of the food placed in the oven. The more mass something has, the more motion its molecules will have. This is why even large items can be cooked in just a little more time than smaller ones.

If you ever made microwave popcorn, you used radiation to speed up the water molecules inside the kernels, causing them to explode. If you don't use a microwave oven to make popcorn, you use conduction. Heat moves up from the bottom of the pan, through the metal, and into the kernels.

SCIENCE Online

Many meals require heat transfer in their preparation. What advantages and disadvantages are there to different methods? Visit Glencoe Science Online at *science.glencoe.com* for a link to the science of food preparation.

Convection

Some energy transfers involve molecules that do not stay in place but move from one place to another. **Convection** (kun VEK shun) transfers thermal energy by the movement of molecules from one place to another in a gas or a liquid. It is because of convection that thermal energy causes changes in the atmosphere that produce weather on Earth. Convection currents are also formed in oceans by cold water from the poles and warm, tropical water, as shown in **Figure 12-15.**

Have you ever seen an eagle or a hawk coasting high in the air? Look at **Figure 12-16.** The bird can fly even though it is not flapping its wings because it is held up by something called a "thermal." A thermal is a column

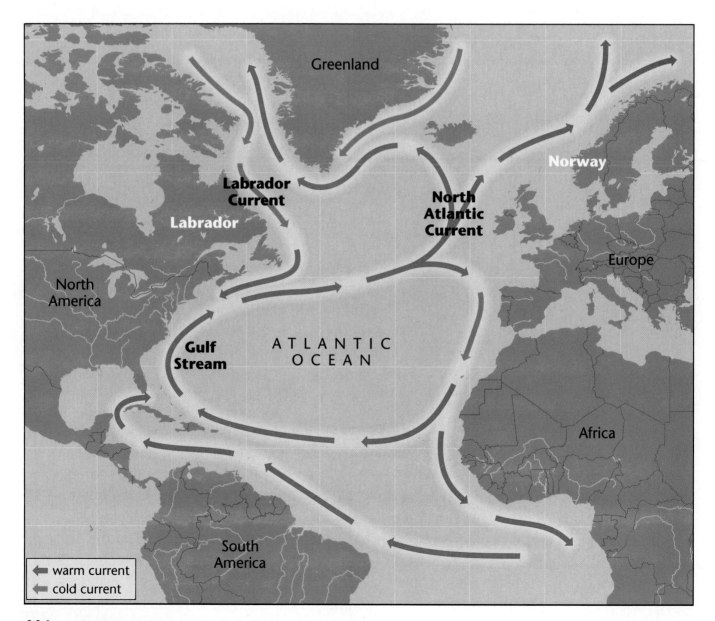

FIGURE 12-15

Even though Labrador is further south than Norway, Norway's ports are kept free of ice by the warm convection current that flows along its coast. Labrador is much colder because of the icy current that flows along its shore.

of warm air that is forced up as cold air around it sinks. A thermal is a convection current in the air.

You can see that thermal energy and its transfer affect you constantly. How many uses of thermal energy can you see around you right now?

FIGURE 12-16

A convection current helps hold this bird in the air. It is the same type of current used by people who are hang gliding. It helps keep them in the air as long as possible.

Section Wrap-up

1. Popcorn can be cooked in a hot-air popper, in a microwave oven, or in a pan on the stove. Identify each method as using either convection, conduction, or radiation.

2. Use a glass of cold tea and a cup of hot tea to explain the differences among temperature, heat, and thermal energy.

3. **Think Critically:** On the Fahrenheit temperature scale, water boils at 212°F and freezes at 32°F. On the Celsius scale, water boils at 100°C and freezes at 0°C. Which scale do you think would be easier to use? Explain.

4. *Skill Builder*
   Making Models A thermos bottle includes a glass container with a shiny coating that is surrounded by a vacuum, which contains no air. This glass is then placed in a container that is the outer part of the thermos bottle. Make a model of what a cross-section of a thermos bottle looks like. If you need help, refer to Making Models on page 557 in the **Skill Handbook.**

USING MATH

To change a temperature from the Fahrenheit scale to the Celsius scale, you subtract 32 from the Fahrenheit temperature and multiply the difference by 5/9. If the temperature is 77°F, what is the Celsius temperature?

12•3 Sun Power

What YOU'LL LEARN

- How using solar energy helps conserve Earth's limited resources
- How to make generalizations

Science Words:
solar energy

Why IT'S IMPORTANT

You will be able to generalize by putting together pieces of information to better understand an issue.

Solar Energy

Much of the energy we use comes from nonrenewable resources. As you learned in Chapter 7, nonrenewable resources cannot be replaced easily or quickly. Coal and oil are nonrenewable resources. Because these resources are limited, many people believe that we should use renewable resources to meet our energy needs. Natural processes can replace renewable resources fairly quickly. Energy from the sun, called **solar energy** (SOH lur • EN ur jee), is an example of a renewable resource.

The sun supplies Earth with a constant source of heat and light. Scientists are exploring ways to gather the sun's energy and transform it into electrical energy. This energy can provide heat and electricity—and even power cars!

One of the most ambitious schemes for providing solar energy on a large scale is to put giant solar collectors, or solar cells, in space. These solar cells, as shown in **Figure 12-17,** would collect energy and transfer it back to Earth using microwave radiation.

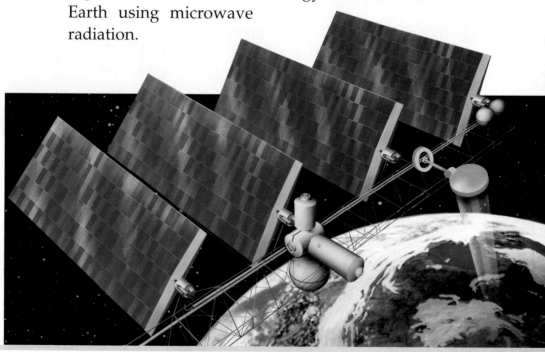

FIGURE 12-17

This is what solar collectors in space will probably look like.

Not all is sunny with this plan, however. To transform energy from the sun into electrical energy, solar cells require materials that are poisonous. In addition, microwaves can potentially harm humans. Solar energy itself can be unreliable. Light from the sun can be blocked by clouds, as in **Figure 12-18,** or air pollution. Even so, we can count on the sun's energy for a long, long time.

As you read material about topics such as solar energy, look for bits and pieces of information that hint at a main meaning—a generalization—about the topic. To make a generalization (jen rool luh ZAY shun), you put together bits of information to arrive at the big picture. For instance, if you knew that a football team was ranked first in the state last year and was unbeaten this season, you would make the generalization that the football team was great. This process will help you to form a better understanding of important issues.

FIGURE 12-18
Unless economical batteries that can store solar energy can be developed, using solar energy depends a lot on how much the sun shines.

 ## Skill Builder: Making Generalizations

LEARNING the SKILL

1. Identify the general topic of what the reading is about.

2. Gather as many facts as you can about the subject. Be sure the information you gather is made up of facts, not opinions.

3. Identify similarities and patterns among the facts. Facts can fit more than one pattern.

4. Use the similarities and patterns to form some general ideas about the subject.

PRACTICING the SKILL

1. What is the main idea of this passage?

2. Which of the following generalizations is supported by the passage?
   a. The use of microwaves with solar cells must be planned and controlled.
   b. The use of microwaves with solar cells does not present a problem.

3. Identify one or two facts that support the correct generalization.

APPLYING the SKILL

Make a general statement that describes your class. Write three facts that support your generalization.

To Build a Fire
by Jack London

He worked slowly and carefully, keenly aware of his danger. Gradually, as the flame grew stronger, he increased the size of the twigs with which he fed it. He squatted in the snow, pulling the twigs out from their entanglement in the brush and feeding directly to the flame. He knew there must be no failure. When it is seventy-five below zero, a man must not fail in his first attempt to build a fire . . .

Science Journal

Read Jack London's short story "To Build a Fire" to find out what happens to the man. What mistakes did he make that eventually put his life in danger? In your Science Journal, make a list of ways that heat, temperature, and thermal energy are important to the outcome of the story.

Jack London's story is set in the Yukon Territory in midwinter. A man hikes through the wilderness along a rarely traveled trail. He is alone except for a dog. It's so cold his spit crackles and freezes before it hits the ground.

The man has never spent a winter in the Arctic before. An old-timer cautioned him not to take this journey, saying that no one should travel alone when the temperature is colder than 50° below zero. But the man ignored the advice. Now, striding along the trail, he is still confident he can handle any problems that come along.

Then it happens. Thin ice covering a spring gives way. The man is soaked up to his knees. Almost instantly, his feet begin to freeze. He must build a fire to thaw his feet and dry his clothes.

But can he? The cold is frighteningly intense. The man's hands grow numb as he struggles with matches and wood. Perhaps he should have listened to the old man. Building a fire is now a matter of life or death.

Read the statements below that review the most important ideas in the chapter. Using what you have learned, answer each question in your Science Journal.

1. Energy is the ability to change your surroundings. Energy is found in many forms. The transformation of energy from one form to another allows humans to both store and use energy. *Compare and contrast potential energy and kinetic energy.*

2. Thermal energy can be transferred from place to place through convection, conduction, and radiation. Heat is thermal energy that is transferred from a warmer object to a cooler one. *What does temperature measure?*

3. Solar energy may be used to reduce use of nonrenewable energy resources, such as coal and petroleum. *Why isn't solar energy used more frequently?*

Using Key Science Words

conduction   potential energy
convection   radiation
energy   solar energy
heat   thermal energy
kinetic energy

Match each phrase with the correct term from the list of Key Science Words.

1. stored energy
2. total kinetic and potential energy in the particles of a material
3. type of transfer of thermal energy that produces weather
4. transfer of thermal energy from a warmer object to a cooler one
5. ability to bring about change

Checking Concepts

Choose the word or phrase that completes the sentence.

6. Solar energy is _____.
   a. a nonrenewable resource
   b. a renewable resource
   c. an inexpensive source of electricity
   d. commonly used as a source of electricity

7. Potential energy is to kinetic energy as _____.
   a. water stored behind a dam is to electrical energy
   b. a flute is to a trumpet
   c. a tractor is to a sports car
   d. an ice cube is to a hot day

8. Thermometers measure _____.
   a. potential energy
   b. molecular motion
   c. mechanical energy
   d. coolness

9. Gravity changes potential energy to kinetic energy when _____.
   a. you push a child in a wagon.
   b. a skier starts down a slope
   c. you throw a ball to a friend
   d. a balloon floats upward

10. When you place a cooler object against a source of heat, you transfer heat by _____.
    a. convection   c. conduction
    b. radiation   d. microwave

Thinking Critically

Answer the following questions in your Science Journal using complete sentences.

11. Energy is the ability to change the temperature, shape, speed, position, or direction of something. Give an example of doing each of these things.

12. How can the chemical energy of your lunch today become the energy you need to deliver newspapers after school tomorrow?

13. How could you use a box of dominoes to represent potential energy, kinetic energy, and transfer of kinetic energy?

14. Use what you know about the movement of thermal energy to explain why you would place cool cloths on a burn.

15. Explain why heat vents placed in the floor of a building would heat the building better than they would cool the building when the air conditioning is used.

Developing Skills

If you need help, refer to the description of each skill in the Skill Handbook.

16. Observing and Inferring: You observe a friend pick up a pot that has been on a lit burner, then quickly set it back down. In terms of thermal energy, infer what happened.

17. Designing an Experiment: Design an experiment that would test the ability of a raw egg to withstand being dropped. Point out the parts of the experiment that would represent each of the following:
a. the controls in the experiment
b. the dependent variable
c. the data to be collected
d. the egg when it has potential energy
e. the egg when it has kinetic energy

18. Comparing and Contrasting: In the activities in this chapter, you made observations and collected data. Data and observations are not the same. Compare and contrast data and observations.

19. Making Models: Draw a diagram that shows the difference between the transfer of heat energy when you make popcorn in a microwave and when you make it on top of a stove.

20. Concept Mapping: Draw a concept map that shows the three ways that thermal energy is transferred. Include an example for each. Be prepared to explain your map to other students and your teacher.

Performance Assessment

1. Designing an Experiment: Use the following items to design an experiment to find out how quickly different materials absorb radiant energy: pockets made out of three different colors of construction paper, a thermometer for each pocket, and a sunny day or a heat lamp.

2. Making Observations and Inferences: Put an empty, glass soft-drink bottle in a freezer for about 15 minutes. Take the bottle out of the freezer. If possible, place it in the sun. Center a dime on top of the mouth of the bottle. Use your observations to infer how thermal energy affects the kinetic energy of molecules in the air.

Chapter Preview

Skills Preview

▶ **Skill Builders**
- recognize cause and effect
- design an experiment to test a hypothesis
- observe

▶ **MiniLABs**
- compare and contrast
- make a model

▶ **Activities**
- hypothesize
- observe and infer
- make and use graphs
- scientific drawing
- make a model

Chapter 13

Electricity and Magnetism

You hear music and feel a breeze on your skin, but you can't see the moving air that causes these things. Magnetism and electricity are like the air. They can't be seen. People learned about magnetic and electric forces by observing how these forces affected the world.

Invisible magnetic force can cause movement, which means magnets can do work. The following activity gives you a chance to test magnetic force.

EXPLORE ACTIVITY

Measure Magnet Motion

1. You'll need a sheet of graph paper and two magnets. Place one magnet at each end of the graph paper. Mark the positions of the magnets.
2. Predict how close the magnets must be to affect each other.
3. Slide one magnet forward one square.
4. Repeat Step 3 until one of the magnets moves without being touched. Measure the distance between the magnets.
5. Turn one magnet a half turn and repeat the activity. Then turn the other magnet and repeat again.
6. Was your hypothesis from Step 2 confirmed?

Science Journal

Imagine you are in a small, metal ship moving around one of the magnets. In your Science Journal, describe how your ship will be affected by the magnet's pull.

13•1 What is an electric charge?

13

What YOU'LL LEARN

- How the structure of the atom helps produce electricity
- Why static electricity is a form of potential energy

Science Words:
electrical energy
static electricity

Why IT'S IMPORTANT

Understanding static electricity will help you understand all the ways you use electricity.

FIGURE 13-1

Lightning is an example of electrical energy.

Static Electricity

Rap, blues, rock, classical—whatever your favorite style of music is, you can hear it any time by popping a compact disc or a cassette into a player. Hearing music from a CD or cassette player wouldn't be possible without electricity. Which type of energy are you using when you play a cassette?

Remember from Chapter 12 the difference between potential energy—stored energy—and kinetic energy? When a skier coasts down a hill, potential energy is transformed to kinetic energy—energy in use. A tape player changes potential electrical energy in the battery to **electrical energy,** the energy of charges in motion.

Where can you find examples of electrical potential energy? Have you ever seen a blinding bolt of lightning slice through a dark sky, as in **Figure 13-1?** What about seeing a spark and hearing a snap after you cross a carpeted floor? These examples of electrical energy are results of static electricity. It is a form of potential electrical energy.

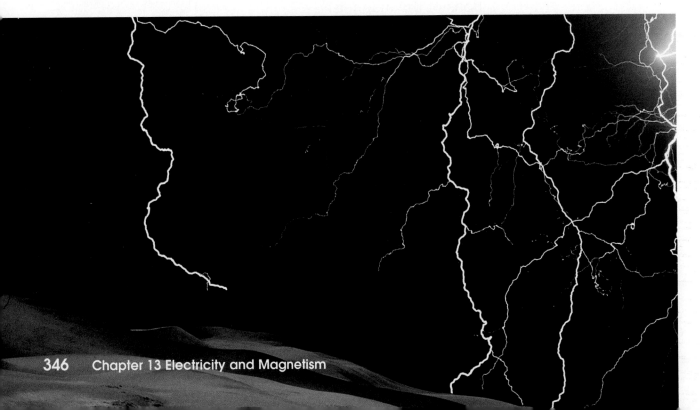

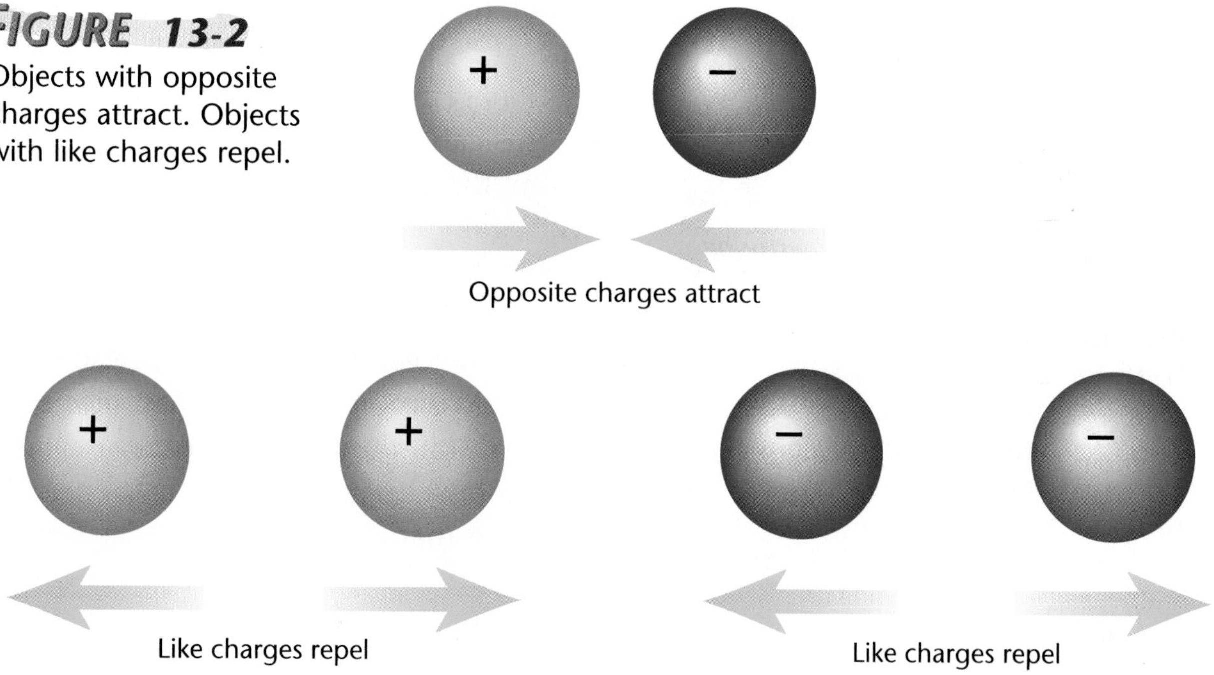

FIGURE 13-2

Objects with opposite charges attract. Objects with like charges repel.

Opposite charges attract

Like charges repel

Like charges repel

To learn more about how you use electrical energy, you first need to understand the differences between electrical potential energy and electrical energy. In Chapter 12, potential and kinetic energy were explained using the example of a skier. The skier carried the energy. To understand electricity, you need to know what carries electrical energy.

Opposites Attract

Does the word *electricity* remind you of any other word? Think back to Chapter 8, where you learned about atoms and electrons, the negatively charged particles of an atom. Electrons and electricity are related. All forms of electricity depend on electrons moving from one place to another. An electron is like the skier that carries the energy.

Positively charged objects attract negatively charged objects, as shown in **Figure 13-2.** There is a pull between the negatively charged electrons around the nucleus and the positively charged protons in the nucleus. This means that electrons closer to the nucleus are held tightly to the atom. Electrons farther from the nucleus are held less tightly. **Figure 13-3** illustrates this.

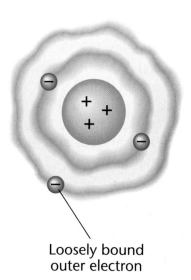

Loosely bound outer electron

FIGURE 13-3

The electron farthest from the nucleus feels less of a pull than the electrons close to the nucleus.

Mini LAB

Contrasting Static Charges

How can potential electrical energy be created and demonstrated?

1. Blow up two rubber balloons and hang them a few inches apart.
2. Rub a glass rod with a square of silk. Bring the glass rod close to one balloon. Observe what happens. Let the rod and balloon touch each another for a few seconds. What happens then?
3. Now, rub one balloon with a wool or silk cloth. Touch that balloon to the other balloon. How does it react? What happens when you touch the first balloon to the glass rod?

Analysis

1. What kinds of charges did the balloons have before rubbing? After?
2. What charge did the glass and cloths have before rubbing? After? How do you know?
3. What made the rod and balloons act as they did? Why do you think so?
4. Why do you think it would be correct to say that electrons are being exchanged, and not protons or neutrons?

Protons and electrons are much too small to see. But you can see evidence of the attraction between these positively and negatively charged particles. Some atoms can lose electrons that are farthest from the atom's nucleus. Other atoms can gain electrons. Materials such as glass and wool are made up of atoms that can lose outer electrons easily. When electrons are lost, some atoms in these materials have more protons than electrons. As a result, they have a positive charge. In contrast, materials such as rubber and silk are made of atoms that easily gain electrons. The added electrons cause atoms to be negatively charged.

The buildup or loss of electrons creates a form of potential electrical energy called **static electricity.** Lightning and sparks result when many electric charges move all at once. This is sometimes called a static discharge, illustrated in **Figure 13-4.** What do you think would happen if you rubbed a balloon with a piece of wool?

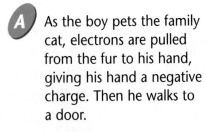

A As the boy pets the family cat, electrons are pulled from the fur to his hand, giving his hand a negative charge. Then he walks to a door.

FIGURE 13-4

Petting the family cat can be a lesson in static electricity.

B The negative charge on his hand repels electrons near the surface of the metal knob. The surface of the knob now has a positive charge; his hand has a negative charge.

C Opposite charges attract. When his hand gets close enough, the extra electrons from his hand leap across the gap. The movement creates a spark and a snap, as electrons transfer energy to the air in the gap.

Problem Solving

Static Image

Photocopying machines work by giving a metal drum a negative charge. Then, a mirror and a bright light are used to put the image to be copied onto the drum. Where the light hits the drum, a positive charge forms, which balances the negative charge. Dark letters and dark parts of pictures don't reflect light. The drum keeps a negative charge where those parts of the image are.

Special dry ink powder called *toner* is brushed onto the drum. Toner sticks only to the charged areas where the letters or pictures were on the image. Then, the paper passes through a heated roller that melts the ink powder. That's why copies are warm when they first come out of the machine.

Solve the Problem:
You have to leave the copier cover up while you copy pages from a big spiral notebook. The edges of the copies are dark and fuzzy. Explain what you think is happening in terms of light and charges. How could you fix the problem?

Think Critically:
The drum has a negative charge. What kind of charge does the toner have? How do you know?

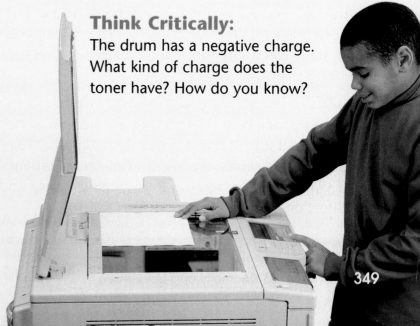

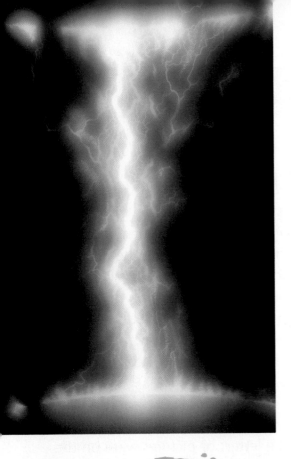

Static Electricity Problems

Lightning is one of the most impressive and dangerous examples of a static electric discharge. Each lightning strike transfers a huge amount of energy. Lightning strikes can start fires and damage trees and buildings. You also may have seen and heard the effect of these huge electrical discharges on your TV or radio.

Smaller static discharges, like that shown in **Figure 13-5,** also can be dangerous. They can harm computers and other electronic devices. Computer chips work on low electric power. The power of a static spark can be enough to overload and permanently burn out the circuits on a chip.

FIGURE 13-5

A spark discharges the static electrical charge built up between two metal objects.

Section Wrap-up

1. Static electrical charges happen in nature. They also can be made by humans. Give one example of natural static electricity and one of human-made static electricity. Explain how each occurs.

2. Electricity does work when it produces motion, heat, light, or sound. Give examples of appliances in your home that do each of those things.

3. **Think Critically:** It is easier for electrons to travel through humid air than through dry air. How does this explain the fact that you see the effects of static electricity more in cool, dry weather?

4. *Skill Builder*
   Recognizing Cause and Effect
   Demonstrate static electricity with a comb, a wool cloth, and bits of tissue paper. Explain what happens with a series of cause-and-effect statements. If you need help, refer to Recognizing Cause and Effect on page 551 in the **Skill Handbook.**

Science Journal

Write a story that tells how your life would change if you had to live without electricity for one day, one week, and one month. How does the length of time without electricity make a difference?

How do circuits work?

Creating a Path for Electrons

If you've watched lightning or touched a doorknob after walking across a carpeted room, you know that a static electric discharge is over in an instant. However, if you want to use electricity to do work, you probably need a nonstop, controlled flow of electrons. A device called a circuit controls the movement of electrons. A **circuit** (SUR kut) is a complete, unbroken path for electrons to follow. Usually, this path is through metal wires. When electrons move in this path, they can do work, as you will find out in the MiniLAB.

A circuit can have one path or many paths. In this book, we will study circuits with only one path. The racetrack in **Figure 13-6** is an example of a circuit with only one path.

What YOU'LL LEARN

- What makes an electric circuit
- The difference between direct and alternating current

Science Words:
circuit
switch

Why IT'S IMPORTANT

Technology that uses electricity is a large part of your day-to-day life.

FIGURE 13-6

This bicycle track is a model of an electrical circuit. The cyclists follow a complete, unbroken path around the track.

Mini LAB

Building a Model Circuit

Even simple circuits have many possible designs.

1. You'll need a battery, a small light-bulb, and one length of wire. Don't use holders.
2. Work with a partner to connect the battery, lightbulb, and wire in such a way that the bulb lights up.
3. Sketch how the circuit looked when you completed it.
4. Find a second way to make the circuit work. (Hint: The circuit can be completed in four possible ways.) Sketch your new circuit.

Analysis

1. How do the sketches of the successful circuits show that electrons move in a circuit?
2. Choose a design that did not work and infer what stopped a circuit from forming.

Controlling Circuits

As electrons move through a circuit, they don't stop at any particular spot and they don't get used up. They are simply moving from a place where extra electrons exist toward a place where there is a shortage of electrons. Electrons can cause movement and operate lights, machines, or appliances when these devices are connected to a circuit, as in **Figure 13-7.** The electrons then move in the devices, producing heat, light, motion, or magnetic signals.

Circuits control the nonstop movement of electrons in two basic ways: (1) they stop or start the movement of electrons, and (2) they have a source of energy. First, let's look at how a circuit stops or starts the movement of electrons. This is done with a **switch,** which is a device that opens or closes a circuit. A switch is like a movable bridge in a model railroad. If the bridge is closed, as in **Figure 13-8A,** the loop is complete. A train on the track could make a complete circle, ending up where it began. But, if the bridge is open, as in **Figure 13-8B,** the train can't go around the track.

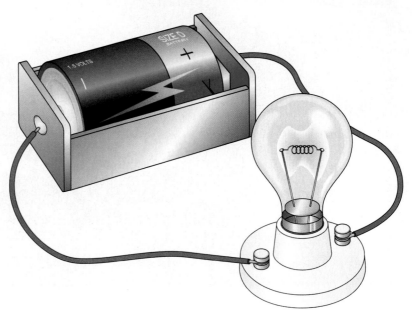

FIGURE 13-7

A simple circuit like this one allows electrons to move and in this case, produce light.

A track is not a complete, closed loop if the bridge is open. Likewise, a circuit is not a complete, closed loop if a switch is open. Electrons cannot travel through the whole circuit if the switch is open.

FIGURE 13-8

A closed electric circuit can be represented by a loop of track.

 When the circuit of track is connected, the model train can run. *What parts of the circuit do the train, track, and bridge represent?*

 An open switch can be represented by a break in the track. The model train cannot run through the open circuit, just as electrons cannot run through an open electric circuit.

Batteries and Circuits

Now you've seen how a switch controls the movement of electrons in a circuit. You also know that circuits use a nonstop, controlled source of electrons. Where does the energy to move those electrons come from? A battery is one source of energy in circuits. A battery works by chemically separating positive and negative charges. This creates extra electrons at one end of the battery and extra positively charged atoms at the other. There is potential electrical energy in the position of the separated charges, just as there is potential energy in an object at the top of a hill, as in **Figure 13-9**. The only way for the negatively charged electrons to combine with the positively charged atoms is by moving outside the battery. If the ends of a battery are connected by a wire, electrons will move in the wire to recombine with the positive charges. This is similar to the uncontrolled way electric charges recombine during a static discharge, such as lightning.

Different types and sizes of batteries add different amounts of energy to electrons. The energy is measured in volts (V). Voltage measures how much potential energy

FIGURE 13-9

A battery can be represented by a hill. Your hand does work to move a model train up the hill. A battery does work to move electrons to one end of the battery. *How could you change the train model to represent a stronger battery? To allow more electrons per second?*

difference there is between the ends of a battery. A 9-V battery adds more energy to electrons than a 1.5-V battery. This is like the skier from Chapter 12 being lifted to different heights on a hill. More height is like more voltage.

Do you have a radio or tape player? How many batteries do they need to operate? Adding another battery to a circuit is like adding more height to the skier's hill. More potential electrical energy is available to be converted to electrical energy. With your parents' permission, compare the batteries needed to operate several different household items: a tape player, a flashlight, a smoke detector, and a television remote control. **Figure 13-10** shows the inside of one of these batteries. In the activity on the next pages, you can test how long these batteries last.

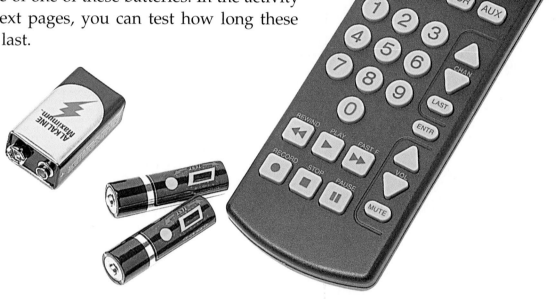

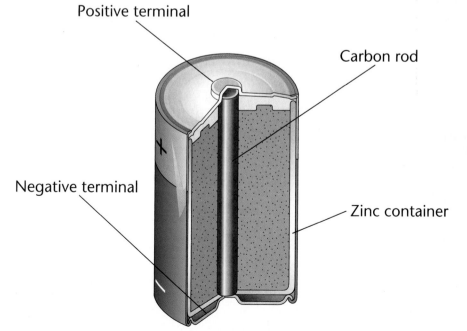

Positive terminal

Carbon rod

Negative terminal

Zinc container

FIGURE 13-10

A battery has two *terminals,* or connection points. When the two terminals of this battery are connected in a circuit, a chemical reaction causes the carbon rod to lose electrons, which accumulate in the zinc. The positively charged carbon rod is the positive terminal. The negatively charged zinc is the negative terminal.

Activity 13-1

Design Your Own Experiment
How long do batteries last?

During a storm, the lights go out and you reach for your flashlight. The batteries are dead! All batteries eventually run down when the chemicals in them are used up. Is this energy used more quickly if the electrons have more work to do?

Possible Materials
- D-cells and holders
- lightbulbs and holders
- buzzers
- bells
- wire
- masking tape

PREPARE

What You'll Investigate
How does adding lightbulbs or machines to a circuit affect the useful life of a battery?

Form a Hypothesis

Think about what you know about circuits and the movement of electrons. **Make a hypothesis** about whether you think adding energy users to the circuit will affect the life of a battery. Give your reasons.

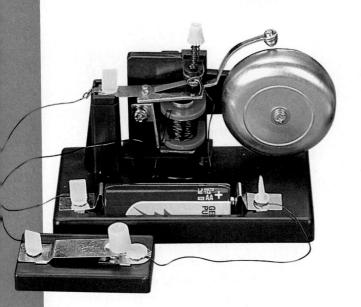

Goals

Design an experiment that tests the life of batteries with machines or lightbulbs connected in a circuit, one after the other.

Observe how many hours each battery will continue to light the lightbulb (or bulbs) to which it is connected. Record your observations in your Science Journal.

Safety Precautions

PLAN

1. **Decide** how to set up the circuit.
2. **List** the things that you will not change during the experiment. List the things that you will change.
3. How will you measure the time each circuit works if you must be out of the classroom?
4. **Plan** how you will record data.

DO

1. Make sure your teacher has approved your plan, including how you will record data.
2. Carry out the experiment.
3. Keep careful records, and post the results where they can be shared with the other groups.

CONCLUDE AND APPLY

1. **Compare** the results of the experiment with your hypothesis. Was your hypothesis supported by the results? Why or why not?
2. **Graph** your experiment results on a class chart with the results of the other groups' experiments.
3. **APPLY** Discuss with the rest of the class the results of the experiment. Agree on how the work required of a battery affects it.

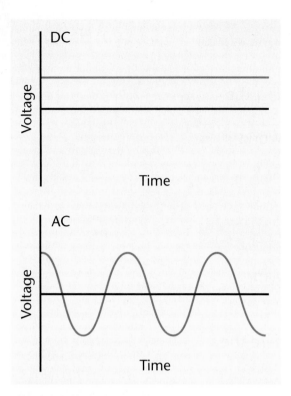

FIGURE 13-11

These graphs compare direct current (top) with alternating current (bottom). The voltage of alternating current changes in a regular pattern.

Alternating Current

You've learned about circuits that use electricity from batteries. But you also know that your TV, microwave oven, hair dryer, and many other appliances don't run on batteries. What kind of electricity are you using when you use these devices? Household devices use a kind of electricity called *alternating current.* With alternating current (AC), electricity travels one way for a while, then it reverses its direction. The direction alternates in a regular pattern. Battery-powered electricity is called *direct current* (DC) because the electricity travels in only one direction. See **Figure 13-11.**

The AC in your home (the electricity from wall sockets) can be dangerous. The voltage is almost 100 times greater than that of a D-cell battery. Most of the AC electricity used in homes in the United States is 120 V and changes direction 60 times each second. AC electricity can be generated at home, but usually comes from a power plant.

Section Wrap-up

1. What part do batteries play in a circuit?

2. What is the difference between a closed and an open circuit?

3. **Think Critically:** Why do electric dryers and ranges require 240 volts?

4. *Skill Builder*
   Designing an Experiment to Test a Hypothesis List the things you controlled in Activity 13-1. What was your dependent variable? Your independent variable? If you need help, refer to Designing an Experiment on page 553 in the **Skill Handbook.**

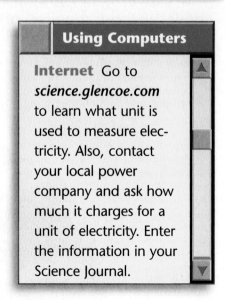

Using Computers

Internet Go to *science.glencoe.com* to learn what unit is used to measure electricity. Also, contact your local power company and ask how much it charges for a unit of electricity. Enter the information in your Science Journal.

What is a magnet? 13•3

Magnets and Electricity

Did a school bell ring to tell you it was time for class? That ringing bell used a magnet and electricity. From school bells to cassette tapes, from your refrigerator to cabinet door latches, magnets are in use all around you. **Figure 13-12** shows a magnetic sculpture you might have experimented with. How does a magnet work?

When you pick up a paper clip with a magnet, one end of the magnet has to be close to the paper clip. The paper clip has to be close enough to the magnet to be affected by its magnetic field. A **magnetic field** is the area around a magnet where magnetic forces act. In a bar magnet, the magnetic field is strong at each end and weaker near the middle of the magnet.

If you have ever played with two bar magnets, you know that like ends repel and opposite ends attract. That means that two N (north) or two S (south) poles push away from each other. On the other hand, an N and an S pole will pull toward each other. In this way, poles (ends) of a magnet are similar to negative or positive charges.

What YOU'LL LEARN

- The relationship between magnetism and electricity
- How a compass works
  Science Words:
  magnetic field
  electromagnet

Why IT'S IMPORTANT

Magnets are part of machines you use every day, such as computers, TVs, and radios.

FIGURE 13-12

The stronger magnet in the base of the sculpture magnetizes the metal diamond shapes. (See Figure 13-13 for an explanation.) When the shapes move out of the base's magnetic field, they lose their magnetism and fall.

FIGURE 13-13

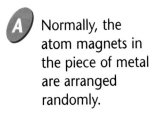

A strong magnet can magnetize a piece of metal.

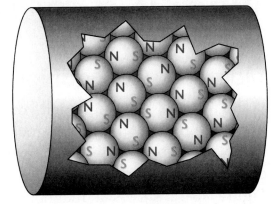

A Normally, the atom magnets in the piece of metal are arranged randomly.

Unmagnetized metal

B A strong south pole attracts all the north poles of the atom magnets, making the metal a magnet.

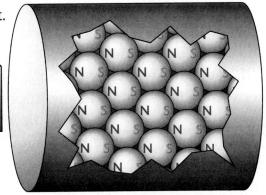

Magnet

Magnetized metal

What makes a magnet?

Why are we talking about magnets in a chapter about electricity? Both electricity and magnetism involve electrons. The atoms that make magnets have certain numbers of electrons arranged in certain ways. Because of this, each atom within a magnetic material is like a small magnet, as shown in **Figure 13-13.** Normally, these atom magnets point in all directions so they cancel each other's magnetism. For a whole piece of material to act as a magnet, most of the atoms in the material must point in the same direction. When they point in the same direction, the individual magnetic forces of each atom add up to a much stronger magnetic force. In the activity on the next page, you can learn more about the relationship between electricity and magnetism.

Lines of Magnetic Force

You usually can't see the lines of magnetic force. In this activity, you'll make the invisible become visible.

What You'll Investigate
How can you illustrate lines of force that show the relationship between magnetism and electricity?

Procedure

1. Put a sheet of graph paper over the bar magnet.
2. Sprinkle iron filings on the paper. **Observe** and **record** the pattern that forms.
3. **Build** a working circuit with a battery and lightbulb. Put a sheet of graph paper over one of the wires.
4. Repeat Step 2.
5. **Wrap** the nail with 20 turns of wire. Connect one end of the wire to a battery. **Touch** the other wire to the other terminal of the battery, and test to see whether the nail can pick up a paper clip. Put a sheet of graph paper over the nail. **CAUTION:** *Do not leave the battery connected unless you are using it. If you do, the wires may become hot enough to cause burns.*
6. Repeat Step 2.

Conclude and Apply

1. **Compare** and **contrast** your data for the bar magnet, circuit, and nail magnet.
2. **Infer** whether magnetic forces exist in all three of the objects that you tested. What is your evidence?
3. **Write** a step-by-step procedure to sketch the lines of attraction for a magnet.

Goals
- Build an electromagnet.
- Observe and compare magnetic fields around different types of magnets.

Materials
- bar magnet
- 20-penny iron nail
- lengths of wire
- D-cell and holder
- lightbulb and holder
- graph paper
- iron filings in a shaker container
- paper clips

FIGURE 13-14

The magnetic fields around bar magnets and electro-magnets have the same shape. *Where would a paper clip be strongly attracted to the electromagnet's iron core? Weakly attracted?*

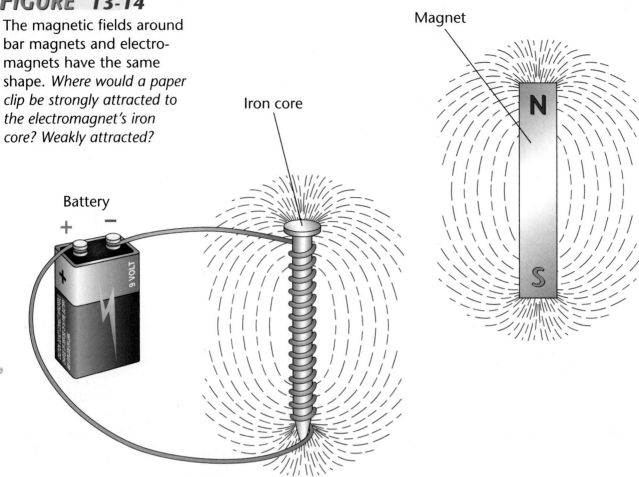

Magnet

Iron core

Battery

Electricity and Magnetism

Magnets and current-carrying wires affect iron filings in much the same way. When a wire is wrapped around a nail, the pattern of iron filings looks more like the pattern made by a bar magnet. This shows an important relationship between electricity and magnetism: A magnetic field always surrounds an electric current.

A nail-and-wire magnet has a special name. An **electromagnet** is a magnet that can be turned on and off. As you can see in **Figure 13-14,** it is made by wrapping a current-carrying wire around an iron core. The strength of an electromagnet can be increased or decreased by changing the strength of the power source. Electromagnets are temporary because their magnetic field stops quickly when the circuit is opened. Electromagnets are used in many everyday devices from doorbells to computer disk drives.

Compasses Are Magnets

You and your parents are in an unfamiliar part of town, and you suddenly realize you're not sure how to get home. You stop someone to ask directions. She tells you to go north and then hurries off before you can ask which way that is. If you only had a compass! How would that help? It's all a matter of magnetism.

In the activity, the iron filings and compass showed the lines of force around a magnet. The filings lined up along the force lines. The compass needle pointed along the lines. The compass works this way because its needle is a magnet.

Left undisturbed, the needle on a compass will point along the lines of force of the nearest or strongest magnet. Usually, this magnet is Earth, and the needle points toward the magnetic north pole of Earth.

FIGURE 13-15

Earth's magnetic field causes magnetic compass needles to line up along north-south lines.

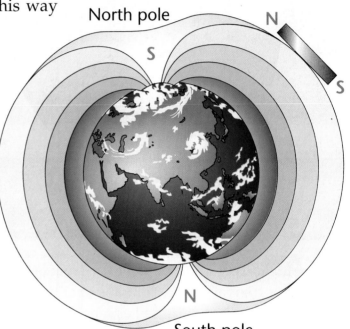

North pole

South pole

A The north magnetic pole is the pole that attracts the north end of magnets. The north magnetic pole corresponds to the south pole of a bar magnet.

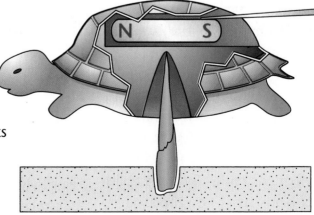

B Small factories in medieval China manufactured magnets for compasses like this one. The turtle's head points to magnetic north, allowing it to be used by travelers.

Because of the connections between electricity and magnetism, they are usually studied together, as *electromagnetism.* Just as electric current can create a magnet, moving magnets can create an electric current.

FIGURE 13-16

A sound engineer uses this equipment to record music. It translates sound from air vibrations to electrical signals to magnetic impulses, and back again.

USING TECHNOLOGY

Magnetic Signals at Work

Audiotapes and videotapes are magnetic messages that are translated into sight and sound by machines. The messages are recorded as tape moves past an electromagnet. As the electric current in the magnet changes, the magnet becomes stronger or weaker. These changes are recorded as stronger and weaker magnetized areas on the tape. When the tape is played, a sensor detects and amplifies the changes. You hear and see the result.

To demonstrate the relationship between magnetism and electronic impulse, get an old videotape from home. *Be sure to get permission to use the tape, because it will be changed during your experiment.* Watch the tape, then rewind. Now, bring a magnet near the exposed part of the tape. The tape will be attracted to the magnet, proving that the tape is magnetic. Now play the tape. What do you hear and see? You have erased part of the tape by scrambling the magnetic code. Try to establish a pattern of blank spaces on the tape.

SCIENCE *Online*

To learn more about magnetic recording, visit Glencoe Science Online at *science.glencoe.com* and follow the link for Chapter 13.

Harnessing Electromagnetic Energy

People first observed the effects of static electricity and magnetized metals thousands of years ago. It was not until the late 1800s that scientists learned to control the flow of electricity in a circuit and make electric batteries and motors. Now electromagnetic energy is a widely used tool. It powers lights and clocks, provides heat or cooling, and is used to record and transmit information, whether your favorite singer on the radio or the computer files of words and pictures used in publishing this book. As with all forms of energy, the more we learn about electromagnetism, the easier it is to use it.

Section Wrap-up

1. A compass needle reacts to what type of field?

2. Magnets have poles that are like positive and negative charges. Describe how these poles act.

3. **Think Critically:** What are some possible uses for a magnet that can be turned on and off?

4. *Skill Builder*
   Observing Watch the behavior of a compass placed on a wire as the wire is connected and disconnected from a battery. Infer what would happen if you used a larger battery. If you need help, refer to Observing and Inferring on page 550 in the **Skill Handbook.**

USING MATH

10-Coil Electromagnet

Batteries	Paper clips picked up
1	3
2	5
3	7

According to the table, how many paper clips would be picked up if four batteries and ten coils were used? How do you know?

13•4 Electricity Sources

What YOU'LL LEARN

- About different ways of generating electricity
- How to make generalizations

Science Words:
generator
turbine

Why IT'S IMPORTANT

You use electricity produced by generators every day.

A Typical Morning

Does your day start like this? You wake up to the bee-dee-BEEP of an electric alarm clock. You switch on the lights and stumble toward the bathroom, where you feel heat from an air duct. You take a hot shower and dress while listening to the radio. Then it's off to the kitchen for a glass of cold juice and some warm, buttered toast. Your day has just begun, and already you've used electricity at least seven times. Did you ever wonder where your electricity comes from?

Power Plants

Most electric power comes from power plants that use huge generators. A **generator** (JEN uh ray tur) is a machine that changes mechanical energy into electric power. One is shown in **Figure 13-17.** The mechanical energy comes from a **turbine** (TUR bun), a large rotating wheel that gets its energy from different sources of energy. As you read about these sources, find facts that hint at general statements about electricity. This will help you to better understand important issues.

Sources of Energy for Electricity Production

Coal, oil, and natural gas—fossil fuels—are the most common sources used to generate electricity in the United States. Fossil fuels are low in cost, but they can be replaced only through slow, natural processes that take millions of years. Plus, when fossil fuels are burned during the production of electricity, they produce pollution.

FIGURE 13-17

Mechanical energy, such as that provided by these windmills, can be changed into electrical power by generators. The large generator above is powered by steam instead of wind.

The sun is another source for electricity production. Solar energy is renewed every day. It does not pollute the air. What are the drawbacks? Places that receive little sunlight are not suited for solar-powered plants. These plants also use thousands of mirrors to reflect sunlight, so they require a lot of space.

Another source is nuclear energy, which harnesses the energy in the nuclei of certain atoms. It does not cause much air pollution, but the waste generated by the process is dangerous. This waste must be stored safely.

Wind and water are sources of power, too. Both are nonpolluting and renewable—but there are drawbacks. It's hard to build a windmill large enough to produce large amounts of electricity. Dams use flowing water to generate electricity. They create large lakes, destroying the habitats of the plants and animals that lived in the area before the dam was built.

Skill Builder: Making Generalizations

LEARNING the SKILL

When you read, find facts that hint at general statements about the subject. This will help you to better understand and explain the reading.

1. To make generalizations (jen rul uh ZAY shuns), you first need to identify the subject matter.

2. Gather as many facts as you can about the subject.

3. Identify similarities and patterns among the facts.

4. Use the similarities and patterns to form some general ideas about the subject.

PRACTICING the SKILL

1. What is the subject matter of this passage?

2. What generalizations can you make about the sources of electricity we use?

3. Identify four facts that support your generalization.

APPLYING the SKILL

Make a general statement about the importance of electricity in your life. Then, identify three facts that support your statement.

Promise Me the Moon
by Joyce Annette Barnes

I do a quick calculation. The total capacity of his system is only 1800 watts on each circuit, which is about enough to run an iron and a vacuum cleaner at the same time. Not nearly enough to supply all the new appliances. That's why the fuses keep blowing, shutting everything down. If lightning strikes his house, the surge of electricity could be too much for the circuit to handle. It's lucky the whole place hasn't gone up in flames . . .

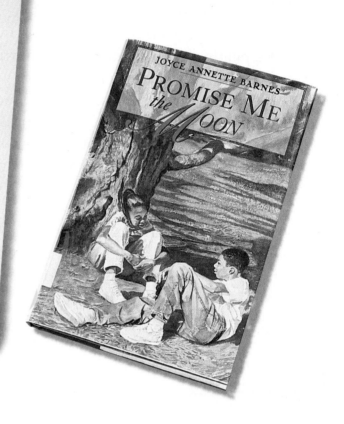

Science Journal

Explore your home like a curious scientist. In your Science Journal, list all the outlets, light fixtures, and electrical appliances you can find in two or three rooms during your investigation. With a parent's permission and help, visit the fuse or circuit breaker box and match your list of outlets and appliances to the circuits there. How many outlets are on a single circuit? Appliances? Lights?

Thirteen-year-old Annie Armstrong, who wants to be an astronaut, can't decide on a topic for her Enriched Science Class project. That is, not until a strange thing happens in her neighborhood. During a thunderstorm, the lights in Mr. Blackstone's stately old house go off. But the lights in the surrounding houses stay on. Annie wonders why. She reads up on electricity, forms a hypothesis, and then goes to investigate the fuse box in the dark, dusty basement of the Blackstone mansion.

Annie's curiosity and hard work pay off. She discovers that the mansion's electrical system is old and unsafe. And she understands what her science teacher means when he says, "Discovery is the essence of science."

Now Annie must discover something else—not about electricity but about herself. Does she have what it takes to get into McAllen High School and its special science and math program—one that can help her realize her dream of becoming an astronaut?

Read the statements below that review the most important ideas in the chapter. Using what you have learned, answer each question in your Science Journal.

1. Static electricity is the buildup of electric charges in one place. *Write a sequence of events that would lead to a discharge of static electricity.*

2. A circuit is a continuous path that electrons can travel. Circuits must be complete (closed) for electrons to move. *Draw a diagram of a complete circuit. Include a source of energy, a pathway, a switch, and evidence of work being done.*

4. Magnetic signals are used in audiotapes and videotapes. *Explain why a large, strong magnet would destroy a videotape.*

3. A magnet has two ends called poles. Opposite poles attract and like poles repel each other. *What would the magnetic field look like around two magnets that were repelling each other? Around two magnets that were attracting each other?*

Using Key Science Words

circuit
electrical energy
electromagnet
generator

magnetic field
static electricity
switch
turbine

Match each phrase with the correct term from the list of Key Science Words.

1. closed path for electrons to follow
2. form of potential electrical energy
3. area where magnetic forces act
4. temporary magnet formed by electricity
5. gate in a circuit

Checking Concepts

Choose the word or phrase that completes the sentence.

6. Lightning represents a discharge of _____.
   a. protons   c. neutrons
   b. electrons   d. atomic nuclei

7. Static electricity moves from place to place because _____.
   a. electrons are attracted to other electrons
   b. potential energy is attracted to kinetic energy
   c. electrons are attracted to Earth's natural magnetic poles
   d. electrons are attracted to positive charges

8. Batteries represent _____.
   a. a form of alternating current
   b. the switch in a circuit
   c. the source of electrical potential energy
   d. a form of atomic energy

9. An electrical circuit requires _____.
   a. a source of electrons, a pathway for the electrons, and a closed switch
   b. a source of protons, a pathway for electrons, and a closed switch
   c. a source of protons, a pathway for protons, and an open switch
   d. a source of electrons, a pathway for electrons, and an open switch

10. All ways of generating electricity _____.
    a. have benefits and drawbacks
    b. are too expensive to use
    c. depend on static
    d. require positive charges

Thinking Critically

Answer the following questions in your Science Journal using complete sentences.

11. Aluminum and glass don't work as cores for electromagnets. Why do you think that is?

12. Playing golf or standing under a tall tree during a thunderstorm is dangerous. Why are these activities dangerous when lightning is present?

13. A circuit has three lightbulbs connected as shown below. If one of the lightbulbs is removed, will the remaining lightbulbs stay lit? Explain.

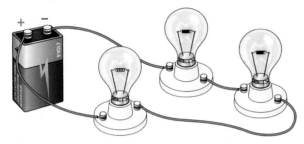

14. A maglev train uses an electromagnet to raise a train above a metal strip and move it along a track. From what you know about magnetic poles, describe how you think a maglev train works.

15. Suppose you needed to establish a new electrical plant in your community. What power source would you use? Why?

Developing Skills

If you need help, refer to the description of each skill in the Skill Handbook.

16. Classifying: Make a list of 12 electrical appliances in your home. Classify them according to whether they are powered by DC or AC.

17. Making Models: Describe how you might make a model of a circuit using a garden hose and marbles.

18. Making and Using a Table: Make a table to compare the properties of five sources of electrical energy. Which source is best? Why? Would your answer change if you had different needs?

19. Measuring: Measure the distance at which a bar magnet will attract a compass needle. Now make the magnet an electromagnet by wrapping it with wire and connecting the wire to a battery. Measure the distance at which it will attract the compass needle now. **CAUTION:** *Disconnect the electromagnet as soon as possible or it will become hot and may cause burns.*

20. Sequencing: List the steps to magnetize a nail.

Performance Assessment

1. Write a Letter: In the Science and Society section, you compared and contrasted ways to produce electricity. Write a letter to your local power company and ask where and how the electricity you use is generated.

2. Calculating: Determine the number of hours in a typical Saturday that you and your family spend using electrical appliances. Pick out three different appliances. Keep a pad and pencil by each one so family members can write down when they are used. Then, total up the time used in just one day.

Physical Science

When the Wright brothers set out to make the first powered airplane, they spent time researching flight and studying designs that had failed, as well as gliders that had already flown. They considered variables such as distance, time, speed, force, and surface area. They recognized the forces involved in flight, such as gravity, lift, and drag (a form of friction). Consider all these variables and forces and help your team design a paper airplane that flies farther or longer than your classmates' airplanes.

A Paper Airplane Contest

Goals
- Research paper airplane design strategies.
- Design and build paper airplanes.
- Measure variables related to the airplane designs.

Researching Flight
Visit Glencoe Science Online at **science.glencoe.com** to find links to paper airplane and flight sites on the web. Learn about the physics of flight, and consider the benefits of different designs.

Procedure
Materials: paper, 50-m tape measure, metric ruler, stopwatch, balance, tape, stapler, paper clips, scissors
1. You may use a single sheet of any type of paper. You may also cut, fold, tape, glue, or staple the paper to form your airplane.

2. Plan one or more designs. What type of paper will your airplane be made of? What shape of wings will you use? Make a sketch of your design and instructions on how to build it. Set up a data table like the one shown in your Science Journal.

3. Build your design. Use a balance to measure your airplane's mass. Record this mass in the data table.

4. Find an indoor testing area. It should be flat and open, such as a cafeteria or gymnasium.

5. Experiment with different ways of flying your airplane. Measure the distance and length of each flight. Record the data in the table in your Science Journal.

6. Make any modifications to your airplane that you think are necessary. Remember to change only one factor at a time. Record each modification in your Science Journal.

7. Tell your teacher when you have finished the airplane that you think will fly as long and as far as possible.

8. Hold a class contest to determine three categories: greatest time in the air, greatest distance flown from starting point, and the greatest overall flight (multiply flight distance and time). Your class will need to decide on the contest rules. Will teams get only one flight, or will they average the results of several? Who will judge flight time? Record the results in your Science Journal.

Conclude and Apply

1. Compare and contrast the designs your class came up with. What features did the winning planes have?

2. How did the planes that did well in the distance category differ from the planes that flew for a long time?

Go Further

Based on the results of your designs, what kind of features would a plane have that is designed to land on a target?

Data Table				
Flight	Mass (g)	Design Change	Flight Distance (m)	Flight Time (s)
1				
2				
3				

UNIT 3

Unit Contents

Earth Science

What's happening here?

In the Toroweap section of the Grand Canyon, you can see the cones of extinct volcanoes. Lava from these volcanoes once spilled over the rim of the canyon, flowing almost a mile to the Colorado River below. The smaller photo is Io, one of Jupiter's moons. Volcanic activity is also changing the surface of Io, as shown in new images from NASA's *Galileo* spacecraft. Materials that make up Earth's structure also are found in meteorites, moons, and planets of our solar system. Could these same materials be found in solar systems of distant galaxies as well?

SCIENCE *Online*

Visit Glencoe Science Online at *science.glencoe.com* for links to information about our solar system. In your Science Journal, write about the materials that scientists hypothesize make up all of the planets and moons in our solar system.

Chapter Preview

Skills Preview

▶ **Skill Builders**
- make a concept map
- develop multimedia presentations
- formulate models

▶ **MiniLABs**
- observe
- model

▶ **Activities**
- observe
- record
- infer
- classify
- draw

376

Chapter 14

The Solar System and Beyond

Here's a view of Earth taken from the moon by astronauts on the *Apollo 17* lunar mission. Do you ever gaze at the night sky? What do you see? On a clear night, it seems like the sky is full of sparkling points of lights. You can see dozens, no hundreds of these sparkles. Just how many stars are there?

EXPLORE ACTIVITY

Estimating the Number of Rice Grains

1. Divide a sheet of black construction paper into two-inch squares. Draw the lines with white crayon or chalk so that they show up clearly.
2. Spill a teaspoonful of rice onto the black paper.
3. Count the number of grains of rice in one square. Repeat this step with a different square. Add the number of grains of rice in the two squares, then divide this number by 2 to calculate the average number of grains of rice in the two squares.
4. Multiply this number by the total number of squares on the paper. This will give you an estimate of the total grains of rice on the paper.

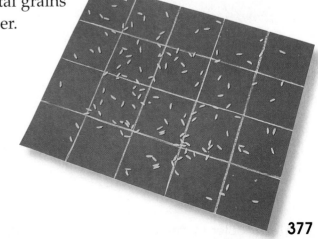

Science Journal

How might scientists use this same method to count the number of stars in the sky? In your Science Journal, describe the process scientists might use.

14•1 Earth's Place in Space

What YOU'LL LEARN

- How seasons are caused by the tilt of Earth's axis
- What causes the phases of the moon

Science Words:
rotation
revolution
eclipse

Why IT'S IMPORTANT

The movements of Earth cause night and day and the seasons.

Earth Moves

You wake up, stretch and yawn, then glance out your window to see the first rays of dawn peeking over the houses. By lunch, the sun is high in the sky. As you sit down to dinner that evening, the sun appears to sink below the horizon. It might seem like the sun moves across the sky. But it is Earth that is really moving.

Earth's Rotation

Earth spins in space like a dog chasing its tail—but not as fast! Our planet spins around an imaginary line called an axis. **Figure 14-1** shows this imaginary axis.

The spinning of Earth on its axis is called Earth's **rotation** (roh TAY shun). Earth rotates once every 24 hours. In the morning, as Earth rotates, the sun comes into view. In the afternoon, Earth continues to rotate, and the sun appears to move across the sky. In the evening, the sun seems to go down because the place where you are on Earth has rotated away from the sun.

You can see how this works by standing and facing the chalkboard. Pretend you are Earth and the chalkboard is the sun. Now turn around slowly in a counterclockwise direction. The chalkboard moves across your vision, then disappears. You rotate until finally you see the chalkboard again. The chalkboard didn't move—you did. When you rotated, you were like Earth, spinning in space so that different parts of the planet face the sun at different times. This movement of Earth, not the movement of the sun, causes night and day.

FIGURE 14-1

The rotation of Earth on its axis causes night and day.

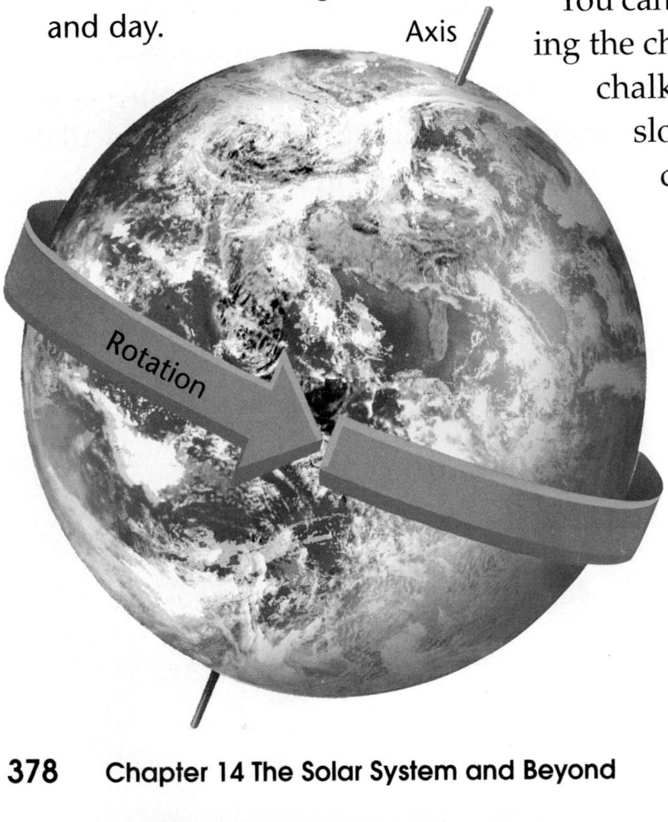

Axis

Rotation

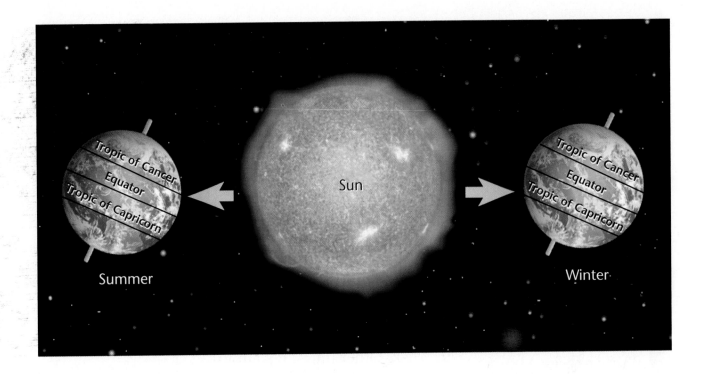

Summer

Sun

Winter

Earth's Revolution

You know that Earth rotates in space. It also moves in other ways. Like an athlete running around a track, Earth moves around the sun in a regular, curved path. This path is called an orbit. The movement of Earth around the sun is known as Earth's **revolution** (rev oh LEW shun). A year on Earth is the time it takes for Earth to revolve around the sun once. How many revolutions old are you?

Seasons

Who doesn't love summer? The days are long and warm. It's a great time to go swimming, ride a bike, or read a book just for fun. Why can't we have summer all year-round? Blame it on Earth's axis. The axis, that imaginary line that our planet spins around, is not straight up and down. It is tilted at an angle. It's because of this tilt that there are seasons in many areas on Earth. Why?

Look at **Figure 14-2.** The part of Earth that is tilted toward the sun receives more direct sunlight and more energy from the sun than the part of Earth that is tilted away from the sun. When the part of Earth that you live on is tilted away from the sun, you have winter. When the part of Earth that you live on is tilted toward the sun, you have summer.

FIGURE 14-2

In the northern hemisphere on June 21 or 22, the sun's rays directly strike the Tropic of Cancer. On December 21 or 22, the sun's direct rays are over the Tropic of Capricorn. *When it's summer in the northern hemisphere, what season is it in the southern hemisphere?*

FIGURE **14-3**

The moon is said to be waxing when it seems to be getting larger night by night. It is said to be waning when it seems to be getting smaller.

B Waxing crescent

A New moon

C First quarter

Movements of the Moon

Imagine a dog running in circles around an athlete who is jogging on a track. That's how you can picture the moon moving around Earth. As Earth revolves around the sun, the moon revolves around Earth. The moon revolves around Earth once every 27.3 days. But, as you have probably noticed, the moon does not always look the same from Earth. Sometimes it looks like a big, glowing disk. Other times, it's a thin sliver.

Moon Phases

How many different moon shapes have you seen? Round shapes? Half-circle shapes? The moon looks different at different times of the month, but it doesn't really change. What does change is the way the moon appears from Earth. We call these changes moon phases. **Figure 14-3** shows the different phases of the moon.

D Waxing gibbous

Light from the Sun

The moon phase you see on any given night depends on the positions of the moon, the sun, and Earth in space. Wait a minute. How can we see the different phases of the moon? Is someone shining a giant flashlight up there? No, the moon receives light from the sun, just as Earth does. And just as half of Earth experiences day while the other half experiences night, one half of the moon is lit by the sun while the other half is dark. As the moon revolves around Earth, we see different parts of the side of the moon that is facing the sun. This makes the moon appear to change shape.

E Full moon

F Waning gibbous

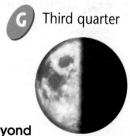

G Third quarter

H Waning crescent

Moon Phases

The moon is our nearest neighbor in space. But the sun, which is much farther away, affects how we see the moon from Earth. In this activity, you'll observe how the positions of the sun, the moon, and Earth cause the different phases of the moon.

What You'll Investigate

How do the positions of the sun, the moon, and Earth affect the phases of the moon?

Procedure

1. **Turn on** the flashlight and darken other lights in the room. **Select** a member of your group to hold the flashlight; this person will be the "sun." **Select** another member of your group to hold up the softball so that the light shines directly on the ball. The softball will be the "moon" in your experiment.
2. The remaining members of your group should sit between the sun and the moon.
3. **Observe** how light shines on the moon. **Draw** the moon, being careful to **shade in** its dark portion.
4. The student who is holding the "moon" should begin to **walk** in a slow circle around the group, stopping at least seven times at different spots. Each time the "moon" stops, **observe** the moon, **draw** the moon, and **shade in** its dark portion.

Conclude and Apply

1. **Compare** and **contrast** your drawings with those of other students. **Discuss** similarities and differences in the drawings.
2. In your own words, **explain** how the positions of the sun, the moon, and Earth affect the phase of the moon we see on Earth.
3. **Compare** your drawings with **Figure 14-3.** Which phase is the moon in for each drawing? **Label** each drawing with the correct moon phase.

Goals

- Observe moon phases.
- Record and label phases of the moon.
- Infer how the positions of the sun, the moon, and Earth affect phases of the moon.

Materials

- drawing paper
- softball
- flashlight

381

FIGURE 14-4
Only a small area of Earth ever experiences a total solar eclipse. *Why?*

Area of partial eclipse

Sun

Moon

Area of total eclipse

Earth

MiniLAB

Observing Distance and Size

1. Place a basketball on a table at the front of the classroom. Then stand at the back of the room.
2. Extend your arm, close one eye, and try to block the ball from sight with your thumb.
3. Slowly move your thumb closer to you until it completely blocks the ball.
4. Repeat the experiment using a golf ball.

Analysis

1. In your Science Journal, describe what you observed. When did your thumb block your view of the basketball? When did your thumb block your view of the golf ball?
2. A small object can sometimes block a larger object from view. Explain how this relates to the moon, the sun, and Earth during a solar eclipse.

Eclipses

Have you ever tried to watch TV with someone standing between you and the screen? You can't see a thing! The light from the screen can't reach your eyes because someone is blocking it. Sometimes, the moon is like that person standing in front of the TV. It moves between the sun and Earth in a position that blocks sunlight from reaching Earth. The moon's shadow travels across parts of Earth. This event, shown in **Figure 14-4,** is called an **eclipse** (ee KLIHPS). Because it is an eclipse of the sun, it is known as a solar eclipse. The moon is much smaller than the sun, so not everywhere on Earth is in the moon's shadow. Sunlight is completely blocked only in the small area of Earth where the moon's shadow falls. In that area, the eclipse is said to be a total solar eclipse.

Lunar Eclipse

Sometimes, Earth can be like a person standing in front of the TV. It gets between the sun and the moon, blocking sunlight from reaching the moon. When

Earth's shadow falls on the moon, we have an eclipse of the moon, which is called a lunar eclipse. **Figure 14-5** shows a lunar eclipse.

Our Neighbors in Space

In this section, you've learned about what causes day and night and the seasons. You've also learned about the moon, Earth's nearest neighbor in space. Next, you'll look at our other neighbors in space—the planets that make up our solar system.

FIGURE 14-5

During a lunar eclipse, Earth moves between the sun and the moon.

Section Wrap-up

1. Explain the difference between Earth's revolution and its rotation.

2. Draw a picture showing the positions of the sun, the moon, and Earth during a solar eclipse.

3. **Think Critically:** Seasons are caused by the tilt of Earth's axis. What do you think seasons would be like if Earth's axis were not tilted?

4. **Skill Builder**
   Concept Mapping Make a cycle map showing the phases of the moon in sequence, beginning with the new moon. If you need help, refer to Concept Mapping on page 544 in the **Skill Handbook.**

USING MATH

Light travels 300 000 km per second. There are 60 seconds in a minute. If it takes eight minutes for the sun's light to reach us, how far is the sun from Earth?

14•2 The Solar System

What YOU'LL LEARN

- About distances in space
- About the objects in our solar system

Science Words:
solar system

Why IT'S IMPORTANT

You'll learn more about how the planets, including Earth, were formed.

Distances in Space

Imagine that you are an astronaut living far in the future, doing research on a space station near the edge of our solar system. You've been working hard for a long time. You need a vacation. Where will you go? How about a tour of the solar system? The **solar system,** shown in **Figure 14-6,** is made up of the nine planets and numerous other objects that orbit the sun. How long do you think it would take you to cross the solar system?

Measuring Space

Distances in space are hard to imagine because space is so vast. Let's get back down to Earth for a minute. Suppose you had to measure your pencil, the hallway outside your classroom, and the distance from your home to school. Would you use the same units for each measurement? Probably not. You'd probably measure

Pluto

Neptune

Uranus

Saturn

Jupiter

your pencil in centimeters. You'd probably use something bigger to measure the length of the hallway, such as meters. You might measure the trip from your home to school in kilometers. We use larger units to measure longer distances. Imagine trying to measure the trip from your home to school in centimeters. If you didn't lose count, you'd end up with a very large number!

Astronomical Unit

Kilometers are fine for measuring long distances on Earth. But we need even bigger units to measure vast distances in space. One such measure is the astronomical (as truh NAHM uh kul) unit. An astronomical unit equals 150 million km, which is the average distance from Earth to the sun. It is abbreviated *AU*. If something is 3 AU away from Earth, it means that the object is three times as far away as Earth is from the sun.

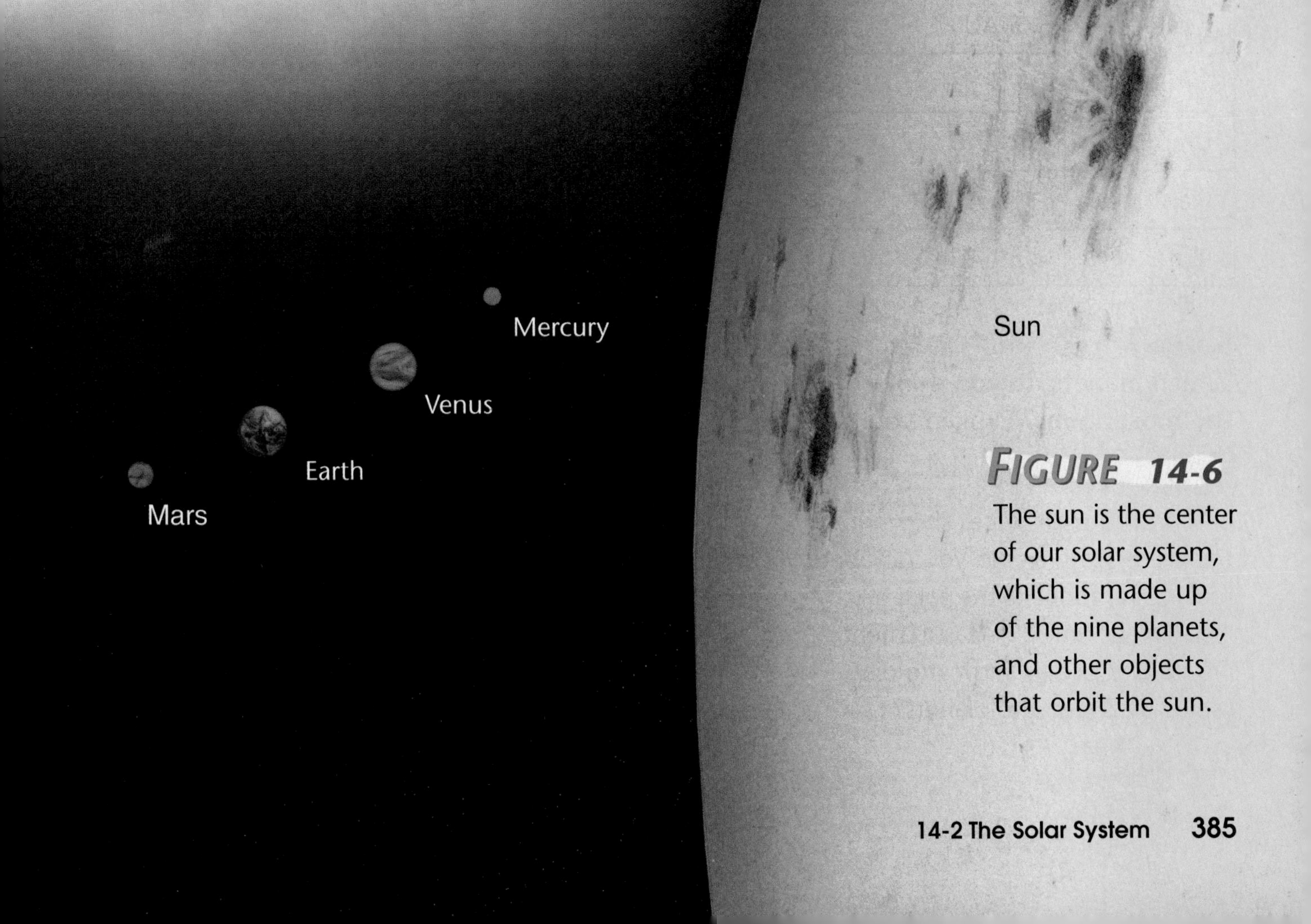

Mercury

Venus

Earth

Mars

Sun

FIGURE 14-6

The sun is the center of our solar system, which is made up of the nine planets, and other objects that orbit the sun.

Problem Solving

Distances in the Solar System

The following table shows the distances in AU between the planets in our solar system and the sun. Study the distances carefully, then answer the questions below.

Planet	Distance from Sun
Mercury	0.38 AU
Venus	0.72 AU
Earth	1 AU
Mars	1.5 AU
Jupiter	5 AU
Saturn	9.5 AU
Uranus	19 AU
Neptune	30 AU
Pluto	39 AU

Solve the Problem:
Which planets do you think scientists know the most about? Explain your answer.

Think Critically:
Based on the distances shown in the table, how would you go about making a model of the solar system? What unit of measurement would you use to show the distances between the planets?

A Tour of the Solar System

Now you know how far you have to travel to tour the solar system, starting from your space station on the outer edge of the solar system. Strap yourself into your spacecraft. It's time to begin your journey. What will you see first as you enter the solar system?

Comets

What's this in **Figure 14-7?** A giant, dirty snowball? No, it's a comet—the first thing you see on your trip. Comets are made up of dust and frozen gases such as ice. From time to time, they swing close to the sun. When they do, the sun's radiation vaporizes some of the material. Gas and dust spurt from the comet, forming bright tails.

FIGURE 14-7

A comet's gas tail and dust tail are blown away from the sun by a combination of solar wind and the pressure of sunlight. Solar wind is a stream of charged particles from the sun.

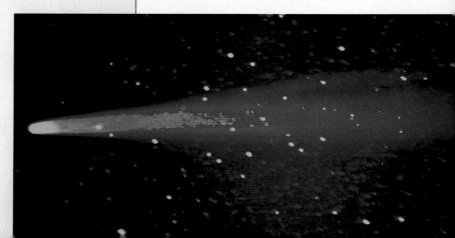

Outer Planets

Moving past the comets, you come to the outer planets. The outer planets are Pluto, Neptune, Uranus, Saturn, and Jupiter. Let's hope you aren't looking for places to stop and rest. Trying to stand on most of these planets would be like trying to stand on a cloud. That's because all of the outer planets except Pluto are huge balls of gas. Each may have a solid core, but none of them has a solid surface. The gas giants have lots of moons, which orbit the planets just like our own moon orbits Earth. They have outer rings made of dust and ice. In fact, the only outer planet that doesn't have rings is Pluto. Pluto isn't a gas giant. What does it look like? You'll soon find out.

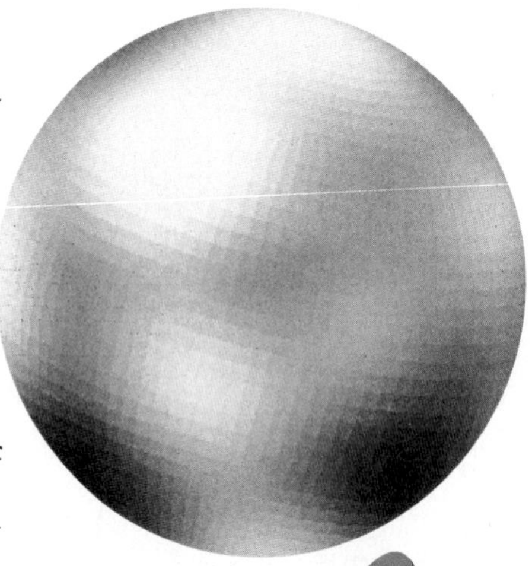

Pluto

The first planet that you come to on your tour is Pluto, a small, rocky planet with a frozen crust. Pluto, the last planet discovered by scientists, is normally farthest from the sun. It is the smallest planet in the solar system, and the one we know the least about. Pluto, shown in **Figure 14-8A,** has no ring system. Its one moon, Charon, is nearly half the size of the planet itself.

Neptune

Neptune is the next stop in your space travel. Neptune, shown in **Figure 14-8B,** is the eighth planet from the sun most of the time. Sometimes, Pluto's orbit crosses inside Neptune's orbit during part of its voyage around the sun. When that happens, Neptune is the ninth planet from the sun. Neptune is the first of the big, gas planets with rings. Neptune's atmosphere is made of a gas called methane. Methane gives the planet a blue-green color.

Uranus

After Neptune, you come to the seventh planet from the sun, Uranus. Uranus needs a careful look because of the interesting way it spins on its axis.

A Pluto

FIGURE 14-8

The outer planets include Pluto and Neptune. Pluto is so small and far away that this is the best image current technology can produce.

B Neptune

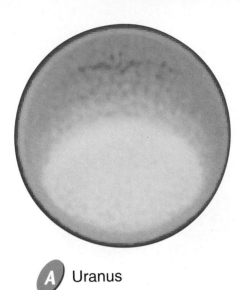

A Uranus

The axis of most planets, including Earth, is tilted just a little, somewhat like the hands of a clock when they are at 1 and 7. Uranus, shown in **Figure 14-9A,** has an axis that is tilted almost even with the plane of its orbit, as if the hands of the clock were at 3 and 9.

Saturn

You thought Uranus was unusual. Wait until you see Saturn, the sixth planet from the sun! You'll be dazzled by its rings, shown in **Figure 14-9B.** Saturn's several broad rings are made up of hundreds of smaller rings, which are made up of pieces of ice and rock. Some of these pieces are like specks of dust. Others are many meters across.

Jupiter

If you're looking for excitement, you'll find it on Jupiter, the largest planet in the solar system and the fifth from the sun. Watch out for a huge, red whirlwind rotating slowly around the middle of the planet. That's the Great Red Spot, a giant storm on Jupiter's surface. Jupiter, shown in **Figure 14-9C,** has 28 moons. Some are larger than Pluto! One of Jupiter's moons, Io, has more active volcanoes than anyplace else in the solar system.

B Saturn

FIGURE 14-9

Uranus, Saturn, and Jupiter are gas giants. *In your Science Journal, list one unique characteristic about each of these planets.*

Asteroid Belt

Look out for asteroids! On the next part of your trip, you must make your way through the asteroid belt that lies between Jupiter and the next planet, Mars. Asteroids are pieces of rock made of minerals similar to those which formed the planets. In fact, asteroids might have become planets if it weren't for that big giant, Jupiter. Jupiter's huge gravitational force probably kept any planets from forming in the area of the asteroid belt.

C Jupiter

Inner Planets

After traveling dozens of astronomical units, you finally reach the inner planets. These planets are solid and rocky. How do we know that? As with all the planets, much of what we know about planets comes from spacecraft that send data back to Earth to help us learn more about space. Look at **Figure 14-10A**. This photograph was taken by a spacecraft.

Mars

Hey! Has someone else been here? You see signs of earlier visits to Mars, the first of the inner planets. Tiny scientific devices have been left behind. But it wasn't a person who left them here. The devices were left by spacecraft sent from Earth to explore Mars's surface. If you stay long enough and look around, you may notice that Mars, shown in **Figure 14-10A,** has seasons and polar ice caps. There are signs that the planet once had liquid water. You'll also notice that the planet looks red. That's because the rocks on its surface contain iron oxide, which is what makes rust look red.

Earth

Home sweet home! You've finally reached Earth, the third planet from the sun. You didn't realize how unusual your home planet was until you saw the other planets. Earth's surface temperatures allow water to exist as a solid, a liquid, and a gas. Also, Earth's atmosphere works like a screen to keep ultraviolet (ul truh VI lut) rays from reaching the planet's surface. Ultraviolet rays are harmful rays from the sun. Because of Earth's atmosphere, life can thrive on the planet. You would like to linger on Earth, shown in **Figure 14-10B,** but you have two more planets to explore.

Venus

Maybe you should have stayed on Earth. You won't be able to see much at your next stop, shown in **Figure 14-10C.** The surface of Venus, the second-closest planet to the sun, is hard to see because it is surrounded by thick clouds.

FIGURE 14-10
Mars, Earth, and Venus are inner planets.

A Mars

C Venus

B Earth

Activity 14-2

Design Your Own Experiment
Space Colony

Have you ever seen a movie or read a book about astronauts from Earth living in space colonies on other planets? Some of these make-believe space colonies look awfully strange! So far, we haven't built a space colony on another planet. But if we did, what do you think it would look like?

PREPARE

Possible Materials
- drawing paper
- markers
- books about the planets

What You'll Investigate
How would conditions on a planet affect the type of space colony that might be built there?

Form a Hypothesis

Research a planet. Review conditions on the surface of the planet. **Make a hypothesis** about the things that would have to be included in a space colony to allow humans to survive on the planet.

Goals

Infer what a space colony might look like on another planet.

Classify planetary surface conditions.

Draw a space colony for a planet.

PLAN

1. **Select** a planet and **study** its surface conditions.
2. **Classify** the surface conditions in the following ways.
   a. solid or gas
   b. hot, cold, or changing temperatures
   c. heavy atmosphere or no atmosphere
   d. bright or dim sunlight
   e. special conditions unlike other planets

3. **List** the things that humans need to survive. For example, humans need air to breathe. Does your planet have air that humans can breathe, or would your space colony have to provide the air?
4. **Make a table** for the planet showing its surface conditions and the features the space colony would have to have so that humans could survive on the planet.
5. **Discuss** your decisions as a group to make sure they make sense.

DO

1. Make sure your teacher has approved your plan and your data table before you proceed.
2. **Draw** a picture of the space colony. **Draw** another picture showing the inside of the space colony. **Label** the parts of the space colony, and **explain** how they help humans to survive.
3. Present your drawing to the class. **Explain** the reasoning behind it.

CONCLUDE AND APPLY

1. **Compare and contrast** your space colony with those of other students who researched the same planet you did. How are they alike? How are they different?
2. Would you change your space colony after seeing other groups' drawings? If so, what changes would you make?
3. What was the most interesting thing you learned about the planet you studied?
4. **APPLY** Was your planet a good choice for a space colony? **Explain** your answer.

FIGURE 14-11

Mercury is the closest planet to the sun. Like our moon, its surface is covered with craters.

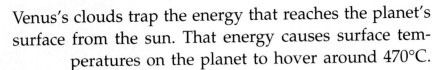

Venus's clouds trap the energy that reaches the planet's surface from the sun. That energy causes surface temperatures on the planet to hover around 470°C. That's hot enough to bake a clay pot, and far too hot for you. You're on to your next stop.

Mercury

The last planet that you visit, and the one closest to the sun, is Mercury. Mercury, shown in **Figure 14-11,** is the second-smallest planet. Its surface is heavily scarred with craters made by meteorites. Meteorites are chunks of rock that fall from the sky when asteroids break up.

Section Wrap-up

1. List the nine planets in order from the sun, beginning with the planet closest to the sun.

2. In general, how are the outer planets different from the inner planets? How are they alike?

3. **Think Critically:** We use larger units of measure to deal with increasingly larger distances. How do you think scientists handle increasingly smaller distances, such as the distances between molecules?

4. *Skill Builder*
   Developing Multimedia Presentations
   Use your knowledge of the solar system to develop a multimedia presentation of the solar system. You may want to begin by drawing a poster that includes the sun, the planets, the asteroid belt, and comets. If you need help, refer to Developing Multimedia Presentations on page 564 in the **Technology Skill Handbook.**

USING MATH

Using the table in the Problem Solving on page 386, calculate the distances of the planets from the sun, using a scale of 10 cm = 1 AU.

Science & Language Arts

Barbary
by Vonda N. McIntyre

. . . But the stars were fantastic. Barbary thought she must be able to see a hundred times as many as on earth, even in the country where sky-glow and smog did not hide them. They spanned the universe, all colors, shining with a steady, cold, remote light. She wanted to write down what they looked like, but every phrase she could think of sounded silly and inadequate.

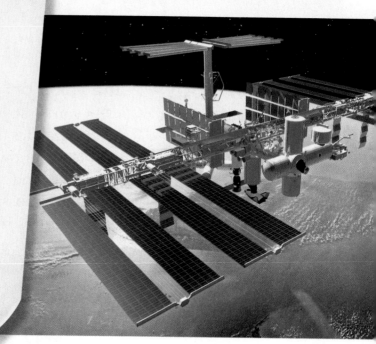

Science Journal

Have you ever dreamed of living on a space station? In your Journal, describe what you think life would be like at a remote outpost in space. What life-support systems would you need? How would you deal with weightlessness? What would you miss most about Earth?

Barbary is hoping to find a home for herself—and her mysterious stowaway—on the space station *Einstein*. On the way, Barbary learns from Jeanne Velory, the astronaut who is taking command of the station, that an alien ship is moving into the solar system. Some people think it may be abandoned; others aren't so sure.

The alien vessel is approaching on a path above the plane of the solar system. *Einstein*, circling Earth in a long orbit, is the ideal spot from which to observe, and hopefully contact, the mysterious ship.

Barbary explores the research station with her new sister, Heather. She learns how to "sly"—move gracefully in zero gravity—pilot a space raft, and do her homework with a very helpful computer. Along the way, she also learns to trust others and her own abilities.

Then without warning, Barbary finds herself using all her newfound skills to save her friends and discover the truth about the alien vessel . . . and its inhabitants.

14•3 Space Exploration: Boom or Bust?

What YOU'LL LEARN

- The costs and benefits of space exploration
- How to recognize bias in the things you read
  Science Words:
  satellite

Why IT'S IMPORTANT

You'll learn to judge the accuracy of what you read.

The Value of Space Exploration

As you sit at your desk reading about the solar system, space satellites and space probes are orbiting our planet or rocketing toward the edge of the solar system. Space **satellites** and space probes are spacecraft that are launched into space to gather information and send it back to Earth. **Figures 14-12** and **14-13** show examples of space technology. These and other kinds of space technology cost our country billions of dollars.

Has the money been well spent? That depends on who you ask. Some people think that the knowledge gained from space exploration has been well worth the cost. They point out that space technology is now used to study everything from the weather to the environment.

Others, though, think that the money would have been better spent elsewhere, and that the risk to human lives—12 astronauts have died—outweighs the benefits of space exploration.

Read the following passages about the space program. The people are fictional, which means they don't really exist. As you read, look for biases in the people's statements. Biases are viewpoints that influence the way people think. Recognizing bias will help you judge the accuracy of what you read.

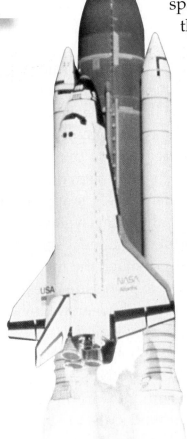

FIGURE 14-12

The space shuttle is a reusable spacecraft that transports astronauts, satellites, and other materials to and from space. *Describe some advantages of using a spacecraft designed to make many trips into space.*

Long Live the Space Program!

"The space program has been the best program this country has ever had. The technology developed for space is now used for many things on Earth. Space satellites are used to track weather patterns and to study the environment. Without the space program, computers would never have been developed."
—Maria Lopez, astronaut

Cut the Space Program!

"Since its beginning, the space program has been nothing but a drain on our country's resources. The billions of dollars spent on this program should have been used to build better roads, improve schools, and feed the hungry. The space program has also taken the lives of several astronauts. It has cost this country billions of dollars."
—Cholena Jones, highway worker

FIGURE 14-13

As part of space shuttle *Discovery's* mission, astronauts repaired the orbiting Hubble Space Telescope.

 ## Skill Builder: Recognizing Bias

LEARNING the SKILL

1. To recognize bias, you need to identify the speakers of the statements and examine their views and possible reasons for saying what they did. Determine how their biases influence the way they view an issue.

2. Sometimes biases are stated as facts, but in reality, they are opinions. Look for words that reflect an opinion, such as *never, best, nothing,* or *should.*

3. Examine the statements to see whether they are balanced. Do they mention other viewpoints?

4. Use your knowledge of the subject or research the subject to identify statements of facts.

PRACTICING the SKILL

1. Based on the material you read in the passages, list three facts about space exploration.

2. Identify the speakers of the statements. What might be some possible reasons for their opposing viewpoints?

3. Identify two facts and two biases in each of the speaker's statements.

APPLYING the SKILL

Look through the letters to the editor in your local newspaper. Write a report analyzing one of the letters for evidence of bias.

14•4 Stars and Galaxies

What YOU'LL LEARN

- How a star is born
- About the galaxies that make up our universe

  Science Words:
  constellation
  galaxy

Why IT'S IMPORTANT

Understanding the vastness of the universe helps us to understand Earth's place in space.

FIGURE 14-14

Find the Big Dipper in the constellation Ursa Major. *Why do you think people call it the Big Dipper?*

Stars

Every night, a whole new world opens to us. The stars come out. The fact is, stars are always in the sky. We just can't see them during the day because the sun's light is brighter than starlight. The sun is a star, too. It is the closest star to Earth. We can't see it at night because as Earth rotates, our part of Earth is facing away from it.

Constellations

Ursa Major, Orion, Taurus. Do these names sound familiar? They are **constellations** (kahn stuh LAY shunz), or groups of stars that form patterns in the sky. **Figure 14-14** shows some constellations.

Constellations are named after animals, objects, and people—real or imaginary. We still use many names that early Greek astronomers gave the constellations. But throughout history, different groups of people have seen different things in the constellations. In early England, people thought the Big Dipper, found in the constellation Ursa Major, looked like a plow. Native Americans saw it as a giant bear. To the Chinese, it looked like a governmental official and his helpers moving on a cloud. What does the Big Dipper look like to you?

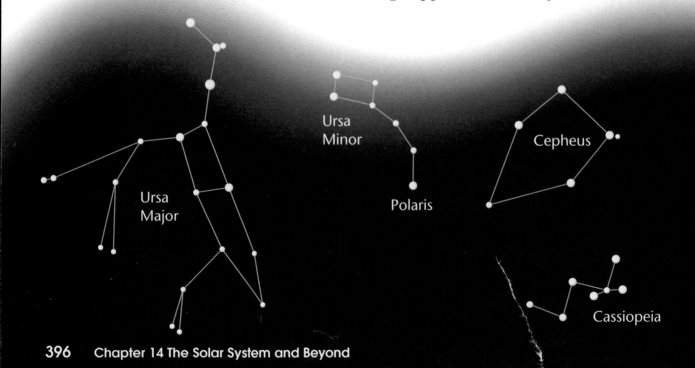

Ursa Major

Ursa Minor

Polaris

Cepheus

Cassiopeia

Starry Colors

When you glance at the sky on a clear night, the stars look like tiny pinpoints of light. It's hard to tell one from another, but stars come in different masses, sizes, and temperatures.

How do you measure a star's temperature? You can't go there with a big thermometer. But you can use a star's color as a clue to its temperature. Red stars are the coolest. Yellow stars are of medium temperature. Bluish-white stars are the hottest. Our sun is a yellow, medium-sized star. The giant red star called Betelgeuse (BEE tul joos) is much bigger than the sun. If this huge star were in the same place as our sun, it would swallow Mercury, Venus, Earth, and Mars.

The Lives of Stars

You've grown up and changed a lot since you were born. You've gone through several stages in your life, and you'll go through many more. Stars go through stages in their lives, just as people do.

Scientists theorize that stars begin their lives as huge clouds of gas and dust. The force of gravity causes the dust and gases to move closer together. When this happens, temperatures within the cloud begin to rise. A star is formed when this cloud gets so dense and hot that it starts producing energy.

The stages a star goes through in its life depend on the star's size. When a medium-sized star like our sun uses up some of the gases in its center, it expands to become a giant. Our sun will become a giant in about 5 billion years. When the remaining gases are used up, our sun will contract to become a black dwarf.

Mini LAB

Modeling Constellations

1. Draw a dot pattern of a constellation on a piece of black construction paper. Choose a constellation from **Figure 14-14** or make your own pattern of stars.
2. With your teacher's help, cut off the end of a round, empty box. You now have a cylinder with both ends open.
3. Place the box over the constellation. Trace around the rim of the box. Cut the paper along the traced circle.
4. Tape the paper to the end of the box. Using a pencil, carefully poke holes through the dots on the paper.
5. Place a flashlight inside the open end of the box. Darken the room and observe your constellation on the ceiling.

Analysis

1. Turn on the overhead light and view your constellation again. Can you still see it? Why or why not?
2. The stars are always in the sky, even during the day. Knowing this, explain how the overhead light is similar to our sun.

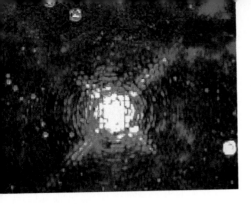

FIGURE 14-15
This photo, taken by the Hubble Space Telescope, shows a massive star in the Pistol Nebula.

Supergiants

When a huge star like the one shown in **Figure 14-15** begins to use up the gases in its core, it becomes a supergiant. Over time, the core of a supergiant collapses. Then the outer part of the star explodes and gets very bright. This is called a supernova. For a few brief days, the supernova might shine more brightly than a whole galaxy. The dust and gases released by this explosion may eventually form other stars.

Meanwhile, the core of the supergiant is still around. It is now called a neutron star. If the core is massive enough, it may rapidly become a black hole. Black holes are so dense that light shone into them would disappear.

USING TECHNOLOGY

Space Probes

Imagine what it would be like to explore a galaxy far away. It will be a long time before we can send astronauts millions of light-years into space. But we've already sent space probes to the outer limits of the solar system—some were launched more than 20 years ago and are still sending back data.

Cassini is one such space probe. It's about the size of a school bus. *Cassini* was launched in October 1997. Its mission is to study the planet Saturn, including Saturn's rings and moons. *Cassini* will reach Saturn in the year 2004. It will orbit the planet for four years, sending back data.

Afterward, if all systems are still working, *Cassini* might explore other planets. And in the future, who knows? Scientists may be able to design a space probe that can "boldly go where no space probe has gone before: to a galaxy far, far away."

SCIENCE Online

Go to Glencoe Science Online at *science.glencoe.com* to learn more about space probes.

Galaxies and the Universe

What do you see when you look at the night sky? If you live in a city, you may not see much. The glare from city lights makes it hard to see the stars. If you go to a dark place, far from the lights of towns and cities, you can see much more. In a dark area, with a powerful telescope, you might see dim clusters of stars grouped together. These clusters are galaxies (GAL uk seez). A **galaxy** is a group of stars, gas, and dust held together by gravity.

Light-Years

Do you remember what you learned earlier about astronomical units or AU? Distances between the planets are measured in AU. But to measure distances between galaxies, we need an even bigger unit of measure. Scientists use light-years to measure distances between galaxies. A light-year is the distance light travels in a year—about 9.5 trillion km. Light travels so fast it could go around Earth seven times in one second!

Have you ever wished that you could travel back in time? In a way, that's what you're doing when you look at a galaxy. The galaxy might be millions of light-years away. So the light that you see started on its journey long ago. You are seeing the galaxy as it was millions of years ago.

The Milky Way Galaxy

What's your address? You have a house number, a street, a city, a state, zip code, and country. Did you know that you have another address? Look at **Figure 14-16.** We all live in the Milky Way Galaxy. There might be a trillion stars in the Milky Way Galaxy, including our sun. Just as Earth revolves around the sun, stars revolve around the centers of galaxies. Our sun revolves around the Milky Way Galaxy about once every 240 million years.

FIGURE 14-16

The Milky Way Galaxy has spiral arms made up of stars, dust, and gas. Its inner region is an area of densely packed stars.

SCIENCE *Online*

Explore Glencoe Science Online at *science.glencoe.com* to learn how telescopes have changed our view of space.

A View from Within

We can see part of the Milky Way Galaxy as a band of light across the sky. But we can't see the whole Milky Way Galaxy. Why not? Think about it. When you're sitting in your classroom, can you see the whole school? No, you are inside the school and can see only parts of it. Our view of the Milky Way Galaxy from Earth is like the view of your school from a classroom. We can see only parts of our galaxy because we are inside it.

The Universe

Each galaxy probably has as many stars as the Milky Way Galaxy. Some may have more. And there may be as many as 100 billion galaxies. All these galaxies, with all their countless stars, make up the universe. Look at **Figure 14-17.** In this great vastness of revolving solar systems, exploding supernovae, and star-filled galaxies is one small planet called Earth. If you think about how huge the universe is, Earth seems like a speck of dust. Yet as far as we know, it's the only place where life exists.

FIGURE 14-17
Stars are forming in the Orion Nebula.

Section Wrap-up

1. What is a constellation? Name three constellations.

2. Describe the life of a medium-sized star such as the sun.

3. **Think Critically:** Some stars may no longer be in existence, but we still see them in the night sky. Why?

4. **Skill Builder**
Formulating Models The Milky Way Galaxy is 100 000 light-years in diameter. How would you build a model of the Milky Way Galaxy? If you need help, refer to Making Models on page 557 in the **Skill Handbook.**

Science Journal

Observe the stars in the night sky. In your Science Journal, draw the stars you observed. Now draw your own constellation based on those stars. Give your constellation a name. Why did you choose that name?

Read the statements below that review the most important ideas in the chapter. Using what you have learned, answer each question in your Science Journal.

1. The light-year and the astronomical unit or AU, are used to measure distances in space. *Why do we need special units of measurement for studying space?*

2. Seasons are created by the tilt of Earth's axis. *Explain why we have our warmest temperatures in summer.*

3. The inner planets and outer planets that make up our solar system revolve around the sun. *How is Pluto different from the other outer planets?*

4. The colors of stars tell us a lot about their temperature. *What color are the hottest stars? The coolest?*

Using Key Science Words

constellation   rotation
eclipse   satellite
galaxy   solar system
revolution

Match each phrase with the correct term from the list of Key Science Words.

1. the shadow produced by the moon or Earth passing in front of the sun
2. the motion of Earth that produces day and night
3. a group of stars, gas, and dust held together by gravity
4. a group of stars that forms a pattern in the sky
5. an object that is launched into space to gather information and send it back to Earth

Checking Concepts

Choose the word or phrase that completes the sentence.

6. The tilt of Earth's _____ causes seasons.
   a. equator   c. moon
   b. oceans   d. axis

7. When the moon is waning, it appears to be _____.
   a. growing larger
   b. growing smaller
   c. a full moon
   d. a new moon

8. An astronomical unit is the distance from _____.
   a. Earth to the moon
   b. Earth to the sun
   c. Pluto to Mercury
   d. Pluto to the sun

9. Earth is the _____ planet from the sun.
   a. first   c. third
   b. second   d. fourth

10. There may be as many as _____ galaxies in the universe.
    a. 1 billion   c. 50 billion
    b. 10 billion   d. 100 billion

Thinking Critically

Answer the following questions in your Science Journal using complete sentences.

11. Describe some advantages and disadvantages of the U.S. space program.

12. What conditions on Earth allow life to thrive?

13. Which of the planets in our solar system seems most like Earth? Which seems most different? Explain your answer, using facts that you've learned about the planets.

14. How might a scientist predict the day and time of a solar eclipse?

15. Throughout history, different groups of people have viewed the constellations in different ways. Infer why this is true.

Developing Skills

If you need help, refer to the description of each skill in the Skill Handbook.

16. **Making and Using Tables:** Research the size, period of rotation, and period of revolution for each planet. Show this information in a table. How do tables help us to better understand information?

17. **Comparing and Contrasting:** Compare the inner planets with the outer planets. How are they alike? How are they different?

18. **Making a Model:** Based on what you have learned about the sun, the moon, and Earth, make a model of a lunar or a solar eclipse.

19. **Sequencing:** Sequence the following terms in order of smallest object to largest group: galaxy, inner planets, solar system, universe, Earth.

20. **Using a CD-ROM:** Choose one of the nine planets and, using a CD-ROM, research information about that planet. Write a report based on your research. In your report, describe the articles, photographs, pictures, or videos found in the CD-ROM.

Performance Assessment

1. **Model:** Based on the information provided in the Problem Solving table on page 386, work as a class to make a three-dimensional model of the solar system. Use the scale you developed from the Using Math activity on page 392 to accurately show the distances between the planets and the sun.

2. **Newspaper Article:** The photo below shows the remains of a supernova explosion. Write a newspaper article describing the life cycle of a star that becomes a supernova. Explain what will happen to the star afterwards.

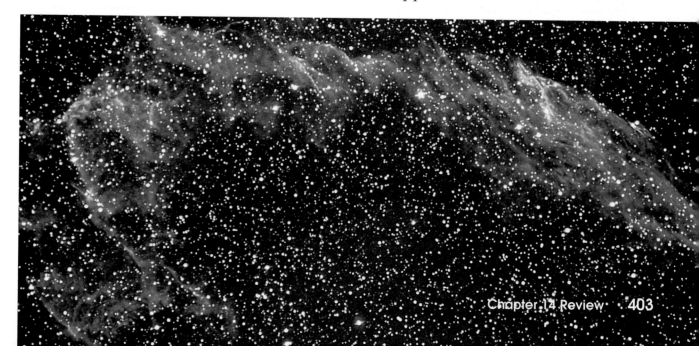

Chapter Preview

Skills Preview

► **Skill Builders**
- use a CD-ROM
- sequence

► **MiniLABs**
- classify
- infer

► **Activities**
- observe
- compare
- hypothesize
- collect data

Chapter 15

Earth Materials

You're at the top! You and a friend have been climbing Zane's Bluff. Now, after all the hard work of the climb, you are treated to a great view. You also take time to get a closer look at the rock you've been climbing. First, you notice that it sparkles in the sun because of the silvery specks that are stuck in the rock. Looking closer, you also see clear, glassy pieces and gray, irregular chunks. What is the rock made of? And how did it get here?

EXPLORE ACTIVITY

Observe a Rock

1. Obtain a sample of granite or basalt from your teacher. You will also need a hand lens.
2. Observe the rock with the hand lens. Your job is to observe and record as many of the features of the rock as you can.
3. Return the rock to your teacher.
4. Try to describe your rock so that other students could pick your rock out of the group of rocks.

Science Journal

In your Science Journal, describe and draw how the parts of the granite or basalt fit together to form the rock. Be sure to label your drawing.

15•1 Minerals: Earth's Jewels

What YOU'LL LEARN

• The difference between a mineral and a rock
• Properties that are used to identify minerals

Science Words:
mineral
rock
crystal
gem
ore

Why IT'S IMPORTANT

You'll learn the value and uses of minerals and rocks.

What is a mineral?

Get ready! You're going on an expedition to find minerals (MIH nuh rools). Where will you look? Do you think you'll have to crawl into a cave or brave the depths of a mine? Well, put away your flashlight and hard hat. You can find minerals right in your own home—in the salt shaker on your table; in the "lead" of your pencil; and in the metal pots and pans, glassware, and dishes in your kitchen. Minerals are everywhere around you, as shown in **Figure 15-1.**

What is a mineral? A **mineral** is an inorganic solid material found in nature. Inorganic substances are things that are not formed by plants or animals. A mineral must always have the same chemical makeup and an orderly pattern of atoms. For example, the mineral calcite always has the same chemical makeup. And the atoms in calcite are always arranged in the same way. Minerals are also the basic building blocks for all rocks. A **rock,** like the granite or basalt used in the Explore activity on page 405, is usually made of two or more minerals. More than 4000 different minerals have been identified. Each mineral has traits or characteristics you can use to identify it.

FIGURE 15-1

You use minerals every day without realizing it. Minerals are used to make many common objects.

A Quartz is melted to form glass.

B The mineral gypsum is used to make drywall, a building material.

How do minerals form?

On your rock-climbing adventure, you started asking questions about how minerals and rocks form. Minerals can form in several ways—from melted rock or from solutions. The first way is from melted rock called magma. As magma cools, atoms combine to form minerals that make up rock. These minerals can form on Earth's surface or deep inside Earth. Minerals also form when solutions evaporate. As water leaves a solution, minerals are left behind. This is how the salt you put on your popcorn was formed. A third way minerals form is also from solution. A solution can hold only so much of the dissolved mineral. If more minerals are added, some of the minerals will drop out as solid materials. Or, if a solution is extremely rich in minerals, the minerals may drop out.

You can tell how a mineral formed by how it looks. If the mineral grains fit together like a puzzle, it cooled from magma. If you see layers of different minerals, the minerals probably formed by evaporation. The beautiful crystal you see in **Figure 15-2** grew from a solution rich in dissolved minerals. To figure out how a mineral forms, you must consider the size of the mineral grain or crystal, the presence of layers, and the texture. Texture is the size of the crystal and how the crystals or mineral grains fit together. Next, you will learn ways to identify different minerals. Knowing how minerals form is interesting, but being able to identify them is more fun.

FIGURE 15-2

This fluorite crystal grew from a solution rich in dissolved minerals.

C These whistles are made from iron ores, such as hematite or magnetite.

D The "lead" in a pencil is not lead, it is the mineral graphite.

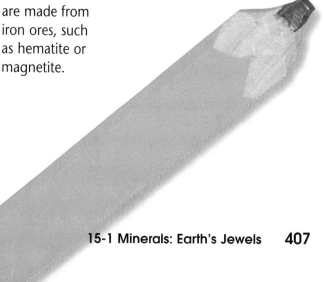

Properties of Minerals

Perched on your rocky outlook, you notice another person on a ledge some distance away. Is it someone you know? She's wearing a yellow shirt and has her long, dark hair in braids, just like a friend you saw this morning. You're only sure it's your friend when she turns, and you recognize her smile. You've identified your friend by physical properties that set her apart from other people—her clothing, hair color and style, and facial features. Each mineral, too, has a set of physical properties that you can use to identify it. No fancy equipment is needed. Most common minerals can be identified with items you have around the house and can carry in your pocket, like a penny and a nail. Now let's take a look at the properties that will help you identify minerals.

Crystals

All minerals have a definite atomic structure. The atoms that make up the mineral are arranged in a repeating pattern. Solid materials that have this pattern of atoms are called **crystals.** Some mineral samples, such as the amethyst pictured in **Figure 15-3,** have beautiful crystals. Crystals have smooth surfaces, sharp edges, and points. Some crystals look like tiny cubes. **Figure 15-4** gives a close look at table salt and reveals that each grain is a little cube. These cubes are crystals of the mineral halite.

FIGURE 15-3

This beautiful amethyst has large crystals. Amethyst is a variety of the mineral quartz.

FIGURE 15-4

Common table salt is the mineral halite, which is called rock salt. Halite crystals are tiny cubes.

Cleavage and Fracture

Another clue to a mineral's identity is the way it breaks. Minerals that split into pieces with smooth, regular edges and surfaces are said to have cleavage (KLEE vuj). The mineral mica in **Figure 15-5A** shows cleavage by splitting into thin sheets. Splitting one of these minerals along a cleavage surface is something like taking pieces of bread from a sliced loaf. Cleavage is caused by weaknesses in the arrangement of the atoms that make up the mineral.

Not all minerals have cleavage. Some break into pieces with jagged or rough edges. Instead of neat slices, these pieces are shaped more like hunks of bread torn from an unsliced loaf. Materials that break this way, such as quartz, have what is called fracture (FRAK chur). **Figure 15-5C** shows the hackly fracture of native copper.

FIGURE 15-5

Cleavage is one of the most useful properties of minerals. Some minerals have one or more directions of cleavage. If minerals do not break in a regular way, they have what is called fracture.

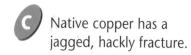

A Mica has one cleavage direction and can be peeled off in sheets.

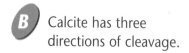

B Calcite has three directions of cleavage.

D The smooth, curved fracture of obsidian is called conchoidal fracture.

C Native copper has a jagged, hackly fracture.

FIGURE 15-6

Luster describes how a mineral looks. If it shines like a metal, it has metallic luster (galena). All other minerals have some type of nonmetallic luster (talc).

Galena

Talc

Streak and Luster

The reddish-gold color of a new penny shows that it's made of copper. The bright yellow color of sulfur is a valuable clue to its identity. Sometimes a mineral's color can help you figure out what it is. But color also can fool you. The common mineral pyrite (PI rite) has a shiny, gold color similar to real gold—close enough to disappoint many prospectors back in the Gold Rush days in the 1800s. Pyrite is called "fool's gold."

A test called the streak test will help identify a mineral, even if it looks a lot like a different mineral. Scratching a mineral sample across an unglazed, white tile (called a streak plate) produces a streak of color, as shown in **Figure 15-7.** The streak is not necessarily the same color as the mineral itself. This streak of powdered mineral is more useful for identification than the mineral's color. Gold prospectors could have saved themselves a lot of heartache if they had known about the streak test. Pyrite makes a greenish-black or brownish-black streak, while gold makes a yellow streak.

Shiny? Dull? Pearly? Words like these describe another property of minerals called luster. Luster is how light is reflected from a mineral's surface. If it shines like a metal, the mineral has metallic (muh TAL lihk) luster. Nonmetallic minerals can be described as having pearly, glassy, dull, or earthy luster, as shown in **Figure 15-6.** Using color, streak, and luster helps identify minerals. What other properties can be used?

FIGURE 15-7

Streak is the color of the powdered mineral. The mineral with a black streak is magnetite, and the mineral with the reddish-brown streak is hematite.

Table 15-1

Mineral Hardness		
Mohs scale		
	softest	Hardness of common objects
Talc	1	
Gypsum	2	fingernail (2.5)
Calcite	3	copper penny (3.5)
Fluorite	4	iron nail (4.5)
Apatite	5	glass (5.5)
Feldspar	6	steel file (6.5)
Quartz	7	streak plate (7)
Topaz	8	
Corundum	9	
Diamond	10	
	hardest	

Hardness

As you investigate different minerals, you'll find that some are harder than others. Some minerals, like talc, are so soft that they can be scratched with a fingernail. Others, like diamond, are so hard that they can be used to cut almost anything else.

In 1822, an Austrian geologist named Friedrich Mohs also noticed this property. He developed a way to classify minerals by their hardness. The Mohs scale, as shown in **Table 15-1,** classifies minerals from 1 (softest) to 10 (hardest). You can determine hardness by trying to scratch one mineral with another to see which is harder. For example, fluorite (4 on the Mohs scale) will scratch calcite (3 on the scale), but calcite will not scratch fluorite. You can also use a homemade mineral identification kit: a penny, a nail, and a glass plate with smooth edges. You just find out what scratches what. Is the mineral hard enough to scratch a penny? Will it scratch glass?

Classifying Minerals

What other properties help to identify minerals?

1. Place your four samples in separate, labeled, sealable, plastic bags, and dip each in a pile of iron filings. Record which mineral(s) attract the filings.
2. Place each sample in a small jar full of vinegar and record what happens.

Analysis

1. Which mineral(s) was magnetic? Which had fizzing properties?
2. Describe, in a data table, the other physical properties of the four minerals.

Problem Solving

Identifying Mineral Origin

At a neighbor's yard sale, you find a dusty old box filled with mineral samples. The neighbor, a former science teacher, says she'll sell you the whole thing for ten dollars. Then she gives you a challenge. If you can come back in a week with the samples organized by the way they formed, she'll give your money back. You rush home to start your detective work.

Solve the Problem:

Think about the traits of mineral crystals that are formed from magma and minerals that settle out of solution. Refer to page 407 for help.

Think Critically:

1. How will you identify and separate the samples by how they formed? What steps will you follow?

2. Determine the characteristics that you will look for in each group.

Gypsum

Dolomite

Biotite

Other Properties

Some minerals have other unusual properties that can help identify them. Magnetism occurs in a few minerals, such as magnetite. The mineral calcite has two other unusual properties. It will fizz when it comes into contact with an acid like vinegar. And if you look through a clear calcite crystal, you will see a double image. Although, because of safety concerns, you should not try it, taste can be used by scientists to identify some minerals. Halite, also called rock salt, has a salty taste. Combinations of all of these properties are used to identify minerals. Learn to use them, and you can be a mineral detective.

Common Minerals

In the chapter opener, the rocks you scrambled up on your climb were made of minerals. But only a few of the more than 4000 minerals make up most rocks. The most common minerals found in rocks include quartz, feldspar, biotite mica, calcite, gypsum, hornblende, fluorite, augite, and hematite, among others. Most rocks on Earth are made up of these common minerals. Most other minerals, such as diamonds and emeralds, are rare.

Gems

Which would you rather win, a diamond ring or a quartz ring? A diamond ring would be more valuable. Why? The diamond in a ring is a kind of mineral called a gem. **Gems** are minerals that are rare and can be cut and polished, giving them a beautiful appearance, as shown in **Figure 15-8.** This makes them ideal for jewelry. Not all diamonds are gems. To be gem quality, a mineral must be clear with no blemishes or cracks. It also must have a beautiful color. Few minerals meet these standards. That's why the ones that do are so rare and valuable.

FIGURE 15-8

Gems such as these are rare and beautiful.

USING TECHNOLOGY

Controlling the Waste from Mining Operations

Mining ore from Earth often takes more than just digging in the ground and finding big gold nuggets. Often, tiny specks of gold too small to see are mixed with the rock. One way of getting gold out of ore is to crush the rock and mix it with chemicals such as cyanide (SI uh nide) and zinc. The major problem with this is that the waste products of this process are poisonous.

Mining operations produce a great deal of waste, which is stored in specially built ponds. To keep the waste from getting into the soil or groundwater, thick layers of clay line the bottoms and sides of the ponds. The clay acts like a barrier, trapping the waste. Scientists must check the water, plants, and soil to make sure the wastes are not leaking out. Mining for gold and other metals means more than just digging the ores out of the ground. Plans must be made for safe disposal of wastes.

SCIENCE *Online*

Visit Glencoe Science Online, *science.glencoe.com*, for a link to a site about protecting wildlife from cyanide.

Ores

A mineral is called an **ore** if it contains something that can be useful and sold for a profit. Many of the metals that we use come from ores. A copper mine is shown in **Figure 15-9.** For example, the iron used to make steel comes from the mineral hematite, lead for batteries is produced from galena, and the magnesium used in vitamins comes from dolomite.

FIGURE 15-9

To be profitable, ores must be found in large deposits or rich veins. Mining is expensive. This mine in Montana is producing copper.

Now you have a better understanding of minerals and their uses. Can you name five things in your classroom that come from minerals? You will find that you use lots of minerals every day. Next, you will look at rocks, which are Earth materials that are made up of combinations of minerals.

Section Wrap-up

1. Explain the difference between a mineral and a rock. Name five common rock-forming minerals.

2. List five properties or traits that are used to identify minerals.

3. **Think Critically:** Would you want to live close to a working gold mine? Explain your answer.

4. **Skill Builder**
   Using a CD-ROM Use an electronic encyclopedia or Earth science CD-ROM to help create your own mineral identification chart. List the traits of at least ten different minerals. If you need help, refer to Using a CD-ROM on page 563 of the **Technology Skill Handbook.**

Science Journal

Select a mineral and write a story in your Journal about how it formed. Be creative, but everything must be scientifically correct. Include how the mineral forms and several of its characteristics. If you need help, refer to Appendix J on pages 538–540.

Igneous and Sedimentary Rocks

Earth's Fire

A rocky cliff, a jagged mountain peak, a huge boulder—they all look about as solid and permanent as anything can be. Rocks seem as if they've always been here and always will be. But things are constantly changing on Earth. New rocks form, and old rocks crack and wear away. These changes produce three main kinds of rocks—igneous, sedimentary, and metamorphic.

If you could travel down into Earth, you would find that the deeper you go, the higher the temperature and the greater the pressure. Deep inside Earth, it is hot enough to melt rock, as seen in **Figure 15-10. Igneous rocks** are produced when melted rock, or magma, from inside Earth cools. Igneous rocks can cool and harden on or under Earth's surface. When magma cools, it makes either an extrusive igneous (EKS trew sihv • IG nee us) rock or an intrusive (IN trew sihv) igneous rock.

What YOU'LL LEARN

- How extrusive and intrusive igneous rocks are different
- How different types of sedimentary rocks form

Science Words:
igneous rock
extrusive
intrusive
sedimentary rock

Why IT'S IMPORTANT

You'll learn about the rocks that form the land around you.

FIGURE 15-10

This erupting volcano is Mount Etna in Italy.

Rocks from Lava

Extrusive igneous rocks form when magma cools on Earth's surface. Magma that reaches Earth's surface is called lava. Lava cools quickly, before large mineral crystals have time to form. That's why extrusive igneous rocks usually have a smooth, sometimes glassy appearance. They have a few or no visible crystals.

Extrusive igneous rocks can form in two ways. In one way, volcanoes erupt and shoot out lava and ash. Also, large cracks in Earth's crust, called fissures (FIHSH urs), can open up. When they do, the lava oozes out onto the ground. Oozing lava from a fissure or a volcano is called a lava flow. It's something like what happens when you put too much batter in the waffle iron. The batter oozes out the sides and flows onto the countertop. In Hawaii, lava flows are so common that you can see a volcanic eruption almost every day.

FIGURE 15-11

Intrusive igneous rocks form when molten rock or magma cools inside Earth. Extrusive igneous rocks form when lava cools at Earth's surface.

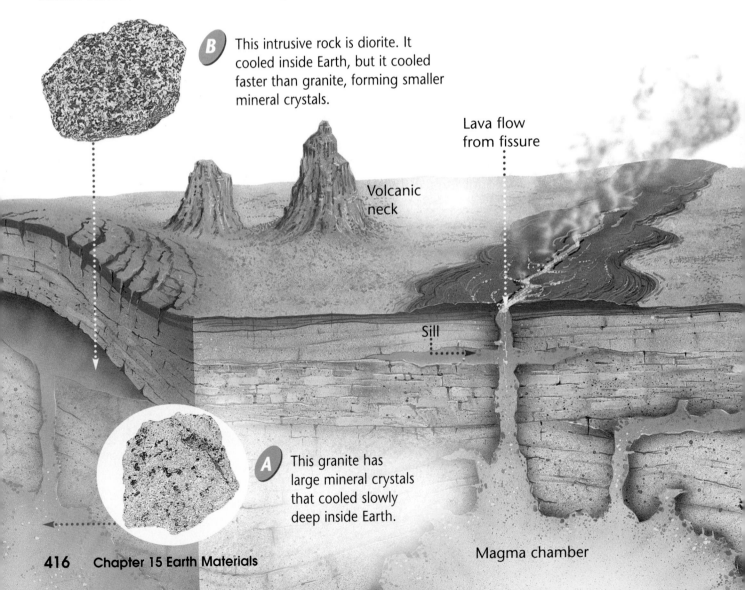

B This intrusive rock is diorite. It cooled inside Earth, but it cooled faster than granite, forming smaller mineral crystals.

Lava flow from fissure

Volcanic neck

Sill

A This granite has large mineral crystals that cooled slowly deep inside Earth.

Magma chamber

Rocks from Magma

What about magma that doesn't reach the surface? Can it form rocks, too? Yes, **intrusive** igneous rocks are produced when magma cools below the crust inside Earth, rather than on the surface, as shown in **Figure 15-11.**

When intrusive igneous rocks form, a huge glob of magma from inside Earth rises toward the surface. But it never quite breaks through to erupt onto the surface. It's similar to a helium balloon that is let loose in a gym and gets stuck on the ceiling. It doesn't make it outside—just to the roof. This hot mass of rock sits right under the surface and cools slowly over thousands of years until it is solid. The cooling takes so long that the minerals in the magma have time to form large crystals. The size of the mineral crystals is the main difference between intrusive and extrusive igneous rocks. *Intrusive* igneous rocks have large crystals that are easy to see. *Extrusive* igneous rocks do not have large crystals that you can see.

SCIENCE Online

What do different igneous rocks look like? Check out the link at Glencoe Science Online. *science.glencoe.com*

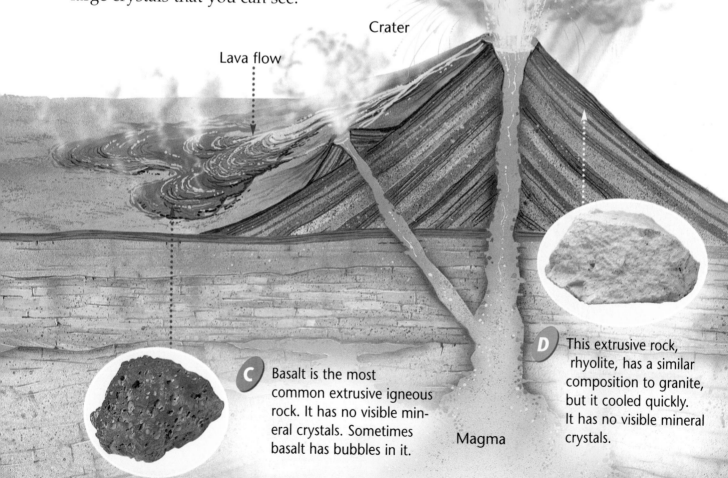

Crater

Lava flow

C Basalt is the most common extrusive igneous rock. It has no visible mineral crystals. Sometimes basalt has bubbles in it.

Magma

D This extrusive rock, rhyolite, has a similar composition to granite, but it cooled quickly. It has no visible mineral crystals.

Design Your Own Experiment
Cool Crystals Versus Hot Crystals

Two types of igneous rocks can look different, even though they are made of the same minerals. Extrusive rocks have an even texture with no visible crystals, while intrusive rocks have large, beautiful crystals. How does this happen?

Possible Materials

- saucepan, 3-quart
- measuring cup
- large wooden spoon
- stove or hot plate
- pint jars or glasses (2)
- paper clips (2)
- pencils (2)
- 3-cm piece of cotton string (2)
- white, granulated sugar (3 cups)
- refrigerator
- igneous rocks (4)

PREPARE

What You'll Investigate

Determine how different-sized crystals are formed from a solution.

Form a Hypothesis

State your **hypothesis** about how you think the rate of cooling affects the different sizes of crystals that can be formed.

Goals

Design an experiment that compares crystal growth in a solution by using different methods of cooling.

Write a general rule about crystal size in igneous rocks based on your observations in this experiment.

Safety Precautions

Never eat or taste anything from a lab, even if you are confident that you know what it is.

PLAN

1. As a group, agree upon and write out your **hypothesis**.
2. As a group, **list** the steps that you will take to test your hypothesis. Be specific, describing exactly how you are going to make different sizes of crystals. If you have questions about how to mix the solution, check with your teacher.
3. **List** all of the materials that you will need to complete your experiment.
4. Before you begin, **make a data table or graph** that will allow you to compare the size of the crystals that form in each solution.
5. **Read** over your entire experiment to make sure that all the steps are in logical order.
6. Be sure to double-check your list of materials and method for creating sizes of crystals.
7. Will you repeat any part of the experiment more than once to allow for human error?
8. Is your data table ready to handle the amount of data that you want to collect?
9. Make sure that your teacher approves your plan and that you have included any changes to the plan.

DO

1. Carry out the lab as planned and approved.
2. Be sure to **record** your observations in the data table as you **complete** each test.

CONCLUDE AND APPLY

1. **Compare** your findings with those of the other student groups.
2. Using your data table, **conclude** how this information could help you to classify igneous rocks.
3. **APPLY** Identify four igneous rock samples as either intrusive or extrusive.

Sedimentary Rocks

The second major group of rocks is called sedimentary (sed uh MEN tree) rocks. **Sedimentary rocks** are made when pieces of other rocks, plant and animal matter, or dissolved minerals collect to form rock layers, as shown in **Figure 15-12**. These pieces of rock and other materials are called sediments (SED uh muntz). Rivers, ocean waves, mudslides, and glaciers can carry sediments. Sediments can also be carried by the wind. When the sediments are dropped by or deposited by wind, ice, gravity, or water, they collect in layers. After the sediments are deposited, they begin the long process of becoming rock. Most sedimentary rocks form over thousands to millions of years. The changes that form sedimentary rocks are always happening in our world, as shown in **Figure 15-13**.

Broken Rocks

When most people talk about sedimentary rocks, they are usually talking about rocks such as sandstone. This type of rock is made up of broken pieces of other rocks. These rocks are deposited by water, ice, gravity, or wind into layers or piles. These piles are then cemented together by other minerals and squeezed or compacted into rock by the weight of sediments on top of them.

FIGURE 15-12

All of these sedimentary rocks are exposed at Cathedral Rocks in Arizona. The layers are different types of sedimentary rocks. *How are sediments carried to form sedimentary rocks?*

A Conglomerate is made up of large pebbles cemented together.

B Sandstone is made up of sand grains cemented together.

These broken sedimentary rocks are identified by the size of the pieces or grains that make up the rock. Rocks made of the smallest grains, clay, are called shale. Silt grains, which are slightly larger than clay, make up siltstone. Sandstone is made of sand. Sand is larger than silt. The largest grains are called pebbles. Pebbles mixed and cemented together with other sediments make up rocks called conglomerates (kon GLAW muh rutz).

Chemical Rocks

Some sedimentary rocks are formed when seawater, loaded with dissolved minerals, evaporates. Chemical sedimentary rock can also form when mineral-rich water from geysers or hot springs evaporates. As the water evaporates, layers of minerals are left behind. If you've ever sat in the sun after swimming in the ocean, you may have noticed salt crystals on your skin. The seawater on your skin evaporated, leaving behind deposits of salt. The salt had been dissolved in the water. Something similar happens to form chemical rocks.

FIGURE 15-13
Waves deposit sand and move sand around on this beach in California.

C Siltstone is made up of grains that are smaller than sand grains.

D Shale is made of the smallest grains, clay.

FIGURE 15-14

There are a great variety of biochemical sedimentary rocks.

A The White Cliffs of Dover, England, are made of chalk.

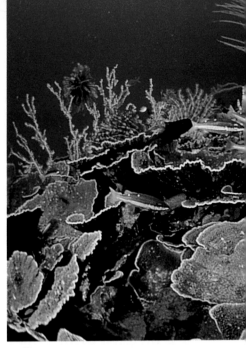

B This coral reef is located in New Guinea. The frame of the reef is limestone.

Mini LAB

How do fossils form rocks?

Create an artificial fossil rock sample of biochemical limestone.

1. Fill a small aluminum pie pan with an assortment of broken macaroni. These represent different kinds of fossils.
2. Mix three tablespoons of white glue into half a cup of water. Pour this solution over the macaroni and set it aside to dry.
3. When your fossil rock sample has set, pop it out of the pan and compare it with a real fossil limestone sample.

Analysis

1. Explain why you used the glue solution and what this represents in nature.
2. Using whole macaroni samples as a guide, match the macaroni pieces in your "rock" to the intact macaroni fossils. Then draw and label them in your Science Journal.

Biochemical Rocks

Would it surprise you to know that the chalk your teacher may be using on the chalkboard may also be a rock? So is the coal that is used as a fuel to produce electricity.

Chalk and coal are examples of the group of sedimentary rocks called biochemical rocks. Biochemical rocks form over millions of years. Living matter dies, piles up, and then is compressed into rock. If the rock is produced from layers of plants piled on top of one another, it is called coal. Biochemical sedimentary rocks can also form in the oceans and are most often classified as limestone. Many different kinds of limestone exist. Chalk is a kind of limestone made from the fossils of millions of tiny animals and algae, as shown in **Figure 15-14.** A fossil is the remains or trace of a once-living plant or animal. A dinosaur bone and footprint are both fossils.

Fossil-rich limestone

C Fossils like these in Dinosaur National Monument are found in sandstones or siltstones.

Camarasaurus skull

Section Wrap-up

1. Describe the different ways that extrusive and intrusive igneous rocks form.

2. Diagram the way that each of the three kinds of sedimentary rocks is formed. List one sample of each kind of rock: broken, chemical, and bio-chemical.

3. **Think Critically:** If someone handed you a sample of an igneous rock and asked you if it was extrusive or intrusive, what would you look for first? Explain.

4. **Skill Builder**
   Sequencing Describe one possible sequence of events that explains how coal is formed. If you need help, refer to Sequencing on page 543 of the **Skill Handbook.**

Science Journal

Select a national park or monument that has had volcanic activity. Read about the park and the features that you'd like to see. Then describe it in your Science Journal.

15•3 Monument or Energy?

What YOU'LL LEARN

- That earth processes produce landscapes and resources
- How land use can change the economy of an area

Science Words:
fossil fuel

Why IT'S IMPORTANT

You will learn that there are consequences to the decisions you make.

In September of 1996, 1.7 million acres of Utah became a national monument. The Grand Staircase-Escalante (es kuh LAWN tay) National Monument is one of the most beautiful areas of the American west, as shown in **Figure 15-15.** For millions of years, running water has carved spectacular canyons into the rocks. Wind has blown bits of sand and silt against other rocks, wearing the rocks away. Water and wind have combined to form mesas (MAY sus), which are flat, tablelike rock structures that rise high above the surrounding land. Wind has also eroded rocks within the monument to form steep cliffs with what look like staircases carved into them.

On the Surface . . .

The monument also contains the remains of ancient cultures. One such group, the Anasazi (ah nuh SAW zee), lived in the caves and shelters that were carved into the rock by wind and water. The rocks of the Grand Staircase-Escalante National Monument also contain many fossils. Earlier, you learned that fossils are the remains of ancient plants and animals. Scientists use fossils to learn about ancient climates, how living things have changed over time, and about Earth's geography.

Beneath Your Feet . . .

Deep beneath the majestic cliffs and canyons of the largest and newest national monument in the United States are other treasures—buried treasures. Experts think that coal, natural gas, and oil valued at billions of dollars lie under the monument, as shown in **Figure 15-16.** These resources are fossil fuels. **Fossil fuels** are the remains of ancient plants and animals that can be burned to produce energy.

FIGURE 15-15

Rock layers eroded to form a series of giant steps, which give the national monument its name.

Preserve or Reserves?

A lot of people are happy about the decision to create the monument. The desert ecosystem is fragile, they say, and should be protected. And, said one scientist, "It's the most beautiful spot in the U.S." But there is a price. Now that the area has been declared a national monument, none of the fossil fuels can be removed.

Because of that, some people worry about the state's economy. These natural resources could provide lots of money and hundreds of jobs for the state of Utah. This money could be used to improve roads and schools, provide career training, and pay for other valuable things. At the same time, the national monument will provide money and jobs because tourists will visit. It is also important to protect and preserve beautiful areas as parks and monuments.

FIGURE 15-16

Large deposits of coal and other fossil fuels are found in this region.

 ## Skill Builder: Predicting Consequences

LEARNING the SKILL

1. Identify the main idea.

2. Identify supporting details.

3. Study any photographs or illustrations that accompany the material.

4. Try to find out about similar decisions or situations. Then ask yourself what the consequences were of a similar action or decision.

PRACTICING the SKILL

1. Identify the main point of the article.

2. Why are some people in Utah worried about the new monument?

3. What points most strongly support the new monument?

APPLYING the SKILL

Former U.S. presidents Bill Clinton, Jimmy Carter, Theodore Roosevelt, and Franklin Roosevelt used the Antiquities Act to establish many monuments and national parks in the United States. Find out more about the act and the resulting consequences.

15•4 Metamorphic Rocks and the Rock Cycle

What YOU'LL LEARN

- The conditions needed for metamorphic rocks to form
- How all rocks are tied together in the rock cycle

Science Words:
metamorphic rock
foliated
non-foliated
rock cycle

Why IT'S IMPORTANT

Learning about metamorphic rocks and the rock cycle will help you see Earth as a constantly changing planet.

New Rocks from Old Rocks

When you wake up in the morning, does the world look different than it did the day before? Usually not. But even if you don't notice, Earth is constantly changing. Layers of sediment are piling up in the bottoms of lakes, landmasses are moving, and rocks are disappearing back under Earth's crust. Some of these changes cause sedimentary and igneous rocks to be heated and squeezed, as shown in **Figure 15-17**. In the process, new rocks form.

FIGURE 15-17

These mountains in the Swiss Alps were formed under great pressures and temperatures. The rocks were squeezed into these spectacular shapes.

FIGURE 15-18

This rock sequence from the state of Washington has been folded and deformed under high pressures and temperatures deep within Earth.

Metamorphic Rocks

Do you recycle your plastic milk jugs? After the jugs are collected, sorted, and cleaned, they are heated and squeezed into pellets. The pellets later can be made into useful products. Or material can be reused and made into art, as shown in **Figure 15-19.** Did you know that rocks get recycled, too? Rocks that form when older rocks are heated or squeezed are called **metamorphic rocks.** The word *metamorphic* (met uh MOR fihk) means "change of form." This is a good description of how some rocks take on a whole new look when under great temperatures and pressures, as shown in **Figure 15-18.**

FIGURE 15-19

This dinosaur statue was made from recycled car parts.

Types of "Changed" Rocks

Metamorphic rocks are divided into two groups: foliated (FOH lee ay tud) and non-foliated, as shown in **Figure 15-20. Foliated** rocks have bands of minerals. These minerals have been heated and squeezed into parallel layers. Many foliated metamorphic rocks have bands of different-colored minerals. Slate, gneiss (NISE), phyllite (FIHL ite), and schist are all examples of foliated rocks. **Non-foliated** metamorphic rocks do not have distinct layers or bands. These rocks, such as quartzite, marble, and soapstone, usually have more even colors than foliated rocks.

Try the next activity to better understand how Earth recycles rocks from one kind to another.

FIGURE 15-20

There are many different types of metamorphic rocks. *What force could cause the parallel layers in foliated rocks?*

A This carved walrus is soapstone, a non-foliated metamorphic rock.

B This schist is a foliated metamorphic rock. The red grains are garnets.

C This statue from a fountain in Italy is made of marble, a non-foliated metamorphic rock.

E Gneiss is a foliated metamorphic rock.

D This roof is made of slate, a foliated metamorphic rock.

Gneiss Rice

You will experiment with pressure to see how it can produce bands of minerals in metamorphic rocks. You will compare real rock samples with clay rocks that you will make.

What You'll Investigate
You will try to re-create conditions that cause an igneous rock to change into a banded rock.

Procedure

1. **CAUTION:** *Don't eat anything from the experiment.*
2. **Sketch** the granite specimen in your Science Journal.
3. **Pour** the rice onto the table. **Roll** the ball of clay in the rice. Some of the rice will stick to the outside of the ball. **Knead** the ball until the rice is spread out fairly evenly inside and out. Roll and knead the ball again, and repeat until your clay sample has lots of "minerals" distributed throughout it.
4. Using the rolling pin, **roll** the clay so that it is about 0.5 cm thick. Don't roll it too hard. The grains of rice should be pointing in different directions. **Draw** a picture of the clay in your Science Journal.
5. Take the edge of the clay closest to you and **fold** it toward the edge farthest from you. **Roll** the clay in the direction you folded it. Fold and roll the clay in the same direction several more times. Flatten the lump to 0.5 cm thickness again. **Draw** what you observe in your "rock" and in the gneiss sample in your Science Journal.

Conclude and Apply
1. What features did the granite and the first lump of clay have in common?
2. What force(s) caused the positions of rice grains in the lump of clay to change? How is this process similar to and also different from what happens in nature?

Goals
- Investigate ways rocks become changed.
- Construct a metamorphic rock.

Materials
- rolling pin
- lump of modeling clay
- uncooked rice (wild rice, if available), half cup
- sample of granite
- sample of gneiss

The Rock Cycle

Rocks can be recycled from one type to another. If you wanted to describe the process to someone, how would you do it? Would you use words or pictures? Scientists have created a diagram called the **rock cycle** to show the process. It shows how different kinds of rock are related to one another and how rocks change from one form to another. Each rock is on a continuing journey through the rock cycle, as shown in **Figure 15-21.**

FIGURE 15-21

This diagram of the rock cycle shows how rocks are constantly recycled from one kind of rock to another.

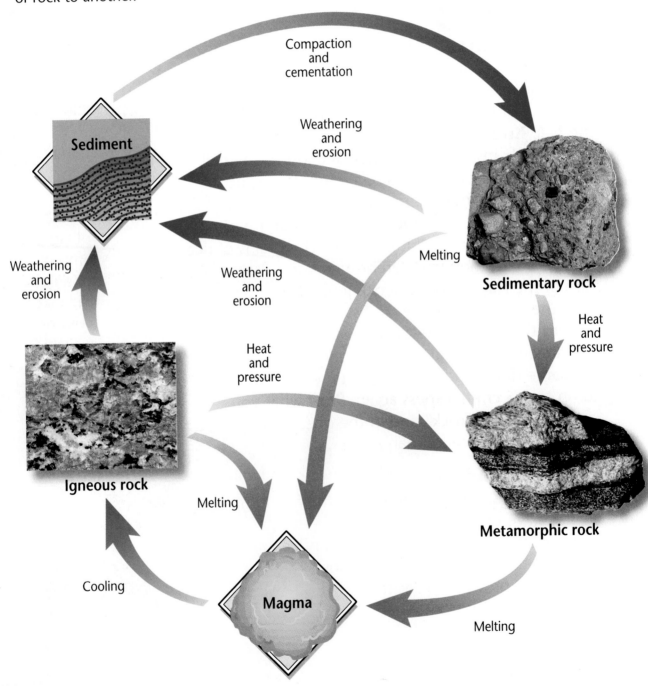

Compaction and cementation

Weathering and erosion

Sediment

Melting

Sedimentary rock

Weathering and erosion

Weathering and erosion

Heat and pressure

Heat and pressure

Heat and pressure

Igneous rock

Melting

Cooling

Magma

Metamorphic rock

Melting

The Journey of a Rock

Pick any point on the diagram of the rock cycle in **Figure 15-21,** and you will see how the rock at that point could become any other kind of rock. Let's start with a blob of lava that oozes to the surface and cools, as shown in **Figure 15-22.** It forms an igneous rock. If that rock happens to fall into a stream, the water will gently wear off small pieces. These pieces of rock are now called sediment and will be washed downstream. In time, this sediment is piled up and cemented together. It becomes a sedimentary rock. But, pressure and heat inside Earth may change the sedimentary rock into a metamorphic rock. In this way, all rocks on Earth are reused and recycled over millions and millions of years. This process is happening right now.

FIGURE 15-22

This lava in Hawaii is flowing into the ocean and cooling rapidly, causing the steam.

Section Wrap-up

1. Identify two factors that combine to produce metamorphic rocks.

2. Name examples of foliated and non-foliated rock samples, and explain the difference between the two types of metamorphic rocks.

3. **Think Critically:** Trace the journey of a granite through the rock cycle. Explain how this rock could be changed from an igneous rock to a sedimentary and metamorphic rock.

4. *Skill Builder*
   Observing Describe a part of the rock cycle you can observe occurring around you or see on television news. If you need help, refer to Observing and Inferring on page 550 in the **Skill Handbook.**

Using Computers

Using a spreadsheet program, create a data table to list and compare the properties of different rocks and minerals that you have studied in this chapter.

People & Science

Alice Derby Erickson, Pipestone Carver

Q **Ms. Erickson, tell us about yourself.**

A I work as a park ranger at the Pipestone National Monument in southwestern Minnesota. I'm a full-blooded Dakota Sioux and a pipestone carver.

Q **What is pipestone?**

A Pipestone is a soft, red stone made of silica and aluminum, also known as catlinite. It's about as hard as your fingernails.

Q **Why is pipestone important to Native Americans?**

A Many Native Americans believe that pipestone is the flesh and blood of our ancestors. For centuries, this sacred stone has been quarried here and carved into small figures and, of course, pipes. The pipe—called *channunpa* in Dakota—is used in most of our important ceremonies.

Q **How is the stone quarried?**

A Only Native Americans are allowed to quarry pipestone. All the work is done by hand. We use sledgehammers, picks, and shovels to remove the hard rock that lies over the pipestone. When we reach the pipestone layer, pieces of it are carefully pried out.

Q **Would you describe the carving process?**

A We first cut out a rough form, then shape it using various files, and smooth the surface with sandpaper. Finally, we heat the stone, apply beeswax, and put it in cold water. That sets the wax and brings out the rich, red color.

Q **What are your earliest memories of carving?**

A I started carving at about age ten. I helped my mother carve turtles. Turtles, to the Dakota people, are a symbol of long life. I'd cut the rough shapes of the turtles and then help my mother sand and finish them. The craft of pipestone carving passes from generation to generation. In my family, we are the fourth generation of carvers.

Career Connection

Think about careers that deal with natural materials. Interview another artist who uses rocks or minerals, and make a presentation to the class about him or her.

Read the statements below that review the most important ideas in the chapter. Using what you have learned, answer each question in your Science Journal.

1. All minerals occur naturally, are inorganic solids, and have a definite pattern of atoms and makeup. *List five properties that can be used to help identify a mineral.*

2. Intrusive igneous rocks form below Earth's surface and have large mineral crystals. *Why don't extrusive igneous rocks have large mineral crystals?*

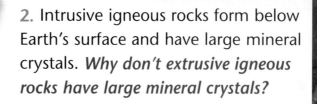

3. Sedimentary rocks can be formed from pieces of other rocks, organic material, and minerals that evaporate or settle out of solution. *How could the grains in a sedimentary rock like sandstone be moved and then deposited?*

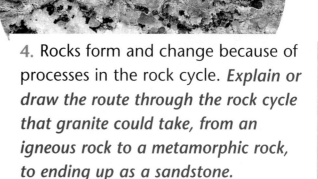

4. Rocks form and change because of processes in the rock cycle. *Explain or draw the route through the rock cycle that granite could take, from an igneous rock to a metamorphic rock, to ending up as a sandstone.*

Using Key Science Words

crystal	metamorphic rock
extrusive	mineral
foliated	non-foliated
fossil fuel	ore
gem	rock
igneous rock	rock cycle
intrusive	sedimentary rock

Match each phrase with the correct term from the list of Key Science Words.

1. type of igneous rocks that are cooled below Earth's crust
2. kind of rock made of pieces of other rocks
3. a rare, precious mineral that can be cut and polished
4. type of metamorphic rocks with layers of different minerals
5. diagram that shows how the formation of igneous, sedimentary, and metamorphic rocks can be interrelated

Checking Concepts

Choose the word or phrase that completes the sentence.

6. Rocks are usually composed of two or more _____.
   a. pieces   c. fossil fuels
   b. minerals   d. foliations

7. Metamorphic rocks are formed when _____.
   a. volcanoes erupt
   b. fissures ooze lava
   c. globs of lava cool underground
   d. heat and pressure change rocks

8. Sedimentary rocks can be classified as _____.
   a. foliated or non-foliated
   b. biochemical, chemical, or broken
   c. extrusive or intrusive
   d. gems or ores

9. _____ rocks are composed of broken pieces of rock.
   a. Sedimentary   c. Old
   b. Banded   d. Extrusive

10. Natural gas, coal, and oil are all _____.
    a. minerals
    b. metamorphic rocks
    c. fossil fuels
    d. igneous rocks

Thinking Critically

Answer the following questions in your Science Journal using complete sentences.

11. Is a sugar crystal a mineral? Why or why not?

12. Metal deposits in Antarctica are not considered to be profitable. List some reasons for this.

13. How could pieces of gneiss, granite, and basalt all be found in one conglomerate?

14. Would you expect to find a well-preserved dinosaur bone in a metamorphic rock like schist? Explain.

15. Explain how the mineral quartz could be in an igneous rock and in a sedimentary rock.

Developing Skills

If you need help, refer to the description of each skill in the Skill Handbook.

16. **Making and Using Tables:** Using the information that you collected in the MiniLAB on page 411, create a data table to compare and contrast the four minerals that you tested. Be sure to include color and luster in addition to the two tests that you performed.

17. **Comparing and Contrasting:** Compare and contrast the differences between intrusive and extrusive rock appearance.

18. **Interpreting Data:** Clem has several mineral samples and is doing his best to identify them. He has tested the hardness, streak, luster, and color of each of the samples but still can't figure out what the samples are. Suggest two other tests that may help. Explain why these tests would be useful.

19. **Observing and Inferring:** You are hiking in the mountains and as you cross a shallow stream, you spy an unusual rock. When you pick it up, you notice it is full of fossil shells. Your dad asks you what it is. What do you tell him and why?

Schist   Conglomerate   Granite

20. **Interpreting Scientific Pictures:** Review the pictures above and determine whether each is a sedimentary, igneous, or metamorphic rock.

Performance Assessment

1. **Display:** Create a display that shows the rock cycle using real rock specimens.

2. **Designing an Experiment:** You were able to make a model for metamorphism using clay and rice. Now experiment with baking soda and vinegar and design an experiment using these two materials to demonstrate how a volcano may erupt and produce igneous rocks. You may want to add red food coloring.

Chapter Preview

Skills Preview

▶ **Skill Builders**
- make models
- sequence

▶ **MiniLABs**
- observe
- make models

▶ **Activities**
- make models
- observe
- hypothesize
- collect data
- design an experiment
- compare results

Chapter 16

Earth's Structure

Next stop on your summer vacation: a Hollywood movie studio. You feel excited as you board a tram that will take you to see giant apes, spaceships, and shark attacks. As the tram starts, the track begins to shake, and buildings around you collapse. You're doomed, it seems. Just then, the tram starts to move and whisks you away to safety. You have just been through a fake earthquake. Thank goodness it was only make-believe. But for people who live in California and in many places around the world, the possibility of an earthquake is real, as in the case of the Loma Prieta earthquake in 1989 shown here.

EXPLORE ACTIVITY

Modeling Quakes

1. You and your lab partner are going to make two different kinds of waves, similar to those that happen during an earthquake. Stretch out a toy spring along the floor until the coils are about an inch apart. You and your lab partner should each hold one end.
2. Gather several coils together and then let them go. Observe the action of the wave as it travels down the spring.
3. Next, whip the end of the spring quickly to your right, keeping it on the floor. Note the difference in the shape of the wave.

Science Journal

In your Science Journal, draw pictures of the waves that you and your partner made with the spring. Describe the motion of the spring in each case.

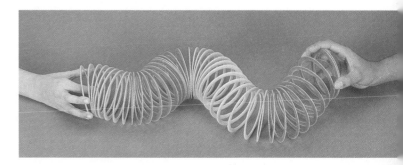

16•1 What's shaking?

What YOU'LL LEARN

- What happens in an earthquake
- How volcanoes erupt and where they occur

Science Words:
earthquake
fault
focus
epicenter
volcano

Why IT'S IMPORTANT

When you see news reports of earthquakes and volcanic eruptions, you'll understand what's causing these events.

Earthquakes

The surface of Earth is always changing. The outer layer of Earth is called the crust. The crust is firm and seems solid, but it is broken up into large pieces. These pieces can move around. Because the pieces do move, energy builds up wherever two pieces meet. The energy builds up in Earth like it does when you bend a stick. As you bend the stick, energy builds until the stick breaks and the energy is released. When that energy is released rapidly, Earth shakes or vibrates.

An **earthquake** is the shaking of Earth produced by the fast release of energy. The energy travels as waves through Earth, as shown in **Figure 16-1.** The amount of damage caused by an earthquake depends upon the size of the waves. Most earthquakes are so small that you can't even feel them.

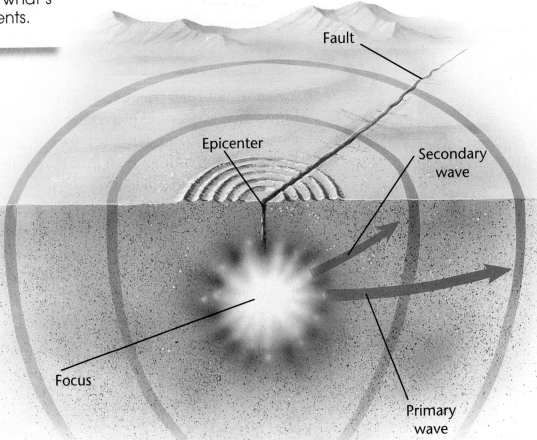

FIGURE 16-1

Earthquakes occur when sudden movement along a fault releases energy. The first earthquake wave is called the primary wave, and the second wave is called the secondary wave.

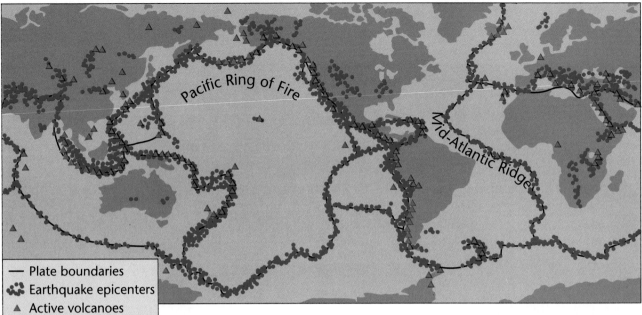

- — Plate boundaries
- ⬡ Earthquake epicenters
- ▲ Active volcanoes

Pacific Ring of Fire

Mid-Atlantic Ridge

FIGURE 16-2

Most of the earth-quakes and active vol-canoes occur along the edge of the Pacific Ocean. This area is known as the Pacific Ring of Fire.

Earthquake Epicenters

Have you noticed that earthquakes seem more common in certain parts of the world than in others? Places such as Japan, China, and Mexico have many earthquakes. The pattern of where earthquakes happen has to do with how Earth's crust fits together. The outer layers of Earth, which include the crust, are broken into large pieces or sections called plates. As shown in **Figure 16-2,** earthquakes are more common where the edges of the giant plates meet. Scientists believe that these plates have been moving for millions and millions of years and continue to move today.

Faults

When the plates rub against one another, energy builds up along areas called faults. A **fault** is like a crack in the crust between the two pieces, where there has been movement. You can think of it like a crack in the sidewalk, but much bigger.

When there is an earthquake, energy moves away from the break in waves, like ripples on a pond. The exact spot inside Earth where the earthquake starts is called the **focus,** as shown in **Figure 16-1.** The waves travel in every direction including upward. The point where the waves hit the surface of Earth, directly over the focus, is called the **epicenter** (EP uh sen tur). A great deal of damage often occurs at the epicenter in a big earthquake.

FIGURE 16-3

Earthquakes are measured in two ways.

A This model shows how earthquake waves travel through Earth. The primary wave (P-wave) is recorded first because it travels faster. Secondary waves (S-waves) are slower and are recorded later.

Perhaps you've seen pictures of crumbled buildings and broken freeway overpasses in cities where strong earthquakes have happened. The strength or magnitude of an earthquake is usually measured in two ways: by the Richter (RICK tur) scale, and by the Modified Mercalli (mur KAH lee) scale shown in **Figure 16-3**. The energy released by an earthquake is measured by the Richter scale, and the damage caused is measured using the Modified Mercalli scale.

Richter Scale

The first scale measures the size or magnitude of an earthquake. This scale was developed by Charles Richter. Richter's scale measures the size of the waves produced by an earthquake. He used an instrument called a seismograph (SIZE muh graf), shown in **Figure 16-3B**, to measure the wave magnitudes. Stronger earthquakes make larger waves that can

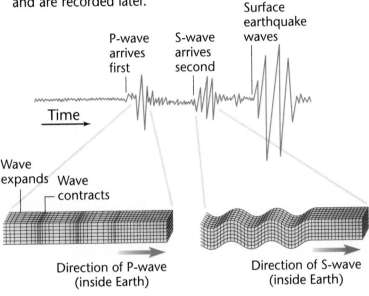

B A seismograph is an instrument that records the wave magnitudes from earthquakes.

C The Modified Mercalli scale is used to measure the damage caused by earthquakes.

Portion of the Modified Mercalli Scale

I. (1) Earth movement is not felt by people.

II. (2) A few people may notice movement if they are sitting still. Hanging objects may sway.

V. (5) Felt by almost everyone. Sleeping people are awakened. Some windows are broken and plaster cracked. Some unstable objects are overturned. Bells ring.

X. (10) Some well-built wooden structures are destroyed. Most masonry structures are destroyed. Ground is badly cracked.

XI. (11) Few, if any, structures remain standing. Broad open cracks appear in the ground.

XII. (12) Destruction is complete. Waves are seen on the ground surface.

be measured by a seismograph. The Richter scale is used worldwide. News reports of earthquakes often mention where the earthquake registered on the Richter scale.

The most obvious way to measure the strength of an earthquake is to look at the damage caused by the quake. Scientists use the Modified Mercalli scale to measure damage. The scale ranges from a low of 1, where you barely notice any damage, to a 12, where buildings are totally destroyed. In an earthquake rated 12 on this scale, the ground moves in rolling waves, and objects on the ground can be sent flying through the air.

Problem Solving

Identifying Epicenters

When an earthquake happens, two types of waves, P (primary) and S (secondary) waves, are produced. P-waves travel faster than the S-waves. P-waves move faster because of how the wave moves through Earth, as seen in **Figure 16-3.** A seismograph records both kinds of waves, but the P-waves arrive first. The amount of time between when the two waves arrive at the station tells how far away the epicenter is. For example, if a P-wave reached the station at 9:00, and the S-wave arrived 4 minutes and 1 second later, from a graph you can calculate that the earthquake epicenter was 2000 km from the station. If you draw a 2000-km circle around that station, the epicenter would lie somewhere along that circle.

Solve the Problem:
Use the information given to find an earthquake epicenter. If the epicenter is 6000 km from station 3, is the epicenter at point A or at point B?

Think Critically:
Explain why you need recordings from three stations to pinpoint the epicenter. Why wouldn't just two stations work? Are four stations better?

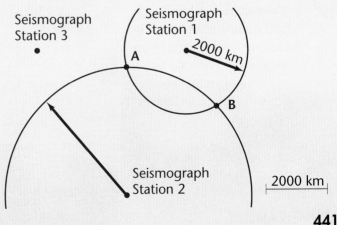

Mini LAB

Constructing a Model Volcano

How is molten rock forced up through Earth's crust?

1. Pour water into a 1-L plastic soda bottle until it is one-third full. Add a large squirt of liquid soap and 3 tablespoons of baking soda. Cap the bottle and shake it for 30 seconds.
2. Place the bottle in the middle of a cookie sheet. Remove the cap, and quickly add 125 mL of vinegar.
3. Observe the reaction.

Analysis

1. The mixture of vinegar and baking soda formed the gas carbon dioxide. Explain why the carbon dioxide gas erupted out of the mouth of the bottle instead of staying inside.
2. Compare and contrast the reaction that you observed with the formation of a real volcano.

FIGURE 16-4

There are three types of volcanoes: composite, shield, and cinder cone. *Which of these three types of volcanoes do you think would be the most dangerous to live near? Explain.*

 Anak Krakatau in Indonesia is a cinder cone volcano. Cinder cone volcanoes have steep sides and are made up of small blocks of solid lava.

Volcanoes

Can you think of another event on Earth that can cause such damage? What about an erupting volcano? Rivers of hot lava rush along, wiping out forests and destroying towns. Clouds of gas and ash spew into the air. Volcanoes, like earthquakes, are signs that Earth is active and moving. **Volcanoes** are places in the world where the hot, liquid magma, which is molten rock from below the crust, breaks through and flows onto the ground. Like earthquakes, volcanoes are also most common along the edges of Earth's plates.

Eruptions

Volcanic eruptions happen when magma is forced up through the surface by pressure inside Earth. The magma collects in a chamber just under the surface. When enough pressure builds up in the chamber, lava, which is what magma is called at the surface, erupts through an opening or vent. This eruption shoots lava, blocks of rock, cinders, ash, and hot gases into the air. These fall down around the vent, forming a small mound. As this happens over and over, a cone builds up around the opening, and a volcano is born.

Different Volcanoes

Scientists use several things to classify volcanoes. Some volcanoes are tall and pointy. Others have long, gentle slopes. Some are made of sticky lava, while some are mostly small chunks of solid lava called cinders. The kind of magma from which the volcano forms and the amount of water in the magma determine the type of volcano. But the main difference among them is how easily they flow. Some magmas are thick and sticky, like cooked oatmeal, and don't flow easily. They produce violent, explosive eruptions. Other magma is more fluid or liquid, like honey, and flows more easily. Magma that flows easily produces quieter lava flows.

Composite and Shield Volcanoes

Figure 16-4 shows three types of volcanoes: composite, shield, and cinder cone. Composite volcanoes—the tall, pointy ones—are made up of layers of thick, sticky lava and ash. Composite volcanoes, such as Mount St. Helens in Washington, are formed by alternating quiet and explosive eruptions. Shield volcanoes, such as Kilauea in Hawaii, are made up of long, sloping layers of rock that formed from quiet eruptions of fluid lava. The third type of volcano is known as a cinder cone. This type is formed by explosive eruptions. Krakatau in Indonesia is an example of a cinder cone volcano. When it erupted in 1883, the explosion was heard 2000 km away in Australia!

C Composite volcanoes like Mount Ruapehu in New Zealand form tall, pointed mountains. They are made of layers of thick, sticky lava and ash.

B Mauna Kea in Hawaii is a shield volcano, with long, sloping sides.

Activity 16-1

Lava Landscapes

In this lab, you will experiment with common, everyday materials and use them to create models of different kinds of lava. You will compare these homemade lavas by racing them down a cookie sheet to help you understand why volcanic eruptions are different.

What You'll Investigate
How do the three kinds of lava flow?

Procedure

1. **Set** the book on your lab table, and **place** the cookie sheet on the book at an angle. You now have a long, gentle slope for your lava to flow down.
2. Each person in your group will make one kind of lava. You will then race the lava down the cookie sheet. Your teacher will provide the lava recipes for the rhyolite (RI uh lite), andesite (AN duh site), or basalt (buh SAWLT) lavas. Each homemade lava will model how real lava flows.
3. **Prepare a data table** where you **record** the distance the flow travels in centimeters. Also time the flow in seconds and **describe** the features of each kind of lava.
4. **Leave** the lavas in the cups until you are ready to race.
5. When all three lavas are prepared, you are ready to experiment.
6. **Place** all three kinds of lava at the top of the tilted cookie sheet. **Compare** the rates that they flow downhill. **Match** the lava with the different kinds of volcanic landforms that can be found around the world.

Conclude and Apply

1. Which lava model—rhyolite, andesite, or basalt—was the most resistant to flow?
2. Which model of lava would be likely to produce a gently sloping lava plain? **Explain** your answer.
3. The amount of silica (liquid glass) in lava helps control how it will flow. The more silica, the stickier the lava. **Sequence** the three kinds of lava from silica-rich to silica-poor based on your **observations**.

Goals
- Model lavas.
- Observe how different lavas flow.

Materials
- cookie sheet
- 6-ounce paper cups (3)
- dehydrated silica gel
- cornstarch
- white glue
- spoon
- thick book
- stopwatch
- meterstick

Where do volcanoes occur?

Geologists have mapped the volcanoes and earthquakes of the world, as shown in **Figure 16-2** on page 439. Looking at the map, do you see any areas with a lot of volcanic activity? One especially active area circles the Pacific Ocean. It has been called the Ring of Fire. Most of the world's active volcanoes are found in this ring around the Pacific Ocean. This area of active volcanoes gives clues to the arrangement of Earth's plates. As with earthquakes, volcanoes occur more commonly where two of Earth's plates meet. What else can scientists learn from volcanoes? Volcanoes can also give information about the inside of Earth. The lava that erupts out of volcanoes, as shown in **Figure 16-5**, came from rock that melted deep inside Earth.

FIGURE 16-5

This eruption in the Galápagos Islands occurred along the Pacific Ring of Fire.

Section Wrap-up

1. Draw a diagram of an imaginary earthquake. Label the focus, epicenter, and fault on your illustration.

2. List the three kinds of volcanoes, and briefly describe why they look different.

3. **Think Critically:** A strong earthquake happens in an area without tall buildings or freeway structures. What other kinds of damage could you look for in order to rank the quake on the Modified Mercalli scale?

4. *Skill Builder*
   Building Models Make a model that shows a slice through a volcano. Explain how the volcano was formed and which type of magma was the likely building material. If you need help, refer to Making Models on page 557 in the **Skill Handbook.**

USING MATH

Mount Mazama exploded about 6900 years ago, creating Crater Lake in Oregon. It is estimated that the eruption produced 1 275 000 tons of ash and rock. Assume 1 ton of rock and ash takes up 25 m³ of space. How much total space did the material exploding out of Mount Mazama take up?

16•2 Predicting Earthquakes

What YOU'LL LEARN

- About earthquakes in California
- How to use library and Internet resources

Why IT'S IMPORTANT

You'll be able to use and understand scientific resources.

If you wanted to learn more about earthquakes, where would you go? A library would be a great place to find out about how and where earthquakes happen. In an encyclopedia, you would learn that earthquakes are caused by movement of Earth's crust along faults. The fault you have probably heard the most about is the San Andreas Fault in California, as shown in **Figure 16-6.** This fault is part of a group of hundreds of faults that crisscross much of California. The thousands of earthquakes that happen in California each year are the result of movements along these faults. Scientists think that the next great earthquake in California will happen along this group of faults. When will it happen?

Shake, Rattle, and Roll

Of all the major earthquakes worldwide in the 20th century, about one fifth have taken place in California. In 1906, a great earthquake almost destroyed San Francisco. In 1971, an earthquake in San Fernando, California, left nearly 60 people dead. Another major earthquake struck California in 1994, at Northridge, causing freeway overpasses to collapse. If you wanted to know if any earthquakes had happened last month, where would you find out? Newspapers and magazines are good sources in which to find information about recent events.

Future Earthquakes

Many of the large California earthquakes have caused a lot of damage, as shown in **Figure 16-7.** But scientists and California residents are still waiting for an earthquake as big as the 1906 San Francisco quake. Scientists are trying to predict when and where the next big earthquake will happen. They study the land around the major faults, as well as California's earthquake history.

FIGURE 16-6

This photo shows an aerial view of the San Andreas Fault in California.

They think that a major quake—with a magnitude similar to the 1906 San Francisco quake—will happen in the next 20 to 30 years.

To predict major earthquakes, scientists are using various technologies. They use lasers and satellites to study small movements along faults. Scientists need to know where the faults are located and which ones are the most likely to cause major damage if they move. It is important to know whether hospitals, police departments, and fire stations will be able to operate after a big quake. This information will make people more prepared for a big earthquake.

Earthquake Information

Where could you get information on recent earthquakes directly—from city officials or from scientists? You can use electronic mail to talk with scientists, or use the Internet to learn about where the faults are located and about past quakes. You may be able to find out about a quake anywhere in the world on the same day it happens!

FIGURE 16-7

This damage was caused by the Loma Prieta earthquake in 1989.

 Skill Builder: **Using Resources**

LEARNING the SKILL

1. Encyclopedias and other reference books in libraries have general information on most subjects.

2. Magazines, journals, and newspapers usually cover fewer topics in depth.

3. The Internet has search engines that locate information you identify.

PRACTICING the SKILL

1. Which of the resources listed would probably be a good source for information about faults?

2. Which resource—an encyclopedia or a newspaper—would have the most recent information about earthquakes? Explain.

3. Where would you look for details about the 1906 San Francisco quake?

APPLYING the SKILL

Find out more about California earthquakes using the Internet and at least two different types of library resources.

16•3 A Journey to Earth's Center

16•3 A Journey to Earth's Center

What YOU'LL LEARN

- What the inside of Earth looks like
- About the evidence that supports the theory of continental drift

Science Words:
mantle
inner core
outer core
continental drift

Why IT'S IMPORTANT

You'll increase your knowledge about what's inside Earth.

FIGURE 16-8

This scientist is studying the rocks from a volcano to learn more about Earth's interior. *Why can't scientists study Earth's interior directly?*

What's inside Earth?

Imagine that you could get into an elevator and ride down to Earth's center. What would you see? It's impossible to do such a thing—Earth's interior is much too hot, but you can learn a lot about the inside of Earth by studying earthquake waves. We also gather clues about the inside of Earth by looking at landforms such as volcanoes, deep-sea trenches, and mountains.

But why does anyone need to know what is inside Earth? Scientists study Earth's interior for many reasons. A major reason to study what's inside the planet is to learn more about earthquakes and volcanoes. Why do they happen and why do they happen where they do? Scientists hope to accurately predict earthquakes and volcanic eruptions to help save lives. Scientists also learn about how Earth was formed and why it looks the way it does today.

Using Volcanoes and Earthquakes

Scientists use information and materials from volcanoes and earthquakes to help them make a model of Earth's interior. Material shot out of volcanoes may have come from deep within Earth. It gives scientists clues to

what the inside of the planet is made of, as shown in **Figure 16-8.** When an earthquake happens, scientists everywhere know about it because the waves are sent all over the world. Seismographs all over the world record the waves and give scientists information about the earthquake. The people studying earthquakes discovered that these waves do not travel through Earth in a regular way or even at the same speeds. At certain places inside Earth, the waves are bent. This is similar to how light waves are bent when they go from air into water. The bending of the earthquake waves gave scientists the first clue that Earth is made up of layers that have different thicknesses and physical properties.

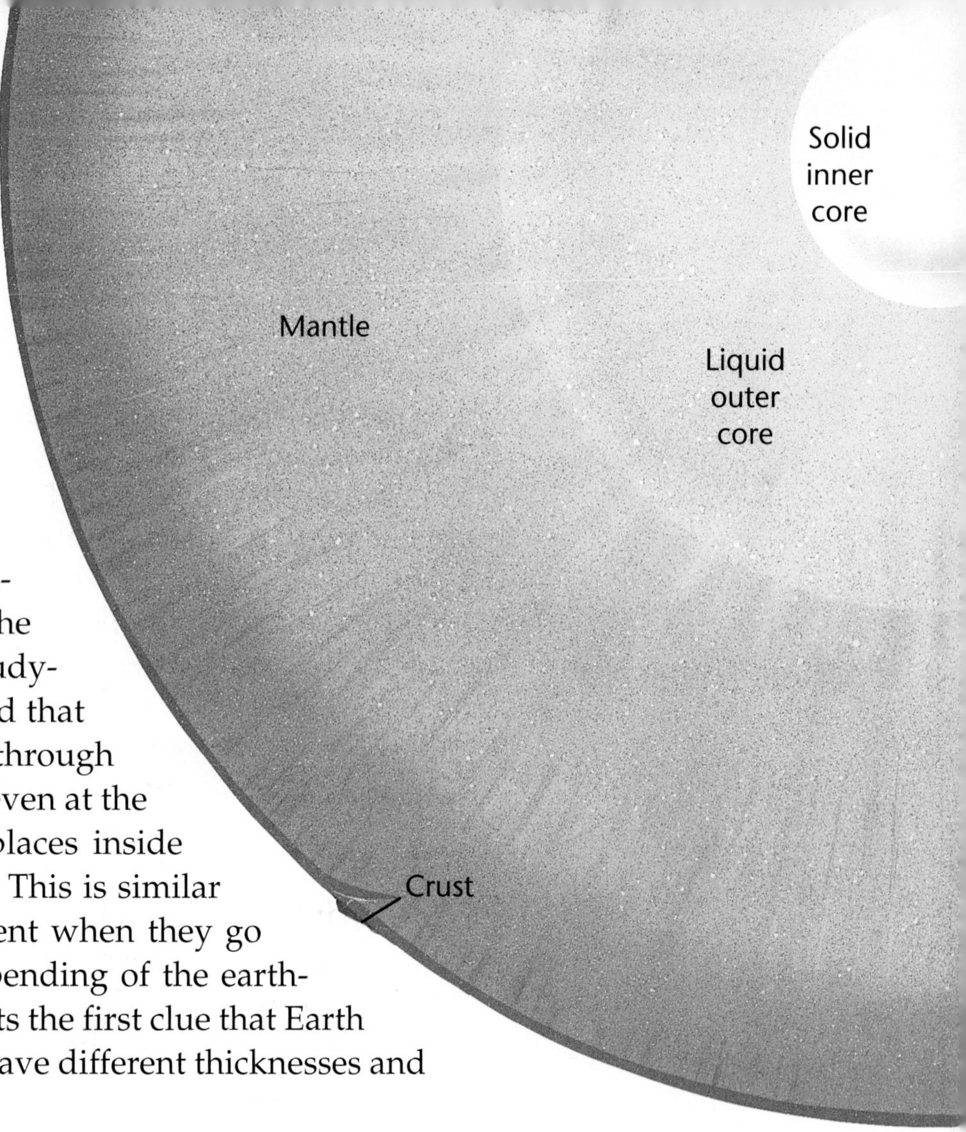

Mantle

Crust

Solid inner core

Liquid outer core

Earth's Layers

Primary (P) waves and secondary (S) waves, as shown in **Figure 16-3,** from earthquakes travel at different speeds through Earth. Both kinds of waves slow down in a plastic-like layer under the crust. This layer is weak and can move or flow like warm tar. This plasticlike layer is part of the thickest layer of Earth called the **mantle,** as shown in **Figure 16-9.** When earthquake waves are bent or slowed down, scientists know that the waves have hit a layer within Earth that has different properties. From these data, scientists are able to infer that Earth must have a solid **inner core** surrounded by a liquid **outer core.** Scientists can tell the outer core is liquid because S-waves are completely stopped at this layer. They already knew S-waves could not travel through fluids. It has taken many years to get a clear picture of the inside of Earth, and scientists learn more every day.

FIGURE 16-9

Earth's interior is made of four major layers: crust, mantle, outer core, and inner core.

SCIENCE
Online

Go to Glencoe Science Online, *science.glencoe.com,* for a link to volcanoes.

Design Your Own Experiment
Making a Seismograph

An earthquake is the vibrations caused by sudden movements of Earth. The sudden movements occur as energy that has built up in Earth's crust is released. The more energy that builds up, the greater the magnitude of the earthquake that occurs. A measure of the vibrations caused by the release of energy can be used to measure the size of the earthquake.

Possible Materials

- ring stand with ring
- wire hook from coat hanger
- masking tape
- sheet of paper
- piece of string
- 2 rubber bands
- fine-tip marker
- metric ruler

PREPARE

What You'll Investigate
How can you measure the size of vibrations?

Form a Hypothesis

If you pound on the top of your desk, you can feel vibrations in the desk. On a much larger scale, this is what happens when an earthquake occurs. If you tried to draw a straight line while a classmate pounded on the table, your line would end up wavy. Using this analogy, create a model that can measure vibrations, and **form a hypothesis** about how your model instrument could be used to measure these vibrations.

Goals

Design an instrument and use it in an experiment that models how vibrations produced by earthquakes are measured.

Safety Precautions

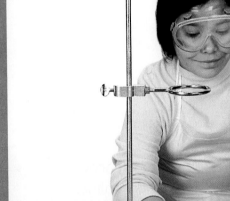

PLAN

1. Agree upon and design a **model** that will measure vibrations.
2. Agree upon and write out your **hypothesis** statement.
3. List the steps needed to build your model and to test your hypothesis. Be specific. Describe exactly what you will do.
4. **Prepare a data table** in your Science Journal.
5. **Read** over your plan to make sure your model can be used in your experiment to test your hypothesis.
6. What **variable** will you be testing? What **variable** will you need to control?
7. Make sure you **measure** and **record** carefully.
8. **Read** over the concluding questions to see if you have included all the necessary steps in your experiment.

DO

1. Make sure your teacher approves your plan and data table before you proceed.
2. **Build** your model and carry out the experiment as planned.
3. While doing the experiment, **record** any observations that you make in your Science Journal.

CONCLUDE AND APPLY

1. **Analyze** your data. Which trial resulted in the most wavy line on your model?
2. How did the movement of the suspended marker **compare and contrast** with the movement of the frame of your model instrument?
3. **Infer** whether the results of your experiment support your original **hypothesis.**
4. **Determine** the cause-and-effect relationship between the size of vibrations caused by pounding on the desk and how wavy the line was.

FIGURE *16-10*

The continents have moved over millions of years from their positions in Pangaea to their present-day positions.

Continental Drift

Has anyone ever made fun of you for trying something unusual or for talking about a different idea? The same thing happens to scientists! Sometimes it takes a long time to find evidence to support new ideas. Many great discoveries and famous scientists were thought to be wrong at first. One example comes from the life of a German scientist named Alfred Wegener (VEG nur).

In 1912, he proposed the idea of a supercontinent that he named Pangaea (pan JEE uh). The word *Pangaea* means "all lands." He looked at a map of the continents as they were at that time and noticed that they could be arranged to fit together. Look at the eastern coastline of South America and the western coastline of Africa, as shown in **Figure 16-10C.** Don't they look like they could fit together as puzzle pieces? Wegener recognized this. He proposed that all the continents had once been connected to form Pangaea. Then, he said, Pangaea broke apart. The continents slowly moved to their present positions and are still moving. He called this the theory of continental drift. **Continental drift** is the movement of continents. Most other scientists at that time thought his theory was totally wrong.

A The fossil plant *Glossopteris* has been found on most of the continents that formed Pangaea.

B About 250 million years ago, the continents were joined to form the supercontinent Pangaea.

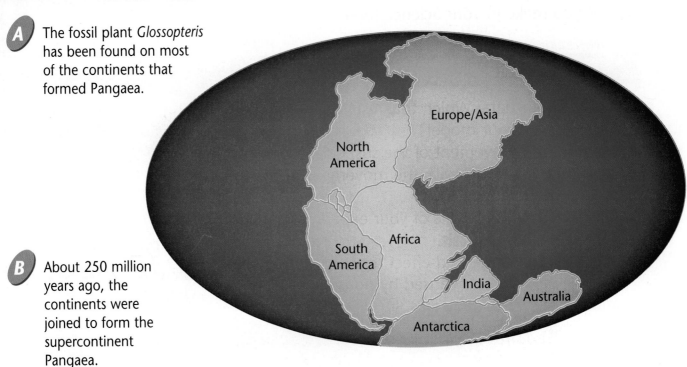

Fossil and Climate Clues

In addition to the puzzlelike fit of the continents, Wegener found other clues. He noticed that many similar fossils were found on the now-separated continents. For instance, fossils of the plant *Glossopteris* (glaws OP trus) were found in Africa, South America, Antarctica, and Australia. And fossils of the reptile *Mesosaurus* were found in Africa and South America. How did these organisms get to these widely separated places? *Mesosaurus* was a freshwater and land animal, so it is unlikely that it swam across oceans. And how did the plant get from one place to another? Using fossils from Antarctica, Wegener also found evidence that it had once had a tropical climate. He found fossils in cold, icy Antarctica of organisms that today live in warm, tropical climates. He also found that glaciers existed long ago in what are now Africa and South America. He thought the best way to explain these things was that the continents had once been together.

C This illustration shows the present-day positions of the continents.

D The fossils of the reptile *Mesosaurus* have been found in South America and Africa, landmasses that are now widely separated.

North America

Europe

Asia

Africa

India

South America

Australia

Antarctica

FIGURE 16-11

Rocks and fossils in the Appalachian Mountains along the eastern coast of North America can be matched with similar rocks, fossils, and mountains across the Atlantic Ocean in Europe and Africa.

USING TECHNOLOGY

Satellite Tracking

Wouldn't it be great to find out exactly where you are anywhere on Earth—from Mount Everest to Miami Beach—at the touch of a button? A network of satellites called the Global Positioning System (GPS) lets you do just that.

No More Getting Lost

GPS has changed forever how people find their way. Hikers use GPS to keep from getting lost. Scientists use GPS to make accurate maps of fossil discoveries, rock outcroppings, and other land features.

Satellite Signals

Twenty-four GPS satellites orbit Earth twice a day. Each satellite sends a nonstop time and location signal. A handheld GPS unit uses signals from at least four satellites to find where a spot is located to within a few meters. It also tells how high above sea level the GPS unit is located.

SCIENCE *Online*

For some other everyday uses of GPS technology, check out the link at Glencoe Science Online.
science.glencoe.com

Rock Clues

Wegener also found similar rocks on the different continents. You can walk along the hills in Scotland following a distinctive layer of rock. Then if you traveled to eastern Canada, you could find a similar rock layer in the hills of Newfoundland. About 250 million years ago, these two areas, which are now separated by the Atlantic Ocean, were joined together. Mountain ranges could be traced across oceans from one continent to another. In all, Wegener found a lot of evidence to support his idea—similar fossils, similar rocks, similar climates, and similar mountain ranges, as shown in **Figure 16-11.** The only problem was that he could not explain how these huge landmasses could have moved. Because he could not explain how it happened, few scientists took his idea seriously. It wasn't until years after his death that scientists using modern technology found new evidence of continental drift to support his ideas.

Today, his ideas form the basis for most of our knowledge of how the continents have changed through time. Next, we'll learn about the new evidence of continental drift.

Section Wrap-up

1. Explain how scientists used earthquakes and volcanoes to infer the structure of Earth's interior.

2. List and discuss the evidence Wegener used to support the theory of continental drift.

3. **Think Critically:** Where would you look in North America to find evidence for Wegener's theory of continental drift?

4. *Skill Builder*
   Making a Model Build a model of the inside of Earth. If you need help, refer to Making Models on page 557 in the **Skill Handbook.**

Science Journal

In your Science Journal, write a short story about what you would see if you could take a trip to the center of Earth. Use your imagination, but be scientifically accurate.

16•4 Crashing Continents

What YOU'LL LEARN

- How new crust forms
- How plate tectonics changes the surface of Earth
- How mountains are formed

Science Words:
plate
seafloor spreading
convection current
plate tectonics

Why IT'S IMPORTANT

You'll understand how the continents move and mountains form.

Earth's Plates

Nearly 90 years after Wegener proposed the theory of continental drift, we have a much better understanding of his ideas. In that time, scientists have gathered more evidence to support his theory.

Our understanding of the upper layers of Earth changed a lot in the late 1900s. Scientists now think that Earth's crust and the upper part of the mantle are made up of 12 large sections called **plates.** The plates can move because they rest on a plasticlike layer within the mantle. The plates can move around on this layer. Imagine standing on pieces of plywood that have been tossed onto a muddy field. As you walk across the pieces of wood, they slip and bump into the pieces next to them. The plates on Earth are doing the same thing, but in slow motion.

FIGURE 16-12

Exploration of the oceans in the 1950s and 1960s revealed a system of ridges and valleys on the ocean floor.

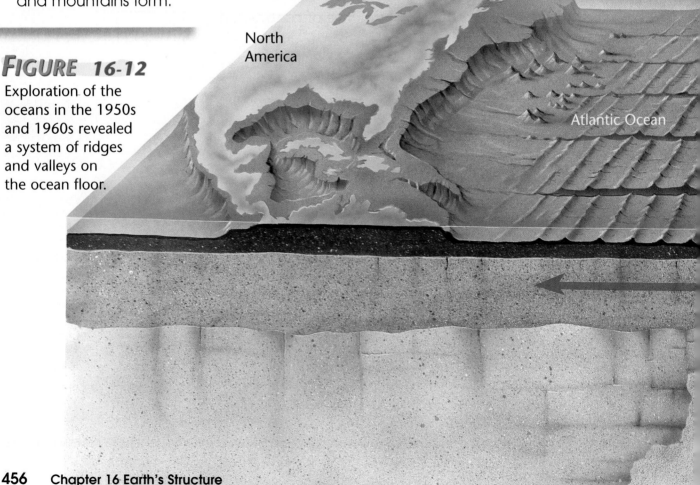

North America

Atlantic Ocean

Seafloor Spreading

Wegener had a problem. He couldn't explain how the continents moved. Scientists have now discovered how the plates move. Exploration of the ocean revealed areas where molten rock or magma oozes out onto the ocean floor. These areas, called mid-ocean ridges, have a gap or valley in the center. If nothing fills the space between the plates, the gap will remain. But as plates move farther apart, magma oozes up and hardens in the gap, adding to the edges of the plates. As more magma moves up onto the surface, the two sides of the ridge are pushed farther apart. In this area, new ocean crust forms as molten rock cools. This process that forms new ocean crust is called **seafloor spreading.** In **Figure 16-12,** you can see that the North American plate is moving slowly away from the European and African plates.

A The North American and South American plates are moving away from the European and African plates. The Atlantic Ocean is slowly getting wider at the rate of about 1.25 cm per year.

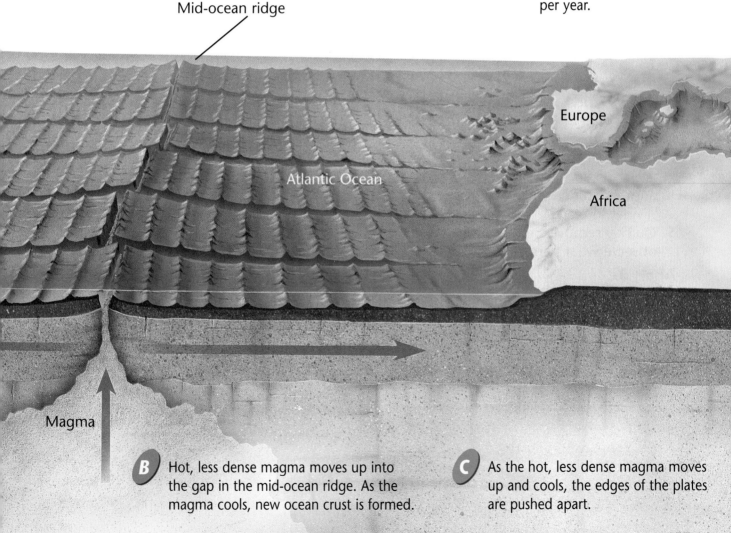

Mid-ocean ridge

Atlantic Ocean

Europe

Africa

Magma

B Hot, less dense magma moves up into the gap in the mid-ocean ridge. As the magma cools, new ocean crust is formed.

C As the hot, less dense magma moves up and cools, the edges of the plates are pushed apart.

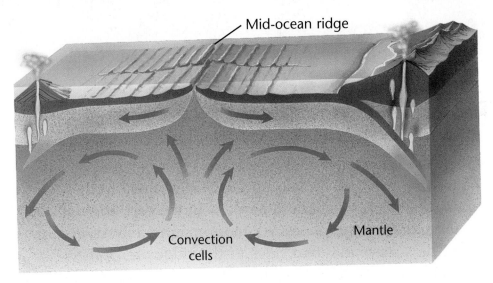

FIGURE 16-13

Convection cells form where hot, less dense magma is forced upward. As the magma cools, it becomes more dense and sinks back down into Earth.

Mid-ocean ridge

Convection cells

Mantle

Mini LAB

Observing Convection

1. Place a plastic divider in a large clear rectangular container or an aquarium.
2. Prepare a separate container with 10 liters of ice water. Add blue food coloring and stir.
3. Add red food coloring to 10 liters of hot water in a second container.
4. With the help of another student, pour both containers of water into the aquarium at the same time. Have another student hold the plastic divider in place.
5. Quickly remove the divider and observe what happens to the water.

Analysis

1. Draw and describe what happens when you remove the divider.
2. Explain why the warm, red water and the cooler, blue water moved as they did.
3. Using this lab as a model, explain how magma would rise and possibly produce volcanoes.

Convection Currents

The motion of magma in the plastic-like layer, just under the crust, causes the movement of the plates. Hot magma is forced upward toward the surface, becoming cooler and more dense as it rises. Then the cooler, dense magma starts to sink back down into the mantle. Deep in the mantle, the magma is heated again. This cycle repeats over and over and is called a **convection current**, as shown in **Figure 16-13.** A convection (kun VEK shun) current works just like hot air from a furnace. Hot air in a room gets pushed up by the cool air under it because the warm air is less dense. Then as it cools, it becomes more dense and it sinks to the floor. As convection currents move around in the plasticlike layer, they drag the stiff plates along with them.

Plate Tectonics

As scientists gather new data, they sometimes blend old and new ideas into new theories. That's what happened in the 1960s. Scientists combined the ideas about continental drift that Wegener had discovered with the evidence of seafloor spreading and came up with the theory of **plate tectonics** (tek TAW nihks). Their new understanding of seafloor spreading and convection currents gave scientists an explanation of how the plates move. Volcanoes and earthquakes are the direct result of the movement of Earth's plates as the plates ride on the giant convection currents.

How do mountains form?

You have learned about how plate tectonics changes the ocean floor because of seafloor spreading. But how is the land changed by plate tectonics? Have you ever wondered how mountains are formed? Can we also use the theory of plate tectonics to explain how mountains form over many millions of years? From the rugged Rocky Mountains, as shown in **Figure 16-14,** in western North America to the gentle Appalachians along the eastern side of North America, mountains form where two of Earth's plates come together. The type of mountain that forms depends on what happens when plates meet.

FIGURE 16-14
Mountains form where two of Earth's plates come together. These mountains are part of the Rocky Mountains in Colorado.

FIGURE 16-15

There are four main kinds of mountains: folded, volcanic, fault-block, and upwarped. The kind of mountain depends on how the mountains were formed.

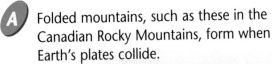

A Folded mountains, such as these in the Canadian Rocky Mountains, form when Earth's plates collide.

B Volcanic mountains, such as Mount Hood in Oregon, can form when one plate gets pushed under another. As the plate melts, the magma rises and volcanoes can form.

C Fault-block mountains form large sections of Earth's crust and are faulted and uplifted, such as the Sierra Nevada Mountains in California.

D Upwarped mountains, such as Mount Princeton in Colorado, form when Earth's crust is stretched and pushed up by forces inside Earth.

Kinds of Mountains

Scientists classify mountains by the way they form. Folded mountains are formed when plates collide. Volcanic mountains arise in areas where one plate gets pushed under another. As a plate gets pushed down into the mantle, the plate starts to melt. Magma is then pushed up through the crust and volcanoes form. This is happening now along the west coast of the state of Washington, forming volcanic mountains like Mount St. Helens. Fault-block mountains arise when whole sections of rock are faulted and broken. When this happens, some of the blocks get pushed up to form mountains. Upwarped mountains, as shown in **Figure 16-15D,** are formed when forces inside Earth, such as rising magma, cause the crust to bulge. Almost all mountains are found at or near the boundaries between two different plates. As you've learned, heat from within Earth drives the movements of the plates. This heat also forms mountains, opens oceans, and causes volcanic eruptions and earthquakes. Earth really is an active planet!

Section Wrap-up

1. What are Earth's plates, and how do they move?

2. List the four kinds of mountains, and describe how they were formed.

3. **Think Critically:** The highest mountain in the world is Mount Everest between Tibet and Nepal. Mount Everest is 8846 m high. Estimate the height of your school. How many "schools" high is Mount Everest?

4. *Skill Builder*
   Sequencing Volcanic mountains are produced near the boundary of two colliding plates, with one plate sliding under the other. Draw your own pictures showing the formation of a range of volcanic mountains. Put the pictures in their proper sequence to show the growth of the mountains. If you need help, refer to Sequencing on page 543 in the **Skill Handbook.**

USING MATH

Based on a seafloor-spreading rate of 5 cm/year, show how far apart two continents will move each 1000 years for the next 20 000 years.

Pacific Crossing
by Gary Soto

"Tokyo's like America," Lincoln said, smiling and trying to make conversation. "You know, we even have our own Cherry Blossom Festival. In San Francisco."

"Yes," Mr. Ono said, braking so hard that Lincoln had to hold on to the dashboard. "Yes, yes." A car was stalled in the left lane. Mr. Ono wiggled his steering wheel as he maneuvered dangerously into the next lane.

"I'm from San Francisco," Lincoln continued. "We're right on a bay like Tokyo."

"Yes, but America is very large," Mr. Ono said . . . "It is big as the sky."

Lincoln Mendoza is a 14-year-old Mexican-American boy from San Francisco who travels to Japan on an exchange program to study shorinji kempo, a martial art. For six weeks, Lincoln lives with his host family, Mr. and Mrs. Ono, and their son, Mitsuo, in the small farming village of Atami outside Tokyo. Both San Francisco and Tokyo are cities that have many earthquakes. As far as the number of earthquakes and their magnitude are concerned, Tokyo and San Francisco could be sister cities.

Before crossing the Pacific, Lincoln had imagined Japan would be all snow-capped mountains, women in kimonos, and temple-like dojos where he would practice kempo on neat straw mats. Instead, he finds that Japan is full of surprises. Tokyo is a lot like San Francisco. But beyond the city, Lincoln discovers small farms, public baths, hard-working people . . . and a dojo that looks like a driveway!

The Onos have misconceptions about the United States, too. Eager to share his culture, Lincoln teaches Mitsuo American slang, and succeeds—almost—in making a real Mexican dinner. But can he succeed in mastering kempo before he must cross the Pacific again?

Science *Journal*

Tokyo and San Francisco are known for their frequent earthquakes. Both cities lie along the Pacific Ring of Fire. In your Journal, draw a map of the Pacific, and show the locations of Tokyo and San Francisco. Label other cities in the Ring of Fire on your map.

Read the statements below that review the most important ideas in the chapter. Using what you have learned, answer each question in your Science Journal.

1. Forces inside Earth cause earthquakes and volcanoes. *Where do most earthquakes occur?*

2. Information on earthquakes in California can be found in the library and on the Internet. *How can damage be minimized during an earthquake?*

3. Information from earthquakes and volcanoes helped scientists understand Earth's interior. *What are the four major layers of Earth?*

4. Mountains commonly form along Earth's plate boundaries. *What type of mountains are formed from the collision of two plates?*

Using Key Science Words

continental drift	mantle
convection current	outer core
earthquake	plate
epicenter	plate tectonics
fault	seafloor spreading
focus	volcano
inner core	

Match each phrase with the correct term from the list of Key Science Words.

1. the surface of Earth directly above the focus of an earthquake
2. cone-shaped mountain formed by igneous activity
3. the thickest layer inside Earth
4. shaking of Earth produced by movement along a fault
5. large section of Earth's crust and upper mantle

Checking Concepts

Choose the word or phrase that completes the sentence.

6. The movement of Earth's plates over millions of years is called _____.
   a. a convection current
   b. plate tectonics
   c. an epicenter
   d. a fault

7. The movement of the plates is believed to be caused by _____.
   a. convection currents
   b. volcanoes
   c. earthquakes
   d. lavas

8. Earthquakes are caused by movement along a feature called a(n) _____.
   a. epicenter
   b. river
   c. landfill
   d. fault

9. Earthquake waves are produced at the _____.
   a. epicenter
   b. plasticlike layer
   c. extinct volcanoes
   d. focus

10. Volcanoes that are formed by quiet eruptions of fluid lava are _____ volcanoes.
    a. shield
    b. composite
    c. cinder cone
    d. earthquake

Thinking Critically

Answer the following questions in your Science Journal using complete sentences.

11. Why do volcanoes and earthquakes often occur in the same areas?

12. Would an earthquake be more likely in Nebraska or Alaska? Explain.

13. Why would the fossils of a seabird found on two different continents not be good evidence of continental drift?

14. Why would scientists want to know how many people live around faults, and how close water lines, gas lines, and hospitals are to faults?

15. Would it be safer to live near a shield volcano or a composite volcano? Explain.

Developing Skills

If you need help, refer to the description of each skill in the Skill Handbook.

16. Making and Using Tables: Research data about ten different earthquakes. Create a data table to list and record the magnitudes of the ten earthquakes. Plot the intensity of the earthquakes on a graph, and compare the strength of each earthquake.

17. Comparing and Contrasting: Compare and contrast the different layers of Earth.

18. Interpreting Data: Carlos is looking at a graph produced by a seismograph. He notices that the normally slightly wavy line had several peaks in it sometime early in the morning. What does this tell him?

19. Observing and Inferring: You are hiking in the desert of California, and you come to a wall of exposed rock. When you look at it, half of the wall appears to have shifted downward while the other half stayed in place. What are you probably looking at?

20. Interpreting Scientific Pictures: Review the pictures below and decide whether each is a composite, shield, or cinder cone volcano.

Performance Assessment

1. Display: Create a display about earthquakes. Include information on how and where earthquakes happen, and drawings or photos of earthquake damage.

2. Designing an Experiment: You created an experiment using baking soda and vinegar to make a model of a volcano erupting. Design an experiment where you use the same materials but a different setup to make a lava flow.

Chapter Preview

Skills Preview

▶ **Skill Builders**
- compare and contrast
- develop multimedia presentations
- predict

▶ **MiniLABs**
- compare
- observe

▶ **Activities**
- observe
- record
- hypothesize
- collect data
- compare and contrast
- model

Chapter 17

Water

Dolphins glide by in a graceful water ballet. Playful sea otters splash and twirl. It's fun to spend a day watching the animals in an aquarium. Don't you wonder what it's like to live in a world that's mostly water?

Well, guess what! You *do* live in a world that's mostly water. Some people call Earth the water planet. If you look at Earth from space, you'll see that water covers most of the planet's surface. But less than one percent of Earth's water is freshwater. This small amount of water we can use easily is important. It's the water we need to survive.

EXPLORE ACTIVITY

Observe Water Filtering

1. Mix some "pollutants"—sand, potting soil, ash, and salt—into a 600-mL beaker containing 300 mL of fresh tap water.
2. Carefully pour the polluted water into another beaker through a coffee filter.
3. Observe and record the kinds of and amount of pollutants that are filtered out.

Science Journal

In your Science Journal, discuss whether or not you think clear water is always clean. How do you think you could test to make sure water is clean enough to drink without tasting it?

17•1 Recycling Water

What YOU'LL LEARN

- How water moves through Earth and its atmosphere
- What groundwater is and how it moves

Science Words:
evaporation
condensation
precipitation
water cycle
groundwater

Why IT'S IMPORTANT

You will better understand the source of your water.

The Water Cycle

The sun is out. The sky is clear and blue. It's a great day to meet your friends at the park for a picnic, as shown in **Figure 17-1.** You get there early and spread out a blanket for everyone to sit on. Soon, you notice that the blanket is damp because the grass is wet. You pick up the blanket and lay it on the sun-warmed picnic table to dry. Sure enough, when your friends show up with the food, both the blanket and the grass are dry. Where did the water go? It evaporated. **Evaporation** (ee vap uh RAY shun) is the process that gradually changes water from a liquid to water vapor. If you place a bowl of water in the hot sun all day, by the end of the day some of the water is gone. Where did the water go? It evaporated.

Later in the day, the sky gets cloudy and dark. You'd better pack up and head for home before it starts to rain! As you scramble to gather up everything, you wonder just where those clouds came from.

You see clouds almost every day. Let's look at how clouds are formed. Remember the water vapor that formed when liquid water on the grass and blanket evaporated? Water vapor rises through Earth's atmosphere, cooling as it rises.

FIGURE 17-1

This family is enjoying a day at the park. For them and for us all, water is a vital resource.

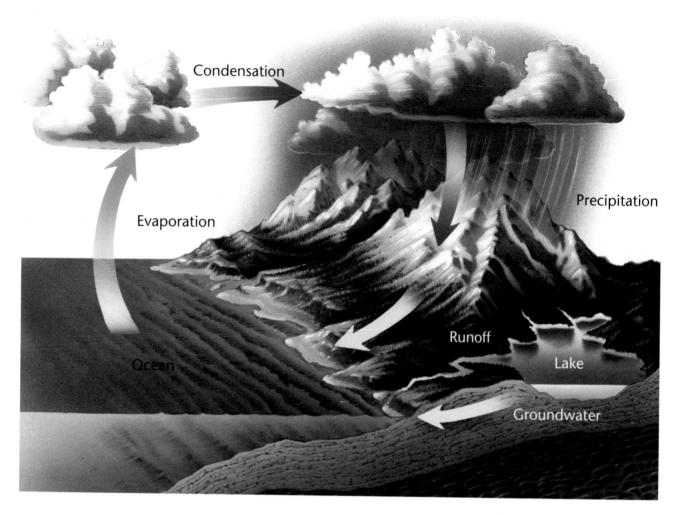

Condensation

Precipitation

Evaporation

Runoff

Ocean

Lake

Groundwater

FIGURE 17-2

The water cycle illustrates how water moves through the atmosphere, on the surface, and under the surface.

During the process of **condensation** (kahn dun SAY shun), the water changes back into liquid form as the vapor cools. As water vapor condenses into tiny water droplets, clouds form.

The tiny droplets of water in the clouds bump into each other. When they do, they form larger drops. When the drops grow so large that they can no longer stay suspended in the clouds, drops of water fall to Earth as **precipitation** (pree sihp uh TAY shun). Rain, snow, hail, and sleet are all kinds of precipitation.

As your day in the park has shown, water constantly moves from Earth's surface into the atmosphere and then back to Earth's surface again. As shown in **Figure 17-2,** this constant movement of water is called the **water cycle.** Even though there is little water in the atmosphere at any one time, this water is important to life on Earth. This small amount of water is recycled through the water cycle to provide the precipitation needed for plants and crops.

How does water affect you?

FIGURE 17-3

You use water all the time for many things. Before you use water, it must be cleaned, tested, and treated to make sure it is safe. *List some other ways you use water.*

A We need lots of clean, safe water to drink.

Where do you get the water you use? You probably just turn on a faucet and water comes out. But where does that water come from? If you live in a city or town, your water may come from a nearby lake, reservoir, or river. In many places, people get water from wells drilled into the ground. If you don't know where the water in your area comes from, find out. Ask your teacher or parents, or check with your city water department.

How many ways have you used water today? The water you drink when you're thirsty is just a small part of all the water you use in a day, as shown in **Figure 17-3**. In the United States, the average family turns on the faucet 70 to 100 times a day. We use gallons and gallons of water to shower, brush our teeth, and flush the toilet. We use still more to cook, wash clothes and dishes, clean house, and water our plants. Can you think of other ways you and your family use water?

B This pool requires a lot of water, and the water must be filtered and treated to be clean and safe.

Where is water found?

Have you ever stood at the edge of a big lake, looking across the water? Some lakes are so big you can't even see the other side. Many rivers wind along for miles and miles. Imagine how much water there must be in all the world's lakes and rivers! Yet the water in lakes and rivers makes up less than one percent of Earth's total freshwater supply. The rest is frozen in glaciers and the polar ice caps or underground.

Underground Water

If you have ever watched as rain falls on the ground, you have seen that sometimes the rain soaks in quickly. Some of the rainwater can run along the surface of the ground to flow into streams and lakes. Other times, the water sits on top of the ground and evaporates or soaks in slowly. The water that soaks into the ground becomes groundwater. **Groundwater** is water that soaks into the ground and collects in the small spaces between bits of soil and rock, as shown in **Figure 17-4.** If the small spaces are connected, then water can flow through layers of rock and soil. People drill down into these layers to make wells. They then pump the water to the surface for use as drinking water, for factories, or for watering crops and animals.

FIGURE 17-4

Groundwater is water that soaks into the ground and collects in the spaces between soil and rock particles.

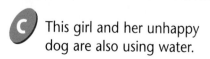

C This girl and her unhappy dog are also using water.

D We use water for our plants, lawns, and gardens.

Design Your Own Experiment
The Ground Is Soaked

Why does water soak into the ground fast after a rainstorm in some places, and in other places it sits on the ground for a long time? Some soils can soak up a lot of water fast, but other soils can only soak up water slowly. In this activity, you will investigate why types of soils soak up water differently, using potting soil, clay, sand, and gravel to model different types of soil.

Possible Materials

- stopwatch or watch with second hand
- 25-mL graduated cylinder
- 600-mL beakers (4)
- potting soil, clay, sand, gravel
- water
- permanent markers
- metric ruler

PREPARE

What You'll Investigate
Compare how water flows through different soils.

Form a Hypothesis

Remember that water sometimes runs along the ground and sometimes soaks into the ground quickly. **Form a hypothesis** about how water will soak into different soils.

Goals

Design and **carry out** an experiment that compares how water flows through three different kinds of soil.

Compare and **contrast** how easily water soaks into different soils.

Safety Precautions

Handle glass beakers with care.

PLAN

1. Agree upon and write out your **hypothesis** statement.
2. **List** the steps needed to test your hypothesis. Be specific. Describe exactly what you will do.
3. **Prepare a data table** in your Science Journal.
4. **Read** over your plan to make sure your experiment tests your hypothesis.
5. What **variable** will you be testing? What variables will you need to control?
6. Make sure you **measure** and **record** carefully.
7. **Read** over the concluding questions to see if you have included all the necessary steps in your experiment.

DO

1. Make sure your teacher approves your plan and data table before you proceed.
2. Carry out the experiment as planned.
3. While doing the experiment, **record** all observations that you make in your Science Journal.

CONCLUDE AND APPLY

1. **Infer** how your data and observations support or do not support your hypothesis. Explain your answer.
2. **Compare** and **contrast** how water soaked into each test soil.
3. **APPLY** If the soil is already soaked with water and you add more water, what will happen to the additional water?

FIGURE 17-5

This artesian well in Michigan flows without a pump. Most cities and towns have water wells that require pumps to bring the water to the surface.

The Importance of Groundwater

In some parts of the world, people rely on wells drilled down into groundwater for their everyday water needs. One type of well, an artesian (ar TEE zhun) well, is shown in **Figure 17-5.** In some places, this water is close to the ground surface. Springs form where the top of the groundwater layer meets Earth's surface. At springs, water flows from the ground.

In the United States, groundwater provides 40 percent of our public water supply. Industries and farms also use groundwater. Groundwater is the only source of water in many agricultural areas. You've learned how people depend on water and where they get the water they use. But water has other important roles on Earth. Next, you'll learn about those roles.

Section Wrap-up

1. Early in the morning, grass and other surfaces are often wet with dew. What happens to the water during the day?

2. Should you be concerned about the safety of your water supply if a large amount of a toxic liquid were spilled on the ground near the town's reservoir? Explain your answer.

3. **Think Critically:** Can a rock layer have small spaces but not allow water to flow through? How might this type of rock layer affect the flow of groundwater?

4. *Skill Builder*
   Comparing and Contrasting Compare and contrast the processes of evaporation and condensation. If you need help, refer to Comparing and Contrasting on page 550 in the **Skill Handbook.**

Using Computers

Model Using computer graphics or drawing software, make a diagram that shows how water circulates through Earth's water cycle.

Earth Shaped by Water

Water Erosion and Deposition

You've gone camping at what seems like a perfect place. It is shady and just a few steps from a clear river. But now, you're getting worried. The level of water in the river is rising. And the water looks muddier than it did earlier. What's going on?

What you don't know is that large amounts of rain upstream from your campsite have made the water level rise. In some places, the river has risen higher than its banks. It has flooded the nearby land. The fast-moving water is stirring up sediment, making the water look muddy. You'd better pack up and come back another day!

The next weekend, you try again. The land around the river isn't flooded anymore. But what a mess! Everything is covered with mud and sand, as shown in **Figure 17-6.** Where did the mud and sand come from? The river left it there. Streams and rivers move a lot of mud, sand, and other sediments (SED uh munts). Moving water lays down sediments in a process called **deposition.** Deposition (dep uh ZIH shun) is one way that streams and rivers change Earth. But how do streams and rivers form in the first place? Let's take a look.

What YOU'LL LEARN

- How water changes Earth's surface
- How rivers and floods move sediment

Science Words:
deposition
runoff
erosion
meander

Why IT'S IMPORTANT

Water helps shape many features of our planet.

FIGURE 17-6

Floods carry mud, sand, and other debris over the area covered by water. This cornfield was flooded in 1993 by the Missouri River.

FIGURE 17-7

The amount of runoff from an area is affected by several factors. These include the amount of vegetation, how steep the land is, the amount of rain, the intensity of the rain, and whether the ground is already wet.

Stream Development

Streams and rivers can be big and mighty. But we're going to start small—with raindrops. Have you ever just sat by the window on a rainy day, watching where the rain goes? Some rainwater sinks right into the soil to become groundwater. Some rain collects on leaves and grass and in puddles. When the sun comes out, this rainwater will evaporate. But some rain doesn't sink in or stay in one place. Because it runs off the surface, it is called runoff. **Runoff** is water that flows over the ground surface and eventually flows into rivers, streams, or lakes.

Runoff is one of the causes of erosion (ee ROH zhun). **Erosion** occurs when soil or rock is loosened and then moved from one place to another. Some parts of the ground are more easily eroded than others. **Figure 17-7** shows some things that affect how much water runs off. The water begins to cut small grooves in some areas. Eventually, the flowing water makes the grooves deeper and wider, forming small valleys called gullies. Water in small gullies flows into streams. Water in streams flows into rivers and eventually into oceans.

Meandering Stream

Deepest part of stream

Sandbar

Sandbar

Steep bank

FIGURE 17-8

When a stream begins to erode its banks more than the channel bottom, wide bends or meanders are formed. The stream cuts into the outside of the meanders and deposits sediment on the inside of the meanders.

Rivers as Earthmovers

In the MiniLAB, you will see that erosion can make a stream deeper or wider. The same thing happens in nature. As a stream is shaped by moving water, erosion cuts its channel deeper and deeper. In steep areas, streams erode the channel bottoms, making the channel deeper. But in flatter areas, streams flow slower and begin to erode the channel sides more than the bottom. When the stream can't get any deeper, it gets wider. Water in the stream moves sediments around, forming wide bends in the stream called **meanders** (mee AN durz). **Figure 17-8** shows a river with meanders.

Mini LAB

Compare Stream Slopes

How does stream slope affect erosion?

1. Fill three long, flat pans with moistened soil.
2. Level the soil in each pan, and gently pack it down.
3. Use a pencil to make a straight or slightly curved groove lengthwise in the soil in each pan.
4. Tilt each pan to a different angle.
5. Using a 250-mL beaker of water, begin pouring a steady stream of water into each pan.

Analysis

1. For each stream, describe where the water eroded the most sediment.
2. In each pan, did the water make the groove deeper or wider? Describe how the slope of the stream affects the type of erosion (deepening or widening). Also, describe how the slope of the pan affects where the most erosion occurs.

Let's Get Settled

You have learned that streams move sediment around. There are different kinds of sediment—some heavy, some light. Streams carry light particles farther than heavy particles. This activity will help you understand how streams move sediments.

What You'll Investigate
How do sediments settle out of water?

Procedure

1. **Label** five cups *A, B, C, D,* and *E.*
2. Use the graduated cylinder to measure 25 mL of gravel. **Pour** the gravel into cup *A.* Do the same for each of the other materials, one type per cup.
3. **Pour** the remaining gravel, salt, sand, clay, and silt (fine sediment that feels like flour) into a 2-L bottle. **Add** water to the bottle, leaving 2-4 cm of space at the top of the bottle. Put the cap on the bottle.
4. **Add** water to each cup until it is about two-thirds full.
5. **Shake** the 2-L bottle and stir each cup thoroughly. **Observe** what happens to the sediment in the bottle. **Record** your observations in a data table.
6. **Observe** each cup for 5 seconds and then at 15-second intervals for 1 minute. Then observe each cup at 2-minute intervals for 16 minutes. **Record** your observations in your data table.
7. After shaking the 2-L bottle, let it settle for 16 minutes. **Observe** the bottle. Use the colored pencils to **draw** and **label** your observations of the 2-L bottle. Use a different color for each layer that forms in the bottle.

Conclude and Apply

1. Which two sediments settled out most quickly? Which three sediments did not settle out? **Explain.**
2. **Describe** the mixture in the 2-L bottle. Are the sediments layered?
3. Based on your observations in this experiment, **explain** what happens to gravel, salt, sand, clay, and silt carried by streams.

Goals
- Observe how sediments settle in water.
- Model how sediments move in streams.

Materials
- graduated cylinder
- clear plastic cups (5)
- 2-L bottle
- water
- spoon
- gravel (50 mL)
- salt (50 mL)
- sand (50 mL)
- clay (50 mL)
- silt (50 mL)
- colored pencils (3)
- stopwatch

Muddy Streams

What have you noticed about streams you have seen? Did they all look the same? Some were clear. Others may have been muddy. The information that you learned in Activity 17-2 should help you understand why the streams you have seen are not all alike. It should also help you understand that water moves different kinds of materials in different ways. Water can push or bounce grains of sediment along the bottom of a stream. Smaller grains of sediment can be carried suspended in the water. Other materials can be carried dissolved in the water. As it constantly moves sediment from one place to another, water truly is an earthmover, as shown in **Figure 17-9** on page 480.

Problem Solving

Interpreting Flooding Problems

The floodplain of a river is the area that is normally flooded during a time of high water. The floodplain is built from sediments deposited by the river. This area makes great farmland but is not the wisest place to build a city. The area of the floodplain is the mostly flat area that exists on both sides of a river. It stretches outward from the river's natural levees or banks. Look at the illustration of the Mississippi River shown here. Try to determine the extent of the river's floodplain.

Solve the Problem:

Interpret the illustration below as to whether any structures exist within the floodplain of this part of the Mississippi River. If you were planning a farm in an area like this, where would you plant crops?

Think Critically:

Based on what you interpreted from the illustration of the Mississippi River and its floodplain, where would you build the farmhouse and barn? Explain why you chose the sites you did for the planted fields, the farmhouse, and the barn.

FIGURE 17-9

This photo was taken during a flash flood of the Colorado River in the Grand Canyon. Notice how muddy the water looks. The river is carrying a lot of sediment. This is how, over millions of years, the relatively small Colorado River eroded the mighty Grand Canyon.

Section Wrap-up

1. During stream erosion, how does the water carry material?

2. Explain how runoff might be affected by the slope of the land and how wet the soil is.

3. **Think Critically:** Why does a stream begin to curve as it develops over time?

4. *Skill Builder*
   Developing Multimedia Presentations
   Use posters, videos, photos, or computers to make a class presentation about how rivers form and develop. Be sure to include information on flooding. If you need help, refer to Developing Multimedia Presentations on page 564 in the **Technology Skill Handbook.**

Science Journal

Write a short story in your Science Journal, describing how a small stream can start in the mountains and flow down to the ocean, forming a wide river valley along the way.

Oceans

Ocean Water

If you were getting ready to sail around the world, what would you take with you? You'd have to take food, clothes, and many supplies. You'd also have to take water. Doesn't that seem kind of strange? You'd be surrounded by water, so why would you need to take your own? The reason is that ocean water is salty. It is not good for people to drink because too much salt will make you sick.

Why is the ocean salty?

How does ocean water get so salty? In the last section, you learned that water flows from streams to rivers and from rivers to the oceans. The water in streams and rivers isn't salty. What happens along the way? Both streams and groundwater pick up elements such as calcium, magnesium, and sodium from rocks and minerals. These elements dissolve in the water and are carried to the oceans. In the oceans, they mix with other elements, such as sulfur and chlorine. These elements have been added to the oceans by volcanic activity. Volcanoes release these elements into the air and they later fall into the oceans. When the elements from the rivers and groundwater combine with the elements already in the ocean, they form salt solutions, as shown in **Figure 17-10.** The most common combination of elements is sodium and chlorine—the same elements that make table salt. Over millions of years, these dissolved salts have built up to make the oceans salty.

What YOU'LL LEARN

- Why oceans are salty
- About currents, waves, and tides

Science Words:
salinity
current
wave
tide

Why IT'S IMPORTANT

You'll better understand the oceans, where most of Earth's water is found.

FIGURE 17-10

Rivers and streams carry many elements in solution. This material is carried to the ocean, where it combines with other elements to form salt solutions.

Ocean Water
To 96.5 parts water
Add 3.5 parts salts

To make salts
Mix 55 parts chloride
30.5 parts sodium
3.7 parts magnesium
7.7 parts sulfate
1.2 parts calcium
1.1 parts potassium
0.7 parts silica
and others

FIGURE 17-11

Many organisms in the ocean use elements in ocean water to build their shells and skeletons. These corals, in the Great Barrier Reef in Australia, have exoskeletons made of calcium carbonate.

A measure of the saltiness of water is called **salinity** (suh LIH nuh tee). The salinity of ocean water has stayed about the same for hundreds of millions of years. That must mean that the amount of salt added to the oceans over time has been balanced by the amount lost over the same time. How do oceans lose salt? Plants and animals that live in the oceans remove some from ocean water. They use elements, such as calcium and silicon, to build their shells and skeletons, as shown in **Figure 17-11.** Some of the elements in the ocean water just settle to the ocean floor. They become part of sedimentary rocks.

Now you know why the oceans are salty—and why you need to take water along when you sail around the world. Something else that could affect your sea trip is ocean motion.

Ocean Motion

Have you ever watched someone stir a big spoonful of sugar into a glass of iced tea? Before the sugar dissolves, you can see it swirling around in the glass. The sugar moves along with the tea because the stirring motion

Surface current

Surface current

Bottom current

forms currents. A **current** is a mass of water (or tea) moving in one direction. Currents happen in the ocean. But it's not because a giant spoon is mixing it up. Let's see why currents form in the ocean.

Surface Currents

You can think of ocean currents as rivers of water flowing through the ocean. Some move along the ocean bottom. Bottom currents are caused mainly by the differences in temperature or density of different water masses. Others, called surface currents, move at or near the ocean surface, as shown in **Figure 17-12.** Wind powers surface currents. Friction between the water in the ocean and the air in the wind makes the water move. You can see how this works by blowing across the top of a bowl of water. Surface currents can carry seeds and plants from one continent to another. Sailors use currents to make sailing easier. People who fish in the ocean can catch more fish if they know about currents. That's because fish and other sea animals often follow currents. Fish follow currents because they eat tiny animals and plants that are carried by the currents.

Currents also can affect the climate. One well-known surface current called the Gulf Stream does just that. The warm water of the Gulf Stream makes the climate of Great Britain much milder than it would be otherwise.

SCIENCE
Online

Do you want to know more about the oceans? Check out Glencoe Science Online, *science. glencoe.com*, for a link to questions and answers from an oceanographic institute.

FIGURE 17-12

Ocean water is always moving. Ocean currents move like rivers flowing through the ocean. Some currents flow near the surface and are caused by winds. Other currents flow along the bottom, or somewhere between the bottom and the surface.

Wind

Surface current

Bottom current

Mini LAB

Observing Waves

Observe how waves move things in the water.

1. Fill a large, flat, plastic box or pan half-way with water.
2. Place some floating items in the water. These things might include a cork, a Ping-Pong ball, or small bits of Styrofoam.
3. Make waves in the pan by moving a wooden rod or large pencil up and down in the pan.

Analysis

1. Observe the objects floating in the pan, and describe how the objects are moved by the water waves.
2. How could you test how individual particles of water are moved by a wave?

FIGURE 17-13

In a wave, water particles move in a circular path. The deeper you go below the water's surface, the smaller the circles. Below a certain depth, little water movement occurs.

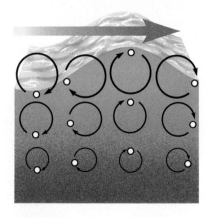

Ocean Waves

One after another, ocean waves crash onto the shore. Couldn't you just sit and watch them forever? No matter how long you sit and watch, the waves would keep coming. What is a wave? A **wave** is a movement of ocean water that makes the water rise and fall. Ocean water is constantly in motion. The movement of waves transmits energy. If you have ever tried to stand up to a big wave at the seashore or in a wave pool, you have felt that energy in the force of the moving water.

But waves in the ocean transmit energy without actually moving the water very far. To understand that, picture what happens when you're floating on a rubber raft in the ocean or at a wave pool. As each wave passes under your raft, the raft moves backward, then up and forward. But it ends up almost where it started. This same type of motion, shown in **Figure 17-13,** makes individual particles of water move in a circular path.

Most of the circular movement of water particles is near the surface of the ocean because that's where most waves are. The wave's energy spreads from one water particle to another through the particles' circular motion. Have you ever been at a sports stadium where fans did "the wave"? In the stadium, one group of people stands up and sits back down. Then the people right next to them do the same thing. Then the next group of people does it, and so on all around the stadium. If you're watching from far away, it looks like a huge wave of people is moving around and around the stadium. But it's really just lots of people standing up and sitting down, one by one. In the ocean, energy spreads from one water particle to another in a similar way.

What makes waves?

What makes waves in the ocean? Wind blows over the ocean's surface. Energy is transferred from the wind to the ocean water. That transfer of energy forms waves.

If you've watched waves roll in toward a beach, you've noticed that the waves "break" as they get close to the shore. That's because the water is shallower close to shore. As a wave comes into the shallow water, the wave doesn't have as much room to move the water around. Water in the wave piles up. As that happens, the top of the wave moves faster than the bottom, and the wave breaks against the shore. Surfers and boogie boarders like to ride those collapsing or breaking waves, as shown in **Figure 17-14.**

FIGURE **17-14**
Ocean waves "break" as they move close to the shore.

USING TECHNOLOGY

Submersibles

Scientists use submersibles (sub MUR suh bulz) such as ALVIN to study the deep ocean. They can also study life in the oceans and human effects on the ocean environment.

In the past, submersibles have been made to sink or rise in the water by changing their weight. This was done by pumping water into or out of tanks in the sub and by using weights. A new type of submersible may change all of that.

Deep Flight I

The experimental craft now called *Deep Flight I* moves up and down in the water by its own power instead of using weights. This new kind of submersible moves in a way similar to the way an airplane flies in the air or a dolphin swims in the ocean. It can move through the water in any direction. To *Deep Flight I*, there is no up or down. It is as much at home in the water in level "flight" as it is preparing for a steep dive.

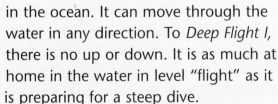

Go to Glencoe Science Online, *science. glencoe.com*, for a link to a site that tells more about ocean research vessels.

Ocean Tides

The waves that constantly crash on the beach are caused by wind, as you just learned. But another kind of wave motion in the ocean is caused by something else. The effects of this water motion are called tides. **Tides** are changes in the level of ocean water during the course of a day. The level of ocean water usually reaches a high and low level twice a day. These high and low levels are called high and low tides, as shown in **Figure 17-15.** At high tide, water is pushed up onto the shore. As the low tide occurs, usually about six hours later, water is pulled away from shore.

What causes the tides? One cause is the pull of gravity between Earth, the moon, and the sun. The moon's gravity exerts a force on Earth. The water in the oceans responds to this force, forming a bulge of ocean water on the side of Earth facing the moon. Another bulge forms

***FIGURE* 17-15**

The pull of gravity between Earth, the moon, and the sun is one cause of the tides in the ocean. In most places, there are two high tides and two low tides each day.

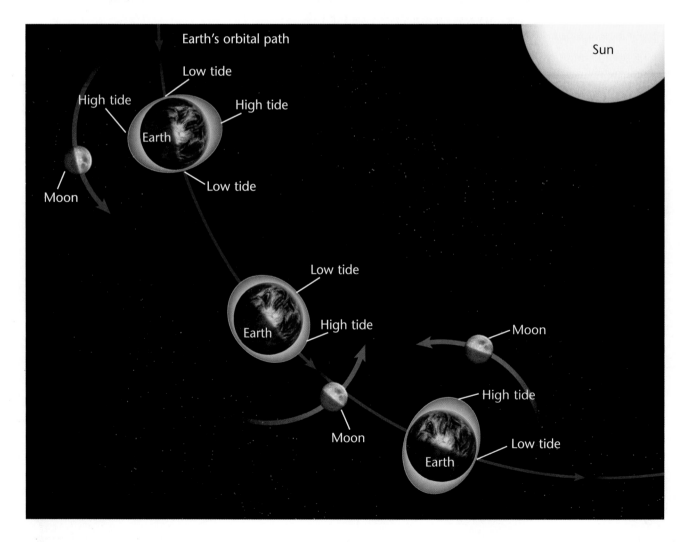

on the opposite side of the planet. The bulge on the side away from the moon also occurs because the water is thrown outward as Earth and the moon revolve around a common point. The water in the oceans tries to travel away from Earth's orbit, but Earth's gravity keeps the water from flying off into space. This is similar to what happens if you spin a bucket of water around in a circle, as shown in **Figure 17-16.** The water in the bucket tries to travel in a straight line and is thrown outward toward the bottom of the bucket. The bottom of the bucket keeps the water contained, and the water can't flow out unless you stop spinning.

The bulges that form in the oceans are the highest parts of the tides. The sun's gravity also has an effect on the tides, but because the sun is much farther away, its effect is less than that of the moon.

You have learned how important water is. Next, you will learn about some of the problems of ocean pollution.

FIGURE 17-16
The bulge of ocean water that forms the tides is thrown outward as Earth and the moon revolve around the sun, like the water in the bucket.

Section Wrap-up

1. Describe how ocean waves get energy from Earth's atmosphere.

2. Describe what causes tides.

3. **Think Critically:** You are swimming in a wave pool at a water park. The wave passes you and moves to the other side of the pool. Why doesn't the wave carry you and the water around you from one side of the pool to the other?

4. **Skill Builder**
   Predicting A surface current moves colder, denser water into an area of warm, less dense water. Predict what will happen to the ocean water in this area. Explain your prediction. If you need help, refer to Predicting on page 560 in the **Skill Handbook.**

Science Journal

In winter, the water of the ocean around Antarctica is saltier and denser than any other ocean water. Find out the reason for this, and write about it in your Science Journal.

17•4 Ocean Pollution

What YOU'LL LEARN

• The importance of oceans as resources
• Various sources of ocean pollution

Why IT'S IMPORTANT

Distinguishing fact from opinion helps you make better decisions.

The Blue Planet

The oceans are the homes for millions of different kinds of life, from tiny floating plants and animals to the largest known creature—the blue whale. And, while humans don't live *in* the oceans, about half of the world's population lives within 100 km of the oceans. Oceans and their beaches give humans some of their food, mineral resources, and recreation.

Sources of Ocean Pollution

Because humans use Earth's oceans for so many things, the oceans can become polluted. The oceans are huge, but they are also fragile resources. Tankers that carry oil have had accidents and spilled oil into the oceans. Beaches and animal and bird populations that live near the shore can suffer greatly from these spills. Oil from other kinds of seagoing ships can also pollute our oceans.

Ships can pollute the oceans by dumping their wastes overboard. Some trash is allowed to be dumped at sea. But things such as glass and metal must first be crushed before they are thrown overboard. The dumping of plastics of any kind is illegal. Along one shore, the Gulf of Mexico, wastes dumped by ships totaled 65 percent of the trash collected during one year.

Ocean beaches also can be polluted by people on the beach, such as the beach shown in **Figure 17-17.** On Padre Island in Texas, during just one of the national beach cleanups that occur several times a year in the United States, volunteers, such as those shown in **Figure 17-18,** collected more than 7 million pieces of trash including 25 000 plastic six-pack holders, 40 000 rubber balloons, 300 000 plastic and glass beverage bottles, and 200 000 metal cans!

FIGURE 17-17

Beaches, like this one in Costa Rica, can become polluted from ocean dumping, shipping accidents, and careless people.

Solutions to Ocean Pollution

For a long time, many people, even some scientists, thought that Earth's oceans were big enough to resist most of the effects of pollution. Now, however, scientists think that the oceans and ocean organisms are suffering from the effects of pollution.

Most people don't realize that many of the things that are poured down drains, thrown away, or dumped out in the backyard can end up in the ocean. Properly disposing of toxic items such as paint, motor oil, and household chemicals can prevent pollution. Reducing and recycling are two other ways to prevent and minimize all forms of pollution. Can you think of other solutions to ocean pollution or any other kind of pollution?

FIGURE 17-18
These students are helping in a beach cleanup.

Skill Builder: Distinguishing Fact from Opinion

LEARNING the SKILL

1. Facts are statements that can be checked for accuracy and proven. Facts often answer the reporter's five questions: *who, what, where, when,* and *why?*

2. Opinions, on the other hand, express personal beliefs and feelings. Opinions can't be proved or disproved. Opinions often include the following words: *I believe, I think, probably, it seems to me, best, worst,* and *greatest,* among others.

PRACTICING the SKILL

1. Reread the first paragraph. Are there any opinions in it? If so, identify them.

2. List at least five facts from magazines or Internet sources on ocean pollution.

APPLYING the SKILL

The statements below were taken from magazine articles on ocean pollution. Identify each as either a fact or an opinion.

1. Fishing should be banned in contaminated areas.

2. About 1 million tons of military equipment were dumped into an ocean trench off the Scottish coast between 1945 and 1976.

3. During 1992 and 1993, pollution caused almost 5000 beach closings.

4. Human activities are killing our beaches.

People & Science

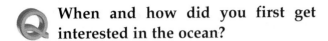

Isidro Bosch, Aquatic Biologist

Q **When and how did you first get interested in the ocean?**

A As a child living in Cuba, some of my earliest memories are of trips with my family to the beach. There was one danger where we went swimming—*los erizos*, the long-spined sea urchins of the Caribbean. As a child, I feared them. Isn't it funny that as a grown-up, I study those same animals?

Q **Dr. Bosch, tell us about your research.**

A I mostly study ocean invertebrates—sea urchins, sea slugs, sponges, and so forth. I'm interested in how these animals live in tough environmental conditions, such as cold polar oceans and the deep sea with its darkness and high pressure.

Q **How have you explored the oceans?**

A In everything from huge research ships to small, inflatable boats. I've dived beneath the sea ice in Antarctica, and explored tropical coral reefs and giant kelp forests. One of my greatest experiences has been exploring the deep sea in tiny submarines that go down into deep parts of the ocean. I've been down to 900 meters in subs.

Q **What's it like to be at those great depths?**

A As the submersible drifts down through the darkness, some organisms have natural luminescence and light up in the water around you. They flash like fireworks on the Fourth of July. On the bottom, you see a world few other people have seen. This world is inhabited by strange animals that have adapted to the crushing pressures, total darkness, and extreme cold.

Q **Recently, you've also been doing research on freshwater lakes. Why?**

A Lakes are threatened habitats, and the biologists who study them have a very important and urgent job. Research in lakes has shown clearly how all life-forms depend on each other—how adding or taking away just one organism can change the entire community of organisms that live there.

Career Connection

Interview someone in your area who works with the oceans, or contact someone at an oceanographic institute on the Internet. Present your findings to the class.

Read the statements below that review the most important ideas in the chapter. Using what you have learned, answer each question in your Science Journal.

1. Water moves through Earth's water cycle by evaporation, condensation, and precipitation. *How does condensation of water vapor in Earth's atmosphere eventually lead to precipitation?*

2. Water that does not run off will soak into the ground. *What is this water that fills and moves through holes in soil and rock called?*

3. Runoff is affected by the amount of rain and its intensity, the slope of the land, the amount of vegetation, and whether the ground is already wet. *Describe how two of these affect runoff.*

4. Streams and groundwater have dissolved elements such as calcium, magnesium, and sodium from rocks and minerals. *How do these elements end up as salts in the oceans?*

Using Key Science Words

condensation   precipitation
current   runoff
deposition   salinity
erosion   tide
evaporation   water cycle
groundwater   wave
meander

Match each phrase with the correct term from the list of Key Science Words.

1. water flowing over the land
2. a wide curve in a stream channel
3. saltiness of ocean water
4. when moving water drops sediment
5. water vapor changing back into a liquid

Checking Concepts

Choose the word or phrase that completes the sentence.

6. The process in which water changes from a liquid to water vapor is

   _____.

   a. evaporation
   b. condensation
   c. runoff
   d. precipitation

7. Earth's _____ is the continual cycling of water between Earth's surface and the atmosphere.
   a. rock cycle   c. water cycle
   b. evaporation   d. runoff

8. Rivers change Earth by dropping sediments in a process called

   _____.

   a. runoff
   b. deposition
   c. condensation
   d. evaporation

9. Currents that move large masses of water across the ocean, at or near the surface, are called _____.
   a. waves   c. hurricanes
   b. tides   d. surface currents

10. The chlorine in ocean water comes from _____.
    a. volcanoes   c. rain
    b. storms   d. rivers

Thinking Critically

Answer the following questions in your Science Journal using complete sentences.

11. Why do you think that floods are often more severe in cities than in the countryside?

12. Would it be better to drill for water in a layer of clay or a layer of sand? Explain.

13. If you look at a map of the United States or the world, you will see that most major cities are located near bodies of water. Explain why you think cities are located near rivers, lakes, and the ocean.

14. If the salinity of ocean water is 3.5 percent, how many grams of solids (salts) are dissolved in 100 kg of water? In 50 kg of water?

15. Why do people worry about mudslides and flooding near hillsides after there has been a forest fire?

Developing Skills

If you need help, refer to the description of each skill in the Skill Handbook.

16. Comparing and Contrasting: Compare and contrast the processes of evaporation and condensation.

17. Recognizing Cause and Effect: What is runoff? How does the amount of vegetation on the ground affect runoff?

18. Hypothesizing: Hypothesize why fine particles are carried farther by a stream than heavier particles.

19. Using Variables, Constants, and Controls: Explain how you could measure undissolved water pollution by comparing how clear the water looks.

20. Comparing and Contrasting: Compare and contrast waves and currents. How are they similar? How are they different?

Performance Assessment

1. Making Observations and Inferences: Experiment to find out the amount of space between particles. Use 100 mL of sand. Measure the amount of water needed to just cover the sand. Repeat your experiment to measure the spaces when you mix 50 mL of gravel and 50 mL of sand. Repeat your experiment to measure the spaces when you mix 50 mL of clay and 50 mL of sand.

2. Observing and Inferring: Study the photo below and infer how water has affected the landscape and organisms shown. Discuss how erosion and deposition have affected the mountains and the land around the river.

Chapter Preview

Skills Preview

▶ **Skill Builders**
- make and use graphs
- compare and contrast
- recognize cause and effect

▶ **MiniLABs**
- observe
- infer

▶ **Activities**
- observe
- model
- infer
- record

18

Earth's Atmosphere

Whoosh! What was that? A big airplane just took off right over your head. Don't you wonder how that giant jet stays in the sky, held up by nothing but air? Air is more than it seems to be. Air is matter. Even though we can't see it, air has mass and takes up space. Air can exert pressure on objects. That's why it's able to provide lift for airplanes. You've felt this pressure when the wind blows. What other properties of air can you observe?

EXPLORE ACTIVITY

Observe the Mass of Air

1. Place an empty balloon on a balance and measure its mass.
2. Blow up the balloon with air. Tie the end so that the air will not get out.
3. Place the blown-up balloon on a balance and determine its mass.

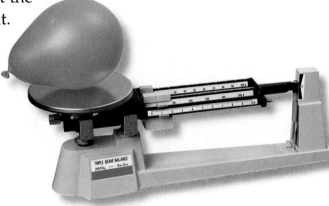

Science
Journal

What was the mass of the empty balloon? What was the mass of the blown-up balloon? In your Science Journal, infer why the two measurements were different. What property of air did you observe?

What YOU'LL LEARN

- What the atmosphere is made of
- The parts of the atmosphere
- The effects of air pressure

Science Words:
atmosphere
troposphere
stratosphere

Why IT'S IMPORTANT

Your life and the lives of all living things on Earth depend upon the atmosphere.

The Air Around You

On a hazy day, you might say that you can see the air. When it's breezy, you feel the air. But most of the time, you can't see the air that is all around you every second of your life. How can you describe something that you can't see? One way is to study what it's made of.

What makes up air?

Take a deep breath. What did you take into your body with that breath? Most people say they breathe oxygen when they take a breath of air. That's partly true. But as you can see from **Figure 18-1,** air is more than oxygen.

Air is a mixture of gases with a tiny amount of solids, such as dust particles, and liquids, like water, in its makeup. A layer of air surrounds Earth like a blanket. This blanket of air is called Earth's **atmosphere** (AT muh sfeer). The most common gas in Earth's atmosphere is nitrogen. Nitrogen makes up about 78 percent of the air we breathe. Oxygen makes up about 21 percent. That leaves about one percent for everything else. The amount of water vapor in the air changes from place to place and from time to time. It can be as low as nearly zero to as high as four percent.

Other Substances in the Atmosphere

We said that the atmosphere is like a blanket. On some days, it looks like a blanket that needs a spin in the washing machine! What could cause the brown layer of air that covers some cities like a dirty blanket?

Look at **Figure 18-2.** Hazy or smoky air is caused by pollutants (puh LEW tuntz) in the atmosphere. Pollutants are materials that harm living things by interfering with life processes. Some pollutants are released by natural

FIGURE 18-1

This graph shows the composition of Earth's atmosphere.

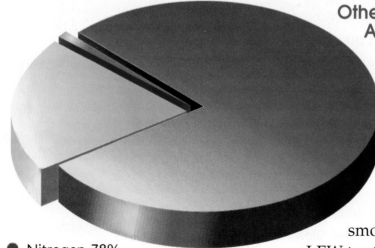

- Nitrogen 78%
- Oxygen 21%
- Other gases 1%

A The burning of fossil fuels, such as coal, releases pollutants into the air.

B Coal scrubbers dissolve the gases in the smoke in water. This is one way that power plants can reduce pollution.

C The exhaust from cars, buses, and trucks adds polluting chemicals to the air.

D All new cars are now equipped with catalytic converters, which change harmful chemicals into harmless compounds. But vehicles are still a major source of air pollution. *Why?*

FIGURE 18-2

Most air pollution is caused by human activities.

events, such as when ashes from volcanic eruptions and forest fires enter the atmosphere. But most pollutants are produced by human activities. Do you do anything that causes air pollution? In one way or another, almost everyone does.

Air Pollution

When you hop in a car or catch a bus, you're usually just thinking about where you're going. But stop and think about how cars and buses affect the air. When cars and buses burn gasoline, they release pollutants into the atmosphere. Pollutants are also given off when coal and other fossil fuels are burned to make electricity. **Figure 18-2** shows how these and other human activities cause air pollution. It also shows some of the things that we use to keep air clean.

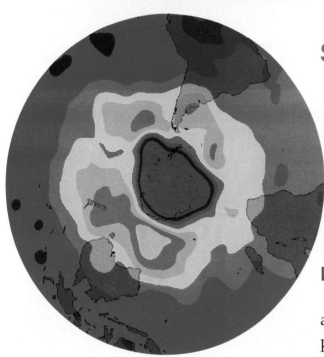

FIGURE 18-3

Every year, the ozone layer thins out over Antarctica by almost 50 percent. The dark pink color shows areas of very low ozone concentration.

Mini LAB

Observing the Effects of Holes in the Ozone Layer

1. Hold a piece of waxed paper a few inches above a globe.
2. Shine a flashlight on the waxed paper from above. Observe how the light shines on the globe.
3. Tear a small hole in the waxed paper. Shine the flashlight on the waxed paper again. Observe how the light shines on the globe.

Analysis

1. How did the amount of light on the globe change after a hole was made in the waxed paper?
2. How is the waxed paper similar to Earth's ozone layer?

Structure of the Atmosphere

In Chapter 14, you went on a tour of the solar system. You learned that Neptune has an atmosphere made of methane and that Saturn is surrounded by rings of rock and ice. But we don't have to go into outer space to find a planet with interesting surroundings. Let's take a trip through Earth's atmosphere to see what's around us.

Layers of the Atmosphere

We'll start close to home. You know the city and state you live in. But do you know what part of the atmosphere you live in? You live in the **troposphere,** the lowest layer of Earth's atmosphere. The troposphere (TROH puh sfeer) is where you'll find weather, clouds, and air pollution. It contains about 75 percent of atmospheric gases.

Going up!

As you rise above the troposphere, you come to the stratosphere (STRAT uh sfeer). The **stratosphere** contains a layer of gas called ozone. This important layer of gas protects life on Earth from ultraviolet rays, which are harmful rays from the sun. Photographs from space satellites, such as the one in **Figure 18-3,** show that the ozone layer is thinning and developing holes. Why? Many scientists think that pollutants in the atmosphere cause the ozone layer to thin. Remember, this is the layer that protects us from the sun's ultraviolet rays. If the ozone layer thins, we could see an increase in skin cancer.

Above the stratosphere are the mesosphere, thermosphere, and exosphere. Beyond the exosphere is space. There is no clear boundary between the atmosphere and space. If you traveled upward through the exosphere, you would encounter fewer and fewer molecules, until eventually, for all practical purposes, you would be out of Earth's atmosphere and in space. **Figure 18-4** gives more information about Earth's atmospheric layers.

Air Pressure

Why doesn't Earth's atmosphere just drift away? What keeps it wrapped around the planet like a cozy blanket? Consider what you already know about air. You know that air is matter. In Chapter 8, you learned that matter has mass. In Chapter 10, you learned that matter has a gravitational attraction to other forms of matter. The gravitational attraction between Earth and the gases in the atmosphere causes the gases to be pulled toward Earth. The weight of the gases at the top of the atmosphere presses down on the gases below so that the molecules that make up air near Earth's surface are squeezed together. It's sort of like stacking a pile of heavy blankets on top of a fluffy quilt. The quilt stuffing gets pressed down and becomes more dense. Air at the bottom of the

FIGURE 18-4

In the mesosphere, air temperature decreases with increasing height. In the thermosphere, air temperature increases with height. Beyond the exosphere is space.

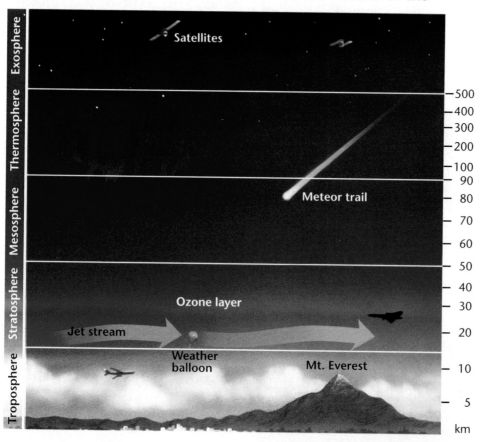

Problem Solving

Air Pressure Egg-zample!

One day, you walk into science class and your teacher announces that she's going to conduct an experiment. Your job is to observe and infer what caused the results of the experiment. Your teacher puts a peeled, hard-boiled egg on the top of an open glass jar. The opening of the jar is smaller than the egg, so the egg won't fit in the jar. Your teacher puts the egg aside, then lights a piece of paper and drops it into the jar. She puts the egg back on top of the jar. The egg wiggles, then, presto! The egg falls into the jar.

Solve the Problem:
Why did the egg fall into the jar? Explain your answer in terms of air pressure.

Think Critically:
What happened to the air in the jar after the paper was set on fire? How is this egg-zample related to your ears popping as you drive up a mountain?

atmosphere is more dense than air at the top. Dense air pushes on objects more than less dense air does. Would the air pressure be greater in the troposphere or exosphere?

Effects of Air Pressure

If you guessed the troposphere, you're right. Overall, air pressure is greater in the troposphere than in upper layers of the atmosphere. Why? In the upper layers, there are fewer molecules exerting less pressure. Would you expect air pressure to be the same everywhere in the troposphere? It isn't. Air pressure changes from place to place. It generally gets lower as you go higher. That's why your eardrums pop when you go up a steep mountain in a car. As you go higher, there is less air pressure around you. The popping is your eardrum trying to balance the higher air pressure inside your ear with the lower air pressure outside your ear.

Changes in air pressure affect humans in other ways, too. Consider a mountain climber. As she climbs higher, she takes in less oxygen with each breath. Why? Because air in high places is less dense and exerts less pressure.

In the following activity, you'll see how temperature affects air pressure.

Air Pressure

Think back to Chapter 12. How does temperature affect the movement of molecules that make up air? What happens to these molecules when they are heated or cooled? You know that air pressure is the force exerted by the molecules that make up air. Do you think differences in temperature have any effect on air pressure?

What You'll Investigate
How does temperature affect air pressure?

Procedure
1. **Place** an open, empty soft-drink bottle in a freezer for two hours. Be careful when you handle the glass bottle. If it breaks, tell your teacher and have your teacher dispose of the glass.
2. **Remove** the bottle. Immediately **stretch** a balloon over the opening of the bottle.
3. **Set** the bottle aside and allow it to come to room temperature for 15 minutes.
4. **Observe** the balloon; **record** your observations.

Conclude and Apply
1. What happened to the balloon after it was stretched over the glass bottle?
2. What property of air was demonstrated by this activity?
3. **Describe** the movement of air in the bottle after the bottle was taken out of the freezer.
4. Does temperature affect air pressure? **Explain** your answer.

Goals
- Observe the effects of temperature changes on air in a container.
- Infer how temperature affects air pressure.

Materials
- empty glass soft-drink bottle
- balloon

The Importance of the Atmosphere

On a cool but sunny day, you head to the beach with a blanket under your arm. You can wrap it around yourself to keep warm and to keep from getting too much sun. Look at **Figure 18-5.** Earth's blanket, the atmosphere, does the same things. It keeps temperatures on Earth just right to support life. It protects living things from the sun's harmful rays. In the next section, you'll learn how the atmosphere affects something that you deal with every day: the weather.

FIGURE 18-5

This photograph, taken from a satellite in space, shows how Earth's atmosphere covers the planet.

Section Wrap-up

1. Compare and contrast the different layers of Earth's atmosphere.

2. Describe some of the human activities that cause air pollution.

3. **Think Critically:** Describe some ways air pressure affects humans.

4. *Skill Builder*
   Making and Using Graphs Make a circle graph that shows the composition of gases in Earth's atmosphere. If you need help, refer to Making and Using Graphs on page 547 in the **Skill Handbook.**

Science Journal

In your Science Journal, write a poem about air. Write about the things that make up air or about the ways that air affects life on Earth.

Weather

What is weather?

Do you like to walk in the rain? Or would you rather watch the clouds in a bright blue sky? We use words like *rain, snow, blue skies,* and *clouds* to describe the weather. But what are we really talking about? **Weather** is what is happening in the atmosphere right now. Take a look at **Figure 18-6**. What's the weather like in these photographs?

Causes of Weather

Look outside. What do you see? Rain or sunshine? Trees blowing in the breeze? What do you think causes the type of weather you see right now? Recall what you learned about Earth's water cycle from Chapter 17. Where does the energy come from that causes the water to evaporate, condense into clouds, and eventually fall to Earth once again? It comes from the sun. Energy from the sun powers Earth's water cycle. The water cycle, in turn, forms the basis of our weather. But there's more to weather than just water.

What YOU'LL LEARN

- The causes of weather
- How to compare and contrast different types of weather

Science Words:
weather
wind
air mass
front

Why IT'S IMPORTANT

Changes in the weather affect you and your family every day.

FIGURE 18-6

Weather is the current state of the atmosphere. *Describe the weather in these three photographs.*

FIGURE 18-7

FIGURE 18-7
The sun's rays strike different parts of Earth at different angles. *Which area receives the most direct sunlight?*

Equator

Other Factors That Affect Weather

If you wanted to vacation in a sunny, warm place, would you head for Alaska or Mexico? Your chances of finding the weather you want would be better in Mexico. You know that some parts of Earth are warmer than other parts. But do you know why? Look at **Figure 18-7.** The round shape of Earth causes the sun's rays to strike different parts of Earth at different angles. This means that some parts of Earth, such as the equator, receive more sun energy—and more heat—than other parts of Earth.

Air Pressure and Wind

Temperature isn't the whole story. A day can be warm or cold. But it can also be cloudy or sunny, rainy or dry. What causes these differences? Look at **Figure 18-7** again. In warmer areas, air pressure is lower because the molecules that make up air are farther apart. In cooler areas, air pressure is higher. The air from high-pressure areas moves into low-pressure areas. This movement of air due to differences in pressure is called **wind.** Wind, air pressure, temperature, and water all work together to cause weather.

Mini LAB

Inferring

What causes wind?

1. Blow up a balloon and hold the end closed with one hand.
2. Place your other hand, palm down, a few inches above the end of the balloon.
3. Slowly let the air out of the balloon.

Analysis

1. Air flows from areas of high pressure to areas of low pressure. Where was the air pressure greater—inside or outside of the inflated balloon?
2. What did you feel on your hand when you opened the end of the balloon? In your Science Journal, explain how this movement of air is similar to the way air moves on a blustery day.

Weather Patterns

The sun was out when you decided to go to an afternoon movie. But now the movie is over, and something has changed. It's still daytime, but the sky is as dark as night. Thunder is crashing. Lightning is flashing. The wind is blowing hard. Heavy raindrops start to splatter on the sidewalk. And of course you didn't bring an umbrella! How did you get into this mess? Let's take a look at what may have caused the change in the weather.

Air Masses

The sudden weather change that left you trapped in the movie theater lobby was caused by movements of air masses. An **air mass** is a large body of air that has the same properties as the area of Earth's surface over which it develops and moves. In other words, if that part of Earth's surface is hot, the air mass will be hot. If the air mass develops over a cold, wet area on Earth, the air mass will be cold and wet. **Figure 18-8** shows the different kinds of air masses that affect weather in the United States.

FIGURE 18-8
Weather in the United States is affected by six major air masses. *Why do air masses have different temperatures and different levels of moisture?*

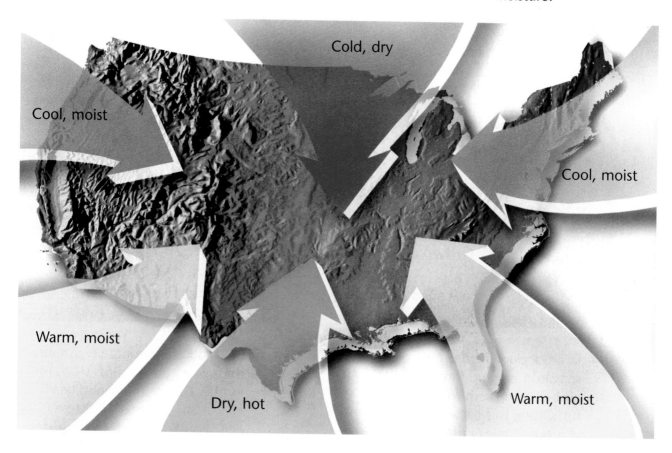

Cold, dry

Cool, moist

Cool, moist

Warm, moist

Dry, hot

Warm, moist

Fronts

Look at **Figure 18-8** again. The air masses come from different directions. Sometimes they bump into each other. They'll slide over one another or push each other out of the way. The place where air masses come together is called a **front.** The weather conditions that occur at a front are caused by interactions between the air masses. **Figure 18-9** shows weather conditions that occur at different kinds of fronts. Look at the picture and tell what kind of weather occurs along a stationary front.

FIGURE **18-9**

There are four different types of fronts: warm, cold, occluded, and stationary.

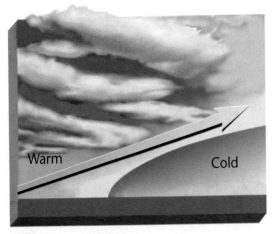

A In a warm front, less dense warm air slides over departing cold air. High, feathery clouds often form along warm fronts.

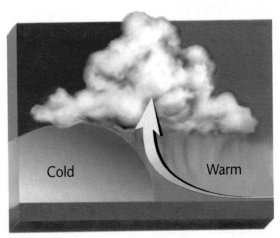

B In a cold front, cold air pushes up warm air, causing a narrow band of violent storms. *What kind of clouds would you expect to see with cold fronts?*

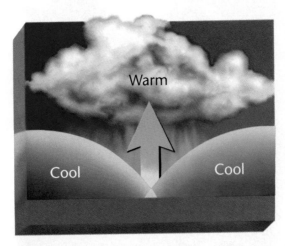

C In an occluded front, two cool air masses meet and force warmer air between them to be pushed up. Often, this causes strong winds and heavy precipitation.

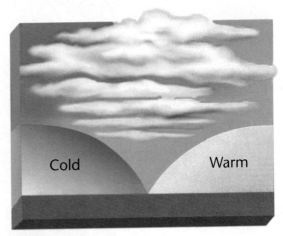

D In a stationary front, a warm air mass or a cold air mass stops moving because of pressure differences. The weather might be rainy and slightly windy for several days.

Severe Weather

We're back in the theater lobby, with thunder and lightning crashing and flashing outside. Should you start walking home in the storm? Not a good idea. Lightning bolts can reach temperatures of around 30 000°C—that's five times hotter than the surface of the sun! Obviously, lightning can be dangerous if it strikes an object or person. What causes the lightning, thunder, and heavy rains of a thunderstorm?

Thunderstorms take place inside warm, moist air masses and at fronts. The warm air is pushed up rapidly, creating an updraft of moist air that forms huge clouds like the one shown in **Figure 18-10.** As the air rises, it cools, and the moisture in the air changes into water droplets. When the droplets become heavier than the air, they begin to fall. The falling droplets collide with other droplets, forming heavy raindrops. The heavy raindrops create downdrafts of air as they fall. This, in turn, creates the strong winds associated with thunderstorms.

Thunderstorms can be dangerous. Lightning can damage property and people. Heavy rains can cause flooding. Strong winds can knock down power lines or tree branches. Sometimes lumps or balls of ice called hail fall from the sky. The hail can damage crops and make dents in cars. Thunderstorms are the most common type of severe weather. But they are not the only type.

FIGURE 18-10

Huge clouds called cumulonimbus clouds can unleash violent thunderstorms. Some cumulonimbus clouds are 18 000 m high.

Activity 18-2

Design Your Own Experiment
Blowing in the Wind

Hold onto your hat! In this activity, you'll learn more about the movement of air called wind. You'll build a simple instrument to measure wind direction. Then you'll compare your measurements to weather conditions in your area to see if there's a relationship between wind direction and weather.

PREPARE

What You'll Investigate
Is there a relationship between wind direction and weather?

Form a Hypothesis
Remember what you've learned about weather patterns and wind. **Make a hypothesis** about how wind direction might affect the weather in your area.

Goals
Design and **construct** an instrument to measure wind direction.

Observe and **record** wind direction and weather conditions.

Infer how wind direction and weather are related.

Safety Precautions
Take care when handling the straight pin and scissors.

Possible Materials
- compass
- drinking straw
- a piece of stiff plastic or poster board
- straight pin
- clear tape or glue
- scissors
- pencil

PLAN

1. **Study** the materials provided by your teacher. Also, study the photograph of the wind vane shown on this page. Wind vanes measure wind direction. As a group, **agree** upon a way to use the materials to **build** a wind vane to measure wind direction.

2. **List** the steps you will take to **build** your wind vane. How will you **cut** and **mount** the plastic so it can **show** wind direction? How will you use the compass to **determine** wind direction?

3. **Make a data table** in your Science Journal to **record** your measurements of wind direction and weather conditions in your area. **Decide** how long your experiment will last. When and where will you make your measurements? Who will **record** the information?

DO

1. Make sure your teacher has approved your plan and your data tables before you proceed.

2. Build the instrument and carry out the experiment as planned.

3. Carefully **measure** and **record** your observations.

CONCLUDE AND APPLY

1. **Explain** how your wind vane works. Why does the arrow of the wind vane point into the wind? Why is the arrow long and narrow?

2. **Compare and contrast** your measurements with other groups' measurements that were taken on the same day. How might you **explain** any differences in the measurements?

3. Does the wind change often in your area? From which direction did the wind blow most often?

4. **APPLY** Is there a relationship between the wind direction and the weather in your area? **Explain** your answer.

Tornadoes

Cows go flying through the air. Cars and trucks are picked up and dropped like toys. Even houses can be whisked off their foundations. Tornadoes are powerful and frightening. What causes them? In very severe thunderstorms, winds blow at different heights and at different speeds. Look at **Figure 18-11.** This difference in wind height and speed can cause a funnel cloud to form. The result? A tornado! Tornadoes are violent, whirling winds that move in a narrow path over land. These twisting winds can spin as fast as 500 km per hour. Tornadoes can destroy lives and property in a matter of seconds.

FIGURE 18-11

Differences in wind height and speed cause a tornado's distinctive funnel cloud.

Hurricanes

Tornadoes are awesome to behold, but they aren't the most powerful storms that develop on Earth. That title goes to hurricanes. Hurricanes are large, swirling, low-pressure systems that form over tropical oceans. They usually begin when several small thunderstorms come together. Strong, high winds that blow in tropical ocean areas cause these thunderstorms to start spinning as one. The heat from the tropical waters gives the storm energy to spin even faster. When the winds of the storm reach at least 120 km per hour, the storm is called a hurricane.

If you watch worldwide weather reports, you can follow the movement of hurricanes. Look at **Figure 18-12.** When hurricanes strike land, they cause high waves and severe thunderstorms. If the wind is strong enough, the thunderstorms formed by hurricanes can cause tornadoes to form! Most hurricanes in the United States hit the Atlantic Coast or the Gulf of Mexico. The National Weather Service tracks these storms so there is enough time to warn people to move out of their way.

The Weather and You

Severe or not, the weather affects you and your family every day. It affects what you wear and what you do. On a larger scale, it affects crops and outdoor events, such as parades or sporting events. When you're trying to make plans, it helps to know what the weather is going to be like a few days in the future. In the next section, you'll learn how scientists predict the weather.

FIGURE 18-12

Hurricanes, the most powerful of storms, have torrential rains and strong winds.

Section Wrap-up

1. What causes wind?

2. How do thunderstorms form?

3. **Think Critically:** Describe what happens when a cold air mass meets a warm air mass. What type of weather is associated with this event?

4. *Skill Builder*
   Compare and Contrast Compare and contrast tornadoes and hurricanes. If you need help, refer to Comparing and Contrasting on page 550 in the **Skill Handbook.**

USING MATH

On the radio, you hear that a tornado is moving toward a nearby town at a speed of 80 km per hour. The storm front is 100 km from the town. How long will it take for the storm front to reach the town?

18•3 Forecasting the Weather

What YOU'LL LEARN

- How scientists forecast the weather
- How to improve your note-taking skills

Science Words:
meteorologist

Why IT'S IMPORTANT

Good note-taking skills will help you improve your study habits.

The Weather and You

Quick—list all the things you do outside: basketball, bike riding, football, kite flying, swimming, running, walking, Rollerblading. Can you do these things if the weather is bad? Probably not. That's one reason people read or listen to weather forecasts in the newspaper or on TV. They want to plan for picnics in the park or trips to the beach. Think back to what you learned about hurricanes. People also need to know about severe weather ahead of time so they can get to safe places.

Who forecasts the weather?

Weather forecasts are made by meteorologists. A **meteorologist** is a scientist who studies weather patterns to predict daily weather. In the past, people used their observations to make the predictions. If they saw a cow lying in a pasture, they might say that rain was on the way. Why? Because people thought that cows could feel moisture in the air and that the cows wanted to lie down in the dry grass before it got wet.

While these kinds of predictions are interesting, they're not very accurate. Today, meteorologists still use observation to predict the weather. But they combine their observations with modern technology to come up with accurate forecasts.

Modern Weather Forecasting

Modern meteorologists use airplanes, weather balloons, weather ships, satellites, computers, and data from weather stations located around the world to make

FIGURE 18-13

This satellite photograph shows weather patterns in the United States. *Why do meteorologists use a combination of technologies to forecast the weather?*

their forecasts. **Figures 18-13** and **18-14** show some examples of weather technology. Why do meteorologists use so many things? They need a complete picture of the weather to make accurate forecasts. They need weather balloons and planes to measure temperature, wind speed, and air pressure from the atmosphere. They need ships to measure ocean currents and water temperatures. They need satellites to send back pictures of shifting clouds. And they need people at weather stations to take local measurements of temperature and rainfall.

All these measurements are constantly updated and recorded by a global network of weather stations. Every three hours, these stations send their data to the world's 13 main weather centers. Meteorologists at these centers enter the data into powerful computers. Your local meteorologist gets his or her information from these computers. Then, based on local weather conditions, your meteorologist makes a forecast.

FIGURE 18-14

This researcher is releasing a weather balloon into the atmosphere.

Skill Builder: Note-Taking Skills

LEARNING the SKILL

1. To take good notes, first read the material to identify the main ideas. The heads and subheads in the material are clues to main ideas.

2. Look for *italicized* or **boldfaced** words in the material. These are also clues to important ideas.

3. Identify details or sentences in the material that support the main ideas.

4. Using the main ideas and the details or sentences that support them, take notes about the material.

PRACTICING the SKILL

1. List three main ideas in this passage.

2. What is a meteorologist? What are some things that meteorologists use to forecast the weather?

3. How did people forecast weather in the past? How is this different from the way we forecast weather now? How is it the same?

APPLYING the SKILL

Choose a subject about the weather that interests you. Research the topic in an encyclopedia, CD-ROM, or other resource. Take notes about the topic, and share what you've learned with the class.

18•4 Climate

Why IT'S IMPORTANT

You'll learn how human activities affect climate.

What is climate?

You've probably noticed that the weather in other parts of the country is different from where you live. If you could see a map of the world's weather, you'd see even greater differences. Some parts of the world always seem rainy. Others are cold every winter. **Figure 18-15** shows different patterns of climate (KLIME ut). **Climate** is the pattern of weather that occurs in a particular area over many years. To determine the climate of a region, scientists figure out the average weather conditions over a period of 30 years or more. They look at the average temperature, precipitation, air pressure, humidity, and the number of days of sunshine in each area.

What causes different climates?

Different parts of Earth receive different amounts of sunlight. The areas nearest the equator receive the most sunlight. They have a tropical climate with warm to hot temperatures year-round. Polar climates are found near the poles, where sunlight strikes Earth at a low angle.

FIGURE 18-15

There are five main types of climate: polar, subarctic, temperate, mild, and tropical. *What type of climate does your area have?*

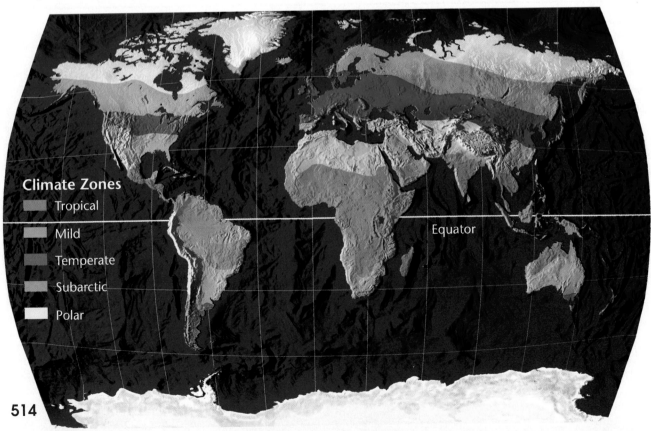

Climate Zones
- Tropical
- Mild
- Temperate
- Subarctic
- Polar

Equator

These places can get very cold. Some are always covered in ice. Subarctic, temperate (TEM prut), and mild climates lie between the tropics and the poles. Temperatures in these areas vary, but on average, they are not as hot as tropical climates and not as cold as polar climates. Is location the only factor that affects climate? No. Oceans, mountains, wind patterns, and even large cities can affect local climate.

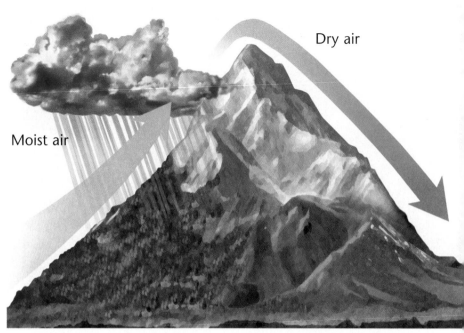

Dry air

Moist air

FIGURE 18-16

The climate on the side of the mountain that faces away from the wind is generally dry and hot. *Why?*

Local Effects on Climate

Imagine you're at the beach on a summer day. You take a moment to enjoy the cool, refreshing breeze that blows in from the ocean. Ocean winds and ocean currents affect the climate along the coasts. These areas are often cooler in the summer and warmer in the winter than areas located just a short distance inland.

As **Figure 18-16** shows, mountains also affect local climate. On the side of the mountain that faces the wind, the climate is generally cool and wet. That's because air cools as it moves up the mountain. As it cools, it drops its moisture as rain or snow. By the time the air reaches the top of the mountain, it is dry. The dry air continues over the mountain, heating up as it goes down the side of the mountain that faces away from the wind.

Cities can affect climate, too, especially in the summer. If you've ever walked barefoot on a hot street, you know that the street absorbs heat. So do parking lots and buildings. Some of the heat they absorb is sent back into the surrounding air. That makes the air in a city hotter.

Location, oceans, mountains, cities—now you know the things that cause different climates. Do the climates of regions ever change? They have in the past. Some people think they might change again in the future.

Climate Changes

Look at **Figure 18-17.** About 20 000 years ago huge sheets of ice called glaciers covered much of Canada and parts of the United States. What caused this type of climate change? Scientists are not sure. Some think that the tilt of Earth's axis or the path of Earth's orbit may change over a long period of time. Others think that a huge volcanic eruption or a meteorite collision with Earth may have caused the climate change.

Some people think we're headed for another big climate change. But this time, the change will be caused by people, not by volcanoes or meteorites.

FIGURE 18-17

This glacier in Argentina is similar to the ones that once covered large parts of North America.

USING TECHNOLOGY

GLOBE Project

Global warming is a big environmental problem—so big that scientists need all the help they can get! That's why more than 4000 schools in 55 countries are part of the GLOBE project. GLOBE stands for Global Learning and Observation to Benefit the Environment. It's a World Wide Web program that helps scientists keep track of changes in the environment. Who's doing the tracking? Students just like you.

In GLOBE, students collect environmental data near their schools. They might measure air temperature or test water samples for pollutants. They carefully record their data, then report their findings on the Internet. Scientists use the data to help them study environmental problems such as global warming.

SCIENCE *Online*

Check out Glencoe Science Online at *science.glencoe.com* for a tour of the GLOBE project.

Will Earth's climate change again?

Look at **Figure 18-18.** Some human activities, such as burning fossil fuels to produce electricity and cutting down rain forests for farmland, increase the amount of carbon dioxide in the atmosphere. Carbon dioxide traps heat in the atmosphere. When there's more carbon dioxide, more heat is trapped, so temperature increases around the world. An increase in average temperatures on Earth is called **global warming.**

Effects of Global Warming

Scientists are not sure what the effects of global warming will be. Some think rising temperatures might cause ice caps to melt. This, in turn, could cause a sudden rise in sea level and flooding along coastal areas. Other scientists aren't convinced that global warming is a problem. All scientists, though, warn against tampering with Earth's climate. We can help protect Earth from harmful climate change by using less electricity and recycling products to reduce our use of fossil fuels.

FIGURE **18-18**

Cutting down rain forests may increase the amount of carbon dioxide in the atmosphere. This, in turn, may lead to global warming.

Section Wrap-up

1. What factors cause climate?

2. List three types of climate. Describe each.

3. **Think Critically:** Can humans affect climate? How?

4. *Skill Builder*
   Recognizing Cause and Effect
   Temperature is one of the weather conditions used to determine a region's climate. Describe how your life might change if the average monthly temperatures in your area decreased by 10°C. If you need help, refer to Recognizing Cause and Effect on page 551 in the **Skill Handbook.**

Science Journal

Using a newspaper, magazine, or the Internet, research global warming. Write about changes that are being made around the world to reduce levels of carbon dioxide and other gases that contribute to global warming.

Science & *History*

SCIENCE *Online*

Other fleets have been destroyed by hurricanes through the years. Visit Glencoe Science Online at *science.glencoe.com* for a link to this history.

How Typhoons Saved Japan

Weather affects our lives every day. It can also change the course of history. One of the most dramatic events caused by weather happened in Japan during the 13th century.

When Kublai Khan, Genghis Khan's successor, came to power in China, he set his sights on conquering Japan. In 1274 and again in 1281, the Khan sent a huge fleet of ships to Japan carrying thousands of soldiers. But on both occasions, the Mongolian forces were utterly destroyed by typhoons. Without warning, these great windstorms swept in from the sea, crushed the ships that had anchored off the Japanese coast, and saved the country from invasion.

Typhoon, hurricane, and *cyclone* are all different names for the same thing—a powerful tropical storm that begins as a low-pressure area over warm seas. Gradually, it grows into a huge, spiraling mass of moisture-rich air, spinning at speeds of more than 120 km per hour. People living near the western Pacific Ocean and the China Sea call such storms typhoons, from *ty fung,* meaning "great wind." People living near the Atlantic Ocean call the storms hurricanes, after *Huracan,* a West Indian storm god. And for people living near the Indian Ocean, the storms are called cyclones, from the Greek word *kuklos,* meaning "circular."

As a typhoon moves across a body of water, it creates waves that can be 20 to 30 meters high. Violent winds and enormous waves make a typhoon one of the most destructive forces on Earth—as Kublai Khan found out.

Reviewing Main Ideas

Read the statements below that review the most important ideas in the chapter. Using what you have learned, answer each question in your Science Journal.

1. Air is mainly a mixture of gases, plus small amounts of solids and liquids. *What percentage of Earth's atmosphere is made of nitrogen? What percentage is made of oxygen?*

2. The sun's rays strike different parts of Earth at different angles. *How does this affect weather?*

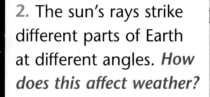

3. Sudden changes in weather are caused by movements of air masses. *What is the most powerful type of storm?*

4. Climate is the pattern of weather that occurs in a particular area over many years. *List three things that affect local climate.*

Using Key Science Words

air mass	meteorologist
atmosphere	stratosphere
climate	troposphere
front	weather
global warming	wind

Match each phrase with the correct term from the list of Key Science Words.

1. the pattern of weather that occurs in an area over many years
2. the place where two air masses meet
3. an increase in average temperatures on Earth
4. the layer of air that surrounds Earth like a blanket
5. the lowest layer of Earth's atmosphere

Checking Concepts

Choose the word or phrase that completes the sentence.

6. The _____ is the outermost layer of Earth's atmosphere.
   a. troposphere
   b. stratosphere
   c. exosphere
   d. thermosphere

7. A(n) _____ front forms when a warm air mass slides up and over a cold air mass.
   a. warm   c. stationary
   b. cold   d. occluded

8. One of the gases that might contribute to global warming is _____.
   a. helium   c. carbon dioxide
   b. hydrogen   d. oxygen

9. The amount of water vapor in the air ranges from _____.
   a. ten percent to 25 percent
   b. zero to four percent
   c. zero to ten percent
   d. 25 percent to 50 percent

10. _____ is a gas found in the stratosphere that protects life on Earth from the sun's ultraviolet rays.
    a. Carbon dioxide   c. Hydrogen
    b. Nitrogen   d. Ozone

Thinking Critically

Answer the following questions in your Science Journal using complete sentences.

11. Would a solar-powered car cause pollution? Explain your answer.

12. Some mountain climbers carry bottles of oxygen when they climb high mountains. Why?

13. Air in warmer places has a lower pressure than air from cooler places. Explain why this is true.

14. Meteorologists use airplanes, weather balloons, satellites, and computers to track weather. Why does the National Weather Bureau track hurricanes?

15. You know that changes in climate can be triggered by human activities. How does cutting down rain forests contribute to global warming?

Developing Skills

If you need help, refer to the description of each skill in the Skill Handbook.

16. **Making and Using Graphs:** Review the composition of Earth's atmosphere shown in **Figure 18-1** on page 496. Make a bar graph showing the amounts of the major gases found in the atmosphere.

17. **Comparing and Contrasting:** Compare and contrast a stationary front and an occluded front.

18. **Recognizing Cause and Effect:** Based on what you know about movement of air masses, can a meteorologist be correct 100 percent of the time? Why or why not?

19. **Interpreting Scientific Illustrations:** What kind of weather is associated with the front shown below? Describe how the air masses move in this type of front.

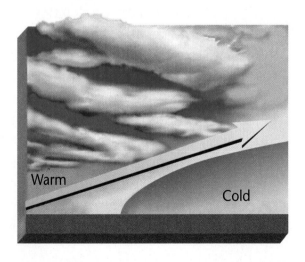

20. **Using a CD-ROM:** Using a CD-ROM, research a weather topic, such as weather forecasting, clouds, or severe storms. Write a report based on your research. In your report, describe the articles, photographs, pictures, or videos found in the CD-ROM.

Performance Assessment

1. **Model:** Design and build a rain gauge, an instrument to measure the amount of rain that falls in a given area. Place the rain gauge near your home or at school. Measure the amount of rain collected over a period of five days. Compare your measurements with those of other students. As a class, discuss some of the things that may have caused different measurements.

2. **Poster:** Review the causes of air pollution on page 497. Then make a poster showing some of the things that you and your family can do to help reduce air pollution.

Internet Project

Earth Science

It's raining cats and dogs!
Red sky at night, sailor's delight.

These sayings are about the weather. You may check out the weather forecast to decide if a concert will be rained out or if you need to wear a coat to the park. Knowing what the weather will be like is important. But how do scientists predict or forecast the weather? They collect weather data every day all over the country and try to find a pattern in the data. Why do they get it wrong sometimes? Let's find out.

Goals

- Collect weather data available on the Internet, in newspapers, and on television.
- Produce a weather forecast based on the data.

Researching Weather

1. Make a data table like the one on the next page.
2. Collect data in your area every day for at least two weeks.
3. Your weather data should include temperature, barometric pressure, wind speed, wind direction, amount and type of precipitation, and cloud cover. Be sure to include the date and your location.

Data Sources

Go to Glencoe Science Online at *science.glencoe.com* to find links to weather data on the Internet. You can post your weather data on the Homepage and collect data from other schools around the country. You can use these data to make your own weather maps. Print the map from the Glencoe

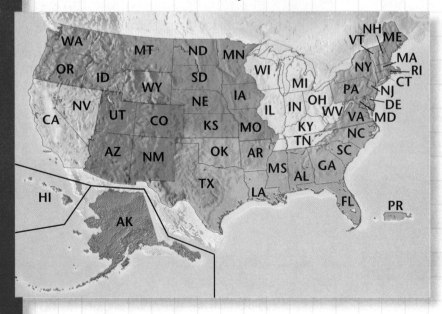

site or post your data on a large map of the United States, using an overlay of tissue paper or plastic. *If you do not have access to the Internet,* you can find weather data on television news shows, in newspapers, or on the radio. You can also make your own weather station. You'll need a thermometer, a barometer, a rain gauge, and a wind vane.

Predicting the Weather

1. After you have collected at least two weeks of weather data for your area, it's time to make your forecast.
2. Using your data, predict what the weather will be like each day for the next two weeks. For each day, make a prediction of the temperature, amount of cloud cover, wind, and whether there will be precipitation. Record your predictions in another data table.
3. Each day, record what the weather was really like and compare it to your prediction for that day.

Go Further

Look up weather data from your area for the last ten years. Find out the record high and low temperatures and amounts of precipitation. How could historical weather data help you make better forecasts?

Conclude and Apply

1. How close did your predictions come to the actual weather? Were your forecasts for the first few days more accurate than the later days' forecasts? Explain.
2. How could you make your predictions more accurate? Would data from other areas help? Explain your answer.

Weather Data Collection Table					
Date					
Location					
Temperature					
Barometric Pressure					
Wind Speed					
Wind Direction					
Type of Precipitation					
Amount of Precipitation					
Cloud Cover					

Appendices

Appendix A

SI Units of Measurement

Table A-1

SI Base Units					
Measurement	**Unit**	**Symbol**	**Measurement**	**Unit**	**Symbol**
length	meter	m	temperature	kelvin	K
mass	kilogram	kg	amount of substance	mole	mol
time	second	s			

Table A-2

Units Derived from SI Base Units		
Measurement	**Unit**	**Symbol**
energy	joule	J
force	newton	N
frequency	hertz	Hz
potential difference	volt	V
power	watt	W
pressure	pascal	Pa

Table A-3

Common SI Prefixes					
Prefix	**Symbol**	**Multiplier**	**Prefix**	**Symbol**	**Multiplier**
Greater than 1			Less than 1		
mega-	M	1 000 000	*deci-*	d	0.1
kilo-	k	1 000	*centi-*	c	0.01
hecto-	h	100	*milli-*	m	0.001
deka-	da	10	*micro-*	µ	0.000 001

Appendix B

SI/Metric to English Conversions

	When you want to convert:	To:	Multiply by:
Length	inches	centimeters	2.54
	centimeters	inches	0.39
	feet	meters	0.30
	meters	feet	3.28
	yards	meters	0.91
	meters	yards	1.09
	miles	kilometers	1.61
	kilometers	miles	0.62
Mass and Weight*	ounces	grams	28.35
	grams	ounces	0.04
	pounds	kilograms	0.45
	kilograms	pounds	2.2
	tons (short)	tonnes (metric tons)	0.91
	tonnes (metric tons)	tons (short)	1.10
	pounds	newtons	4.45
	newtons	pounds	0.23
Volume	cubic inches	cubic centimeters	16.39
	cubic centimeters	cubic inches	0.06
	cubic feet	cubic meters	0.03
	cubic meters	cubic feet	35.30
	liters	quarts	1.06
	liters	gallons	0.26
	gallons	liters	3.78
Area	square inches	square centimeters	6.45
	square centimeters	square inches	0.16
	square feet	square meters	0.09
	square meters	square feet	10.76
	square miles	square kilometers	2.59
	square kilometers	square miles	0.39
	hectares	acres	2.47
	acres	hectares	0.40
Temperature	Fahrenheit	5/9 (°F − 32)	Celsius
	Celsius	9/5°C + 32	Fahrenheit

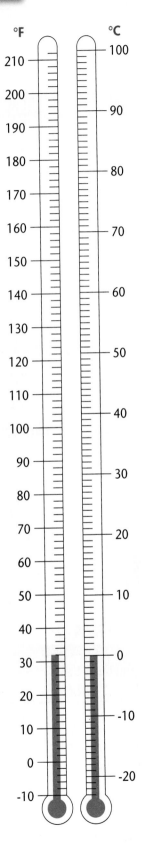

*Weight as measured in standard Earth gravity

Appendix C

Safety in the Classroom

1. Always obtain your teacher's permission to begin an investigation.
2. Study the procedure. If you have questions, ask your teacher. Be sure you understand any safety symbols shown on the page.
3. Use the safety equipment provided for you. Goggles and a safety apron should be worn during an investigation.
4. Always slant test tubes away from yourself and others when heating them.
5. Never eat or drink in the lab, and never use lab glassware as food or drink containers. Never inhale chemicals. Do not taste any substances or draw any material into a tube with your mouth.
6. If you spill any chemical, wash it off immediately with water. Report the spill immediately to your teacher.
7. Know the location and proper use of the fire extinguisher, safety shower, fire blanket, first aid kit, and fire alarm.
8. Keep all materials away from open flames. Tie back long hair and loose clothing.
9. If a fire should break out in the classroom, or if your clothing should catch fire, smother it with the fire blanket or a coat, or get under a safety shower. NEVER RUN.
10. Report any accident or injury, no matter how small, to your teacher.

Follow these procedures as you clean up your work area.
1. Turn off the water and gas. Disconnect electrical devices.
2. Return all materials to their proper places.
3. Dispose of chemicals and other materials as directed by your teacher. Place broken glass and solid substances in the proper containers. Never discard materials in the sink.
4. Clean your work area.
5. Wash your hands thoroughly after working in the laboratory.

Table C-1

First Aid	
Injury	**Safe Response**
Burns	Apply cold water. Call your teacher immediately.
Cuts and bruises	Stop any bleeding by applying direct pressure. Cover cuts with a clean dressing. Apply cold compresses to bruises. Call your teacher immediately.
Fainting	Leave the person lying down. Loosen any tight clothing and keep crowds away. Call your teacher immediately.
Foreign matter in eye	Flush with plenty of water. Use eyewash bottle or fountain.
Poisoning	Note the suspected poisoning agent and call your teacher immediately.
Any spills on skin	Flush with large amounts of water or use safety shower. Call your teacher immediately.

Appendix D

SAFETY SYMBOLS

SAFETY SYMBOLS	HAZARD	PRECAUTION	REMEDY
Disposal	Special disposal required	Dispose of wastes as directed by your teacher.	Ask your teacher how to dispose of laboratory materials.
Biological	Organisms that can harm humans	Avoid breathing in or skin contact with organisms. Wear dust mask or gloves. Wash hands thoroughly.	Notify your teacher if you suspect contact.
Extreme Temperature	Objects that can burn skin by being too cold or too hot	Use proper protection when handling.	Go to your teacher for first aid.
Sharp Object	Use of tools or glassware that can easily puncture or slice skin	Practice common sense behavior and follow guidelines for use of the tool.	Go to your teacher for first aid.
Fumes	Potential danger from smelling fumes	Must have good ventilation and never smell fumes directly.	Leave foul area and notify your teacher immediately.
Electrical	Possible danger from electrical shock or burn	Double-check setup with instructor. Check condition of wires and apparatus.	Do not attempt to fix electrical problems. Notify your teacher immediately.
Irritant	Substances that can irritate your skin or mucous membranes	Wear dust mask or gloves. Practice extra care when handling these materials.	Go to your teacher for first aid.
Chemical	Substances (acids and bases) that can react with and destroy tissue and other materials	Wear goggles and an apron.	Immediately flush with water and notify your teacher.
Toxic	Poisonous substance	Follow your teacher's instructions. Always wash hands thoroughly after use.	Go to your teacher for first aid.
Fire	Flammable and combustible materials may burn if exposed to an open flame or spark	Avoid flames and heat sources. Be aware of locations of fire safety equipment.	Notify your teacher immediately. Use fire safety equipment if necessary.

Eye Safety
This symbol appears when a danger to eyes exists.

Clothing Protection
This symbol appears when substances could stain or burn clothing.

Animal Safety
This symbol appears whenever live animals are studied and the safety of the animals and students must be ensured.

Appendix E

Mitosis

Mitosis is the process by which a nucleus divides into two nuclei, each containing the same number of chromosomes that the original cell had. Usually the cytoplasm then also divides.

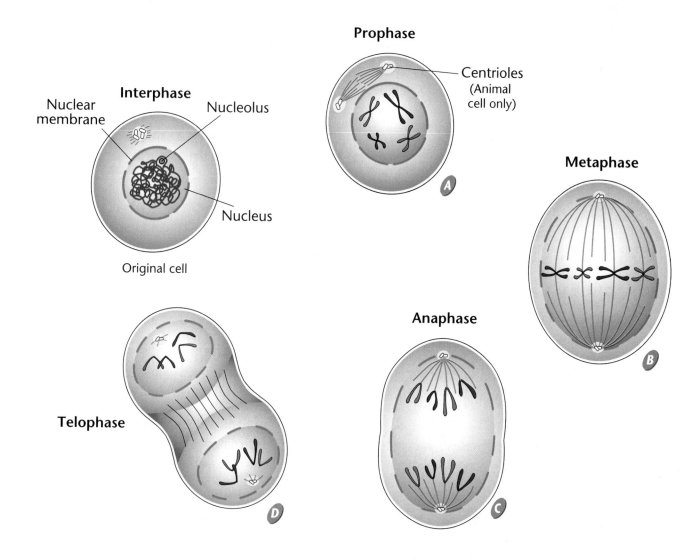

Interphase   Before mitosis begins, the chromosomes duplicate.

A Prophase   Duplicated chromosomes become visible.

B Metaphase   Duplicated chromosomes line up at the equator of the cell.

C Anaphase   Duplicated chromosomes separate.

D Telophase   The cytoplasm separates. Two new cells contain same number of chromosomes as the original cell.

Appendix F

Diversity of Life: Classification of Living Organisms

A six-kingdom system of classification of organisms is used today. Two kingdoms—Kingdom Archaebacteria and Kingdom Eubacteria—contain organisms that do not have a nucleus and that lack membrane-bound structures in the cytoplasm of their cells. The members of the other four kingdoms have a cell or cells that contain a nucleus and structures in the cytoplasm, some of which are surrounded by membranes. These kingdoms are Kingdom Protista, Kingdom Fungi, Kingdom Plantae, and Kingdom Animalia.

Kingdom Archaebacteria

one-celled; some absorb food from their surroundings; some are photosynthetic; some are chemosynthetic; many are found in extremely harsh environments including salt ponds, hot springs, swamps, and deep-sea hydrothermal vents

Kingdom Eubacteria

one-celled; most absorb food from their surroundings; some are photosynthetic; some are chemosynthetic; many are parasites; many are round, spiral, or rod-shaped; some form colonies

Kingdom Protista

Phylum Euglenophyta one-celled; photosynthetic or take in food; most have one flagellum; euglenoids

Kingdom Eubacteria
Bacillus anthracis

Phylum Chlorophyta
Desmids

Phylum Bacillariophyta one-celled; photosynthetic; have unique double shells made of silica; diatoms

Phylum Dinoflagellata one-celled; photosynthetic; contain red pigments; have two flagella; dinoflagellates

Phylum Chlorophyta one-celled, many-celled, or colonies; photosynthetic; contain chlorophyll; live on land, in freshwater, or salt water; green algae

Phylum Rhodophyta most are many-celled; photosynthetic; contain red pigments; most live in deep, saltwater environments; red algae

Phylum Phaeophyta most are many-celled; photosynthetic; contain brown pigments; most live in saltwater environments; brown algae

Phylum Rhizopoda one-celled; take in food; are free-living or parasitic; move by means of pseudopods; amoebas

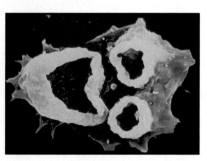

Amoeba

Phylum Zoomastigina one-celled; take in food; free-living or parasitic; have one or more flagella; zoomastigotes

Phylum Ciliophora one-celled; take in food; have large numbers of cilia; ciliates

Phylum Sporozoa one-celled; take in food; have no means of movement; are parasites in animals; sporozoans

Phylum Myxomycota
Slime mold

Phyla Myxomycota and Acrasiomycota one- or many-celled; absorb food; change form during life cycle; cellular and plasmodial slime molds

Phylum Oomycota many-celled; are either parasites or decomposers; live in freshwater or salt water; water molds, rusts and downy mildews

Kingdom Fungi

Phylum Zygomycota many-celled; absorb food; spores are produced in sporangia; zygote fungi; bread mold

Phylum Ascomycota one- and many-celled; absorb food; spores produced in asci; sac fungi; yeast

Phylum Basidiomycota many-celled; absorb food; spores produced in basidia; club fungi; mushrooms

Phylum Deuteromycota members with unknown reproductive structures; imperfect fungi; *Penicillium*

Mycophycota organisms formed by symbiotic relationship between an ascomycote or a basidiomycote and green alga or cyanobacterium; lichens

Phylum Oomycota
Phytophthora infestans

Lichens

Appendix F

Kingdom Plantae

Divisions Bryophyta (mosses), **Anthocerophyta** (hornworts), **Hepatophytal** (liverworts), **Psilophytal** (whisk ferns) many-celled nonvascular plants; reproduce by spores produced in capsules; green; grow in moist, land environments

Division Lycophyta many-celled vascular plants; spores are produced in conelike structures; live on land; are photosynthetic; club mosses

Division Sphenophyta vascular plants; ribbed and jointed stems; scalelike leaves; spores produced in conelike structures; horsetails

Division Pterophyta vascular plants; leaves called fronds; spores produced in clusters of sporangia called sori; live on land or in water; ferns

Division Ginkgophyta deciduous trees; only one living species; have fan-shaped leaves with branching veins and fleshy cones with seeds; ginkgoes

Division Cycadophyta palmlike plants; have large, featherlike leaves; produces seeds in cones; cycads

Division Coniferophyta deciduous or evergreen; trees or shrubs; have needlelike or scalelike leaves; seeds produced in cones; conifers

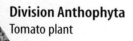

Division Anthophyta
Tomato plant

Division Gnetophyta shrubs or woody vines; seeds are produced in cones; division contains only three genera; gnetum

Division Anthophyta dominant group of plants; flowering plants; have fruits with seeds

Kingdom Animalia

Phylum Porifera aquatic organisms that lack true tissues and organs; are asymmetrical and sessile; sponges

Phylum Cnidaria radially symmetrical organisms; have a digestive cavity with one opening; most have tentacles armed with stinging cells; live in aquatic environments singly or in colonies; includes jellyfish, corals, hydra, and sea anemones

Phylum Platyhelminthes bilaterally symmetrical worms; have flattened bodies; digestive system has one opening; parasitic and free-living species; flatworms

Division Bryophyta
Liverwort

Phylum Platyhelminthes
Flatworm

Appendix F

Phylum Chordata

Phylum Nematoda round, bilaterally symmetrical body; have digestive system with two openings; free-living forms and parasitic forms; roundworms

Phylum Mollusca soft-bodied animals, many with a hard shell and soft foot or footlike appendage; a mantle covers the soft body; aquatic and terrestrial species; includes clams, snails, squid, and octopuses

Phylum Annelida bilaterally symmetrical worms; have round, segmented bodies; terrestrial and aquatic species; includes earthworms, leeches, and marine polychaetes

Phylum Arthropoda largest animal group; have hard exoskeletons, segmented bodies, and pairs of jointed appendages; land and aquatic species; includes insects, crustaceans, and spiders

Phylum Echinodermata marine organisms; have spiny or leathery skin and a water-vascular system with tube feet; are radially symmetrical; includes sea stars, sand dollars, and sea urchins

Phylum Chordata organisms with internal skeletons and specialized body systems; most have paired appendages; all at some time have a notochord, nerve cord, gill slits, and a postanal tail; include fish, amphibians, reptiles, birds, and mammals

Appendix G

PERIODIC TABLE OF THE ELEMENTS

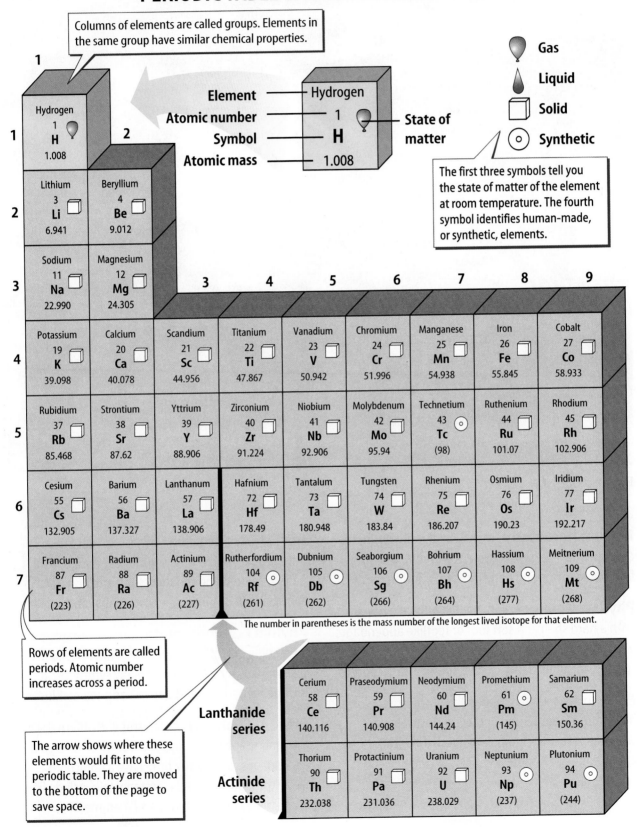

Columns of elements are called groups. Elements in the same group have similar chemical properties.

Element — Hydrogen
Atomic number — 1
Symbol — H
Atomic mass — 1.008

State of matter

Gas
Liquid
Solid
Synthetic

The first three symbols tell you the state of matter of the element at room temperature. The fourth symbol identifies human-made, or synthetic, elements.

1

1	Hydrogen 1 H 1.008
2	Lithium 3 Li 6.941
3	Sodium 11 Na 22.990
4	Potassium 19 K 39.098
5	Rubidium 37 Rb 85.468
6	Cesium 55 Cs 132.905
7	Francium 87 Fr (223)

2

Beryllium 4 Be 9.012
Magnesium 12 Mg 24.305
Calcium 20 Ca 40.078
Strontium 38 Sr 87.62
Barium 56 Ba 137.327
Radium 88 Ra (226)

	3	**4**	**5**	**6**	**7**	**8**	**9**
	Scandium 21 Sc 44.956	Titanium 22 Ti 47.867	Vanadium 23 V 50.942	Chromium 24 Cr 51.996	Manganese 25 Mn 54.938	Iron 26 Fe 55.845	Cobalt 27 Co 58.933
	Yttrium 39 Y 88.906	Zirconium 40 Zr 91.224	Niobium 41 Nb 92.906	Molybdenum 42 Mo 95.94	Technetium 43 Tc (98)	Ruthenium 44 Ru 101.07	Rhodium 45 Rh 102.906
	Lanthanum 57 La 138.906	Hafnium 72 Hf 178.49	Tantalum 73 Ta 180.948	Tungsten 74 W 183.84	Rhenium 75 Re 186.207	Osmium 76 Os 190.23	Iridium 77 Ir 192.217
	Actinium 89 Ac (227)	Rutherfordium 104 Rf (261)	Dubnium 105 Db (262)	Seaborgium 106 Sg (266)	Bohrium 107 Bh (264)	Hassium 108 Hs (277)	Meitnerium 109 Mt (268)

The number in parentheses is the mass number of the longest lived isotope for that element.

Rows of elements are called periods. Atomic number increases across a period.

The arrow shows where these elements would fit into the periodic table. They are moved to the bottom of the page to save space.

Lanthanide series

| Cerium 58 Ce 140.116 | Praseodymium 59 Pr 140.908 | Neodymium 60 Nd 144.24 | Promethium 61 Pm (145) | Samarium 62 Sm 150.36 |

Actinide series

| Thorium 90 Th 232.038 | Protactinium 91 Pa 231.036 | Uranium 92 U 238.029 | Neptunium 93 Np (237) | Plutonium 94 Pu (244) |

Appendix G

Metal

Metalloid

Nonmetal

Recently discovered

The color of an element's block tells you if the element is a metal, nonmetal, metalloid, or has been discovered so recently that more study is needed.

SCIENCE Online
Visit the Glencoe Science Web site at **science.glencoe.com** for updates to the periodic table.

		13	14	15	16	17	18
							Helium 2 He 4.003
		Boron 5 B 10.811	Carbon 6 C 12.011	Nitrogen 7 N 14.007	Oxygen 8 O 15.999	Fluorine 9 F 18.998	Neon 10 Ne 20.180

10	11	12					
			Aluminum 13 Al 26.982	Silicon 14 Si 28.086	Phosphorus 15 P 30.974	Sulfur 16 S 32.065	Chlorine 17 Cl 35.453	Argon 18 Ar 39.948
Nickel 28 Ni 58.693	Copper 29 Cu 63.546	Zinc 30 Zn 65.39	Gallium 31 Ga 69.723	Germanium 32 Ge 72.64	Arsenic 33 As 74.922	Selenium 34 Se 78.96	Bromine 35 Br 79.904	Krypton 36 Kr 83.80
Palladium 46 Pd 106.42	Silver 47 Ag 107.868	Cadmium 48 Cd 112.411	Indium 49 In 114.818	Tin 50 Sn 118.710	Antimony 51 Sb 121.760	Tellurium 52 Te 127.60	Iodine 53 I 126.904	Xenon 54 Xe 131.293
Platinum 78 Pt 195.078	Gold 79 Au 196.967	Mercury 80 Hg 200.59	Thallium 81 Tl 204.383	Lead 82 Pb 207.2	Bismuth 83 Bi 208.980	Polonium 84 Po (209)	Astatine 85 At (210)	Radon 86 Rn (222)
Ununnilium * 110 Uun (281)	Unununium * 111 Uuu (272)	Ununbium * 112 Uub (285)		Ununquadium * 114 Uuq (289)		Ununhexium * 116 Uuh (289)		Ununoctium * 118 Uuo (293)

* Names not officially assigned. Discovery of elements 114, 116, and 118 recently reported. Further information not yet available.

Europium 63 Eu 151.964	Gadolinium 64 Gd 157.25	Terbium 65 Tb 158.925	Dysprosium 66 Dy 162.50	Holmium 67 Ho 164.930	Erbium 68 Er 167.259	Thulium 69 Tm 168.934	Ytterbium 70 Yb 173.04	Lutetium 71 Lu 174.967
Americium 95 Am (243)	Curium 96 Cm (247)	Berkelium 97 Bk (247)	Californium 98 Cf (251)	Einsteinium 99 Es (252)	Fermium 100 Fm (257)	Mendelevium 101 Md (258)	Nobelium 102 No (259)	Lawrencium 103 Lr (262)

Care and Use of a Microscope

Eyepiece Contains a magnifying lens you look through

Arm Supports the body tube

Low-power objective Contains the lens with low-power magnification

Stage clips Hold the microscope slide in place

Coarse Adjustment Focuses the image under low power

Fine Adjustment Sharpens the image under high and low magnification

Base Provides support for the microscope

Body tube Connects the eyepiece to the revolving nosepiece

Revolving nosepiece Holds and turns the objectives into viewing position

High-power objective Contains the lens with the highest magnification

Stage Supports the microscope slide

Light source Allows light to reflect upward through the diaphragm, the specimen, and the lenses

Care of a Microscope

1. Always carry the microscope holding the arm with one hand and supporting the base with the other hand.
2. Don't touch the lenses with your fingers.
3. Never lower the coarse adjustment knob when looking through the eyepiece lens.
4. Always focus first with the low-power objective.
5. Don't use the coarse adjustment knob when the high-power objective is in place.
6. Store the microscope covered.

Using a Microscope

1. Place the microscope on a flat surface that is clear of objects. The arm should be toward you.
2. Look through the eyepiece. Adjust the diaphragm so that light comes through the opening in the stage.
3. Place a slide on the stage so that the specimen is in the field of view. Hold it firmly in place by using the stage clips.

4. Always focus first with the coarse adjustment and the low-power objective lens. Once the object is in focus on low power, turn the nosepiece until the high-power objective is in place. Use ONLY the fine adjustment to focus with the high-power objective lens.

Making a Wet-Mount Slide

1. Carefully place the item you want to look at in the center of a clean glass slide. Make sure the sample is thin enough for light to pass through.
2. Use a dropper to place one or two drops of water on the sample.
3. Hold a clean coverslip by the edges and place it at one edge of the drop of water. Slowly lower the coverslip onto the drop of water until it lies flat.
4. If you have too much water or a lot of air bubbles, touch the edge of a paper towel to the edge of the coverslip to draw off extra water and force out air.

Appendix I

Rocks

Rock Type	Rock Name	Characteristics
Igneous (intrusive) Photos on pages 416, 430.	Granite	Large mineral grains of quartz, feldspar, hornblende, and mica. Usually light in color.
	Diorite	Large mineral grains of feldspar, hornblende, mica. Less quartz than granite. Intermediate in color.
	Gabbro	Large mineral grains of feldspar, hornblende, and mica. No quartz. Dark in color.
Igneous (extrusive) Photos on page 417.	Rhyolite	Small mineral grains of quartz, feldspar, hornblende, and mica or no visible grains. Light in color.
	Andesite	Small mineral grains of feldspar, hornblende, mica or no visible grains. Less quartz than rhyolite. Intermediate in color.
	Basalt	Small mineral grains of feldspar, hornblende, mica or no visible grains. No quartz. Dark in color.
	Obsidian	Glassy texture. No visible grains. Volcanic glass. Fracture looks like broken glass.
	Pumice	Frothy texture. Floats.
Sedimentary (broken, detrital) Photos on pages 420-421, 430.	Conglomerate	Coarse-grained. Gravel or pebble-sized grains.
	Sandstone	Sand-sized grains 1/16 to 2 mm in size.
	Siltstone	Grains are smaller than sand but larger than clay.
	Shale	Smallest grains. Usually dark in color.
Sedimentary (biochemical) Photos on pages 422-423.	Limestone	Major mineral is calcite. Usually forms in oceans, lakes, and rivers. Often contains fossils.
	Coal	Occurs in swampy, low-lying areas. Compacted layers of organic material, mainly plant remains.
Sedimentary (chemical)	Rock Salt	Commonly formed by the evaporation of seawater.
Metamorphic (foliated) Photos on pages 427-429.	Gneiss	Well-developed banding because of alternating layers of different minerals, usually of different colors. Common parent rock is granite.
	Schist	Well-defined parallel arrangement of flat, sheet-like minerals, mainly micas. Common parent rocks are shale, phyllite.
	Phyllite	Shiny or silky appearance. Looks wrinkled. Common parent rocks are shale, slate.
	Slate	Harder, denser, and shinier than shale. Common parent rock is shale.
Metamorphic (non-foliated) Photos on pages 428, 430.	Marble	Interlocking calcite or dolomite crystals. Common parent rock is limestone.
	Soapstone	Composed mainly of the mineral talc. Soft with a greasy feel.
	Quartzite	Hard and well cemented with interlocking quartz crystals. Common parent rock is sandstone.

Appendix J

Minerals

Mineral (formula)	Color	Streak	Hardness	Breakage pattern	Uses and other properties
graphite (C)	black to gray	black to gray	1-2	basal cleavage (scales)	pencil lead, lubricants for locks, rods to control some small nuclear reactions, battery poles
silver (Ag)	silvery white, tarnishes to black	light gray to silver	2.5	hackly	coins, fillings for teeth, jewelry, silverplate, wires; malleable and ductile
galena (PbS)	gray	gray to black	2.5	cubic cleavage perfect	source of lead, used in pipes, shields for X rays, fishing equipment sinkers
gold (Au)	pale to golden yellow	yellow	2.5-3	hackly	jewelry, money, gold leaf, fillings for teeth, medicines; does not tarnish
bornite (Cu_5FeS_4)	bronze, tarnishes to dark blue, purple	gray-black	3	uneven fracture	source of copper; called "peacock ore" because of the purple shine when it tarnishes
copper (Cu)	copper red	copper red	3	hackly	coins, pipes, gutters, wire, cooking utensils, jewelry, decorative plaques; malleable and ductile
chalcopyrite ($CuFeS_2$)	brassy to golden yellow	greenish black	3.5-4	uneven fracture	main ore of copper
chromite ($FeCr_2O_4$)	black or brown	brown to black	5.5	irregular fracture	ore of chromium, stainless steel, metallurgical bricks
pyrrhotite (FeS)	bronze	gray-black	4	uneven fracture	often found with pentlandite, an ore of nickel; may be magnetic
hematite (specular) (Fe_2O_3)	black or reddish brown	red or reddish brown	6	irregular fracture	source of iron; roasted in a blast furnace, converted to "pig" iron, made into steel
magnetite (Fe_3O_4)	black	black	6	conchoidal fracture	source of iron, naturally magnetic, called lodestone
pyrite (FeS_2)	light, brassy, yellow	greenish black	6.5	uneven fracture	source of iron, "fool's gold," alters to limonite

Appendix J

Minerals

Mineral (formula)	Color	Streak	Hardness	Breakage pattern	Uses and other properties
talc $(Mg_3(OH)_2Si_4O_{10})$	white, greenish	white	1	cleavage in one direction	easily cut with fingernail; used for talcum powder; soapstone; is used in paper and for tabletops
kaolinite $(Al_2Si_2O_5(OH)_4)$	white, red, reddish, brown, black	white	2	basal cleavage	clays; used in ceramics and in china dishes; common in most soils; often microscopic-sized particles
gypsum $(CaSO_4 \cdot 2H_2O)$	colorless, gray, white, brown	white	2	basal cleavage	used extensively in the preparation of plaster of paris, alabaster, and dry wall for building construction
sphalerite (ZnS)	brown	pale yellow	3.5-4	cleavage in six directions	main ore of zinc; used in paints, dyes, and medicine
sulfur (S)	yellow	yellow to white	2	conchoidal fracture	used in medicine, fungicides for plants, vulcanization of rubber, production of sulfuric acid
muscovite $(KAl_3Si_3O_{10}(OH)_2)$	white, light gray, yellow, rose, green	colorless	2.5	basal cleavage	occurs in large flexible plates; used as an insulator in electrical equipment, lubricant
biotite $(K(Mg, Fe)_3AlSi_3O_{10}(OH)_2)$	black to dark brown	colorless	2.5	basal cleavage	occurs in large flexible plates
halite $(NaCl)$	colorless, red, white, blue	colorless	2.5	cubic cleavage	salt; very soluble in water; a preservative
calcite $(CaCO_3)$	colorless, white, pale blue	colorless, white	3	cleavage in three directions	fizzes when HCl is added; used in cements and other building materials
dolomite $(CaMg(CO_3)_2)$	colorless, white, pink, green, gray, black	white	3.5-4	cleavage in three directions	concrete and cement; used as an ornamental building stone

Appendix J

Minerals

Mineral (formula)	Color	Streak	Hardness	Breakage pattern	Uses and other properties
fluorite (CaF_2)	colorless, white, blue, green, red, yellow, purple	colorless	4	cleavage	used in the manufacture of optical equipment; glows under ultraviolet light
limonite (hydrous iron oxides)	yellow, brown, black	yellow, brown	5.5	conchoidal fracture	source of iron; weathers easily, coloring matter of soils
hornblende ($CaNa(Mg, Al,Fe)_5(Al,Si)_2 Si_6O_{22}(OH)_2$)	green to black	gray to white	5-6	cleavage in two directions	will transmit light on thin edges; 6-sided cross section
feldspar (orthoclase) ($KAlSi_3O_8$)	colorless, white to gray, green and yellow	colorless	6	two cleavage planes meet at 90° angle	insoluble in acids; used in the manufacture of porcelain
feldspar (plagioclase) ($NaAlSi_3O_8$) ($CaAl_2Si_2O_8$)	gray, green, white	colorless	6	two cleavage planes meet at 86° angle	used in ceramics; striations present on some faces
augite ($(Ca, Na) (Mg, Fe, Al) (Al, Si)_2O_6$)	black	colorless	6	2-directional cleavage	square or 8-sided cross section
olivine ($(Mg, Fe)_2 SiO_4$)	olive green	colorless	6.5	conchoidal fracture	gemstones, refractory sand
quartz (SiO_2)	colorless, various colors	colorless	7	conchoidal fracture	used in glass manufacture, electronic equipment, radios, computers, watches, gemstones
garnet ($(Mg, Fe,Ca)_3 (Al_2Si_3O_{12})$)	deep yellow-red, green, black	colorless	7.5	conchoidal fracture	used in jewelry; also used as an abrasive
topaz ($Al_2SiO_4 (F, OH)_2$)	white, pink, yellow, pale blue, colorless	colorless	8	basal cleavage	valuable gemstone
corundum (Al_2O_3)	colorless, blue, brown, green, white, pink, red	colorless	9	fracture	gemstones: ruby is red, sapphire is blue; industrial abrasive

Skill Handbook

Table of Contents

Organizing Information

Communicating

Being able to explain ideas to other people is an important part of our everyday lives. Whether reading a book, writing a letter, or watching a television program, people everywhere are giving their opinions and sharing information with each other.

Science Journal One way to record information, and express how you think about a topic in science is by writing in your Science Journal. It also lets you show how much you know about a subject.

There are many different kinds of Science Journal assignments. You may be asked to pretend you are a scientist, a TV reporter, or a committee member of a local environmental group and write from that point of view. Maybe you will be communicating your opinions to a member of Congress, a doctor, or to the editor of your local newspaper. Sometimes, you will summarize information, make an outline or a diagram, or write a letter or a paragraph in your Science Journal.

FIGURE 1

A Science Journal entry

FIGURE 2

Classifying dishes

Classifying

You may not realize it, but you make things orderly in the world around you. If you hang all your shirts together in the closet or if your favorite CDs are stacked together by performer, you have used the skill of classifying. Classifying is the process of sorting objects or events into groups based on things they have in common. When classifying, first look closely at the objects or events to be classified. Then, select one characteristic or feature that is shared by some members in the group but not by all. Place those members that share that feature into their own group. You can classify members into smaller and smaller groups based on similar characteristics.

What do you do with the dishes after they are washed? You classify them as you put them away, as shown in **Figure 2.** You separate the silverware from the plates and glasses. Forks, spoons, and knives each have their own place in the drawer. You would keep separating the dishes until all are classified and put away. Remember that all the smaller groups still share the common feature of being an eating utensil.

Sequencing

A sequence is an arrangement of things or events in a certain order. When you are asked to sequence objects or events, decide what comes first, then think about what should come next. Continue to choose objects or events until all are in order. Then, go back over the sequence to make sure each thing or event in your sequence logically leads to the next.

A sequence you are familiar with is alphabetical order. Another example of sequence would be the steps in a recipe, as shown in **Figure 3.** Think about following a recipe. Steps in a recipe for chocolate-chip cookies have to be followed in order for the cookies to turn out right.

FIGURE 3

A recipe for chocolate-chip cookies contains a sequence of steps.

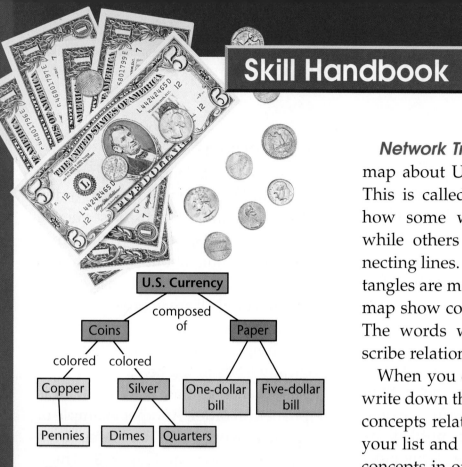

FIGURE 4

Network tree describing U.S. currency

Concept Mapping

If you were taking a trip in a car, you would probably take along a road map. The road map shows you where you are, where you are going, and other places along the way.

A concept map is similar to a road map. But a concept map shows relationships among ideas (or concepts) rather than places. A concept map is a diagram that shows how concepts are related visually. Because the concept map shows relationships among ideas, it can make the meanings of ideas and terms clear and help you understand better what you are studying.

Three types of concept maps are described here: a network tree, an events chain, and a spider map.

Network Tree Look at the concept map about U.S. currency in **Figure 4.** This is called a network tree. Notice how some words are in rectangles while others are written across connecting lines. The words inside the rectangles are main ideas. The lines in the map show connections between ideas. The words written on the lines describe relationships between concepts.

When you construct a network tree, write down the topic and list the major concepts related to that topic. Look at your list and begin to put the ideas or concepts in order from general to specific. Branch the related concepts from the major concept and describe the relationships on the lines.

Events Chain An events chain map is used to describe concepts in order.

FIGURE 5

Events chain of a typical morning routine.

In science, an events chain can be used to describe a sequence of events, the steps in a procedure, or the stages of a process.

When making an events chain, first find the one event that starts the chain. This event is called the initiating event. Then, find the next event in the chain and continue until you reach an outcome. Suppose you are asked to describe what happens when your alarm clock rings. An events chain map describing the steps might look like **Figure 5.**

Cycle Map A cycle concept map is a special type of events chain map. In a cycle concept map, the series of events does not produce a final outcome. The last event in the chain relates back to

the initiating event. Because there is no outcome and the last event relates back to the initiating event, the cycle repeats itself. Look at the cycle map describing the relationship between day and night in **Figure 6.**

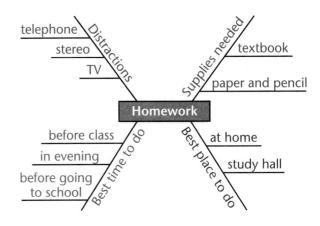

FIGURE 7

Spider map about homework.

Spider Map A fourth type of concept map is the spider map. This is a map that you can use for brainstorming. Once you brainstorm ideas from a central idea, you may find you have a jumble of ideas. Many of these ideas are related to the central idea but are not necessarily clearly related to each other. As illustrated by the spider map in **Figure 7,** you may begin to separate and group unrelated terms so that they become more useful by writing them outside the main concept.

FIGURE 6

Cycle map of day and night.

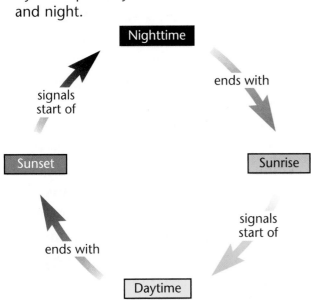

Making and Using Tables

Browse through your textbook and you will notice tables in the text and in the activities. In a table, data or information is arranged in a way that makes it easier for you to understand. Activity tables help organize and interpret the data you collect during an activity.

Most tables have a title. The title tells you what the table is about. A table is divided into columns and rows. The first column lists items to be compared. In **Figure 8,** a collection of recy-

Recycled Materials			
Day of Week	Paper (kg)	Aluminum (kg)	Plastic (kg)
Mon.	4.0	2.0	0.5
Wed.	3.5	1.5	0.5
Fri.	3.0	1.0	1.5

FIGURE 8

Table of recycled materials.

clable materials is being compared in a table. The row across the top lists the specific characteristics being compared. Collected data are recorded within the grid of the table. To make a table, list the items to be compared down in columns and the characteristics to be compared across in rows.

The title of the table in **Figure 8** is "Recycled Materials." What is being compared? This table shows the different materials being recycled and on which days they are recycled. To find out how much plastic, in kilograms, is being recycled on Wednesday, locate the column labeled "Plastic (kg)" and the row "Wed." The datum in the box where the column and row intersect gives the answer. Did you answer "0.5"? How much aluminum, in kilograms, is being recycled on Friday? If you answered "1.0," you understand how to use the parts of the table.

Making and Using Graphs

After scientists organize data in tables, they often show the data in a graph. A graph is a diagram that shows the relationship of one item or variable to another. A graph makes interpretation and analysis of data easier. There are three basic types of graphs used in science: the line graph, the bar graph, and the circle graph.

Line Graphs A line graph is used to show the relationship between two variables. The variables being compared go on two axes of the graph. The independent variable always goes on the horizontal axis, called the *x*-axis. The independent variable is the condition that is being changed. The dependent variable always goes on the vertical axis, called the *y*-axis. The dependent variable is any change that results from the changes in the independent variable.

Suppose your class started to record the amount of materials they collected in one week for their school to recycle. The collected information is shown in **Figure 9.**

Materials Collected During Week		
Day of Week	Paper (kg)	During Week (kg)
Mon.	5.0	4.0
Wed.	4.0	1.0
Fri.	4.0	2.0

FIGURE 9

Amount of recyclable materials collected during one week.

You could make a graph of the materials collected over the three days of the school week. The three weekdays are the independent variables and are placed on the *x*-axis of your graph. The amount of materials collected is the dependent variable and would go on the *y*-axis. After drawing your axes, label each with a scale. The *x*-axis lists the three weekdays. To make a scale of the amount of materials collected on the *y*-axis, look at the data values. Because the lowest amount collected was 1.0 and the highest was 5.0, you will have to start numbering at least at 1.0 and go through 5.0.

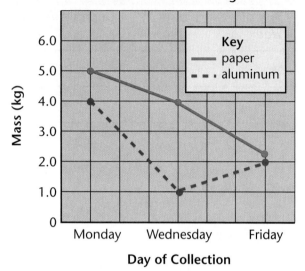

Material Collected During Week

Mass (kg)

Key
— paper
- - - aluminum

Day of Collection

FIGURE 10

Line graph of materials collected during week.

Next, plot the data points for collected paper. The first pair of data you want to plot is Monday and 5.0 kg of paper. Locate "Monday" on the x-axis and locate "5.0" on the y-axis. Where an imaginary vertical line from the x-axis and an imaginary horizontal line from the y-axis would meet, place the first data point. Place the other data points the same way. After all the points are plotted, connect them with the best smooth curve. Repeat this procedure for the data points for aluminum. Use continuous and dashed lines to distinguish the two line graphs. The resulting graph should look like **Figure 10.**

Bar Graphs Bar graphs are similar to line graphs. They compare data that do not continuously change. In a bar graph, vertical bars show the relationships among data.

To make a bar graph, set up the x-axis and y-axis as you did for the line graph. The data are plotted by drawing vertical bars from the x-axis up to a point where the y-axis would meet the bar if it were extended.

Look at the bar graph in **Figure 11** comparing the mass of aluminum collected over three weekdays. The x-axis is the days on which the aluminum was collected. The y-axis is the mass of aluminum collected, in kilograms.

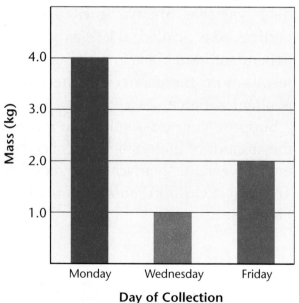

Aluminum Collected During Week

Mass (kg)

Day of Collection

FIGURE 11

Bar graph of aluminum collected during week.

Circle Graphs A circle graph or pie graph uses a circle divided into sections to show data. Each section represents part of the whole. All the sections together equal 100 percent.

Suppose you wanted to make a circle graph to show the number of seeds that sprouted or grew from a package of seeds. You count the total number of seeds. You find that there are 143 seeds in the package. This represents 100 percent, the whole circle.

You plant the seeds, and 129 seeds sprout. The seeds that sprouted will make up one section of the circle graph, and the seeds that did not sprout will make up the remaining section.

To find out how much of the circle each section should take, divide the number of seeds in each section by the total number of seeds. Then multiply your answer by 360, the number of degrees in a circle. Round to the nearest whole number. The section of the circle graph in degrees that represents the seeds sprouted is figured below.

$$\frac{129}{143} \times 360 = 324.75 \text{ or } 325 \text{ degrees}$$

To plot these data on a circle graph, you need a compass and a protractor. Use the compass to draw a circle. It will be easier to measure the part of the circle representing the seeds that did not sprout, so subtract 325° from 360° to get 35°. Draw a straight line from the center of the circle to the edge of the circle. Place your protractor on this line and use it to mark a point at 35°.

Use this point to draw a straight line from the center of the circle to the edge. This is the section for the group of seeds that did not grow. The other section represents the group of 129 seeds that did grow. Label the sections and title the graph, as shown in **Figure 12.**

FIGURE 12

Circle graph of
seed germination

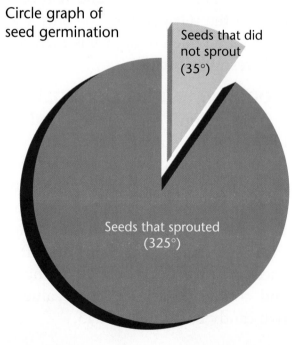

Seeds that did
not sprout
(35°)

Seeds that sprouted
(325°)

Thinking Critically

Observing and Inferring

Observing Scientists try to make careful and accurate observations. When possible, they use instruments such as microscopes, thermometers, and balances to make observations. Measurements with a balance or thermometer provide numerical data that can be checked and repeated. Other observations are made using your senses. The basis of all scientific inquiry is observation.

Inferring Scientists often make inferences based on their observations. An inference is a conclusion about what was observed.

When making an inference, be certain to use correct data and observations. Analyze all of the data that you've collected. Then, based on everything you know, draw a conclusion about what you've observed. If possible, investigate further to find out if your inference was correct.

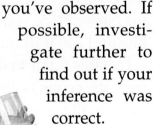

Nutritional Value		
	Candy A	Candy B
Serving Size	103 g	105 g
Calories	220	160
Total fat	10 g	10 g
Protein	25 g	2.6 g
Total carbohydrate	30 g	15 g

FIGURE 13

Table comparing the nutritional value of *Candy A* and *Candy B*.

When you drank a glass of orange juice after the volleyball game, you observed that the orange juice was cold. You might infer or conclude that the juice was cold because it had been made earlier in the day and had been kept in the refrigerator. The only way to be sure which inference is correct is to investigate further.

Comparing and Contrasting

Observations can be analyzed by looking at the similarities and the differences between two or more objects or events that you observe. When you look at objects or events to see how they are similar, you are comparing them. Contrasting is looking for differences in objects or events. Compare and contrast the nutritional value of two candy bars in **Figure 13.**

Recognizing Cause and Effect

Have you ever watched something happen and tried to figure out how or why it came about? If so, you have observed an effect and inferred a reason for the event. The event is an effect, and the reason for the event is the cause.

Suppose that every time your teacher fed the fish in a classroom aquarium, he or she tapped the food container on the edge of the aquarium. Then, one day your teacher just happened to tap the edge of the aquarium with a pencil. You observed the fish swim to the surface of the aquarium to feed, as shown in **Figure 14.** What is the effect, and what would you infer to be the cause? The effect is the fish swimming to the surface of the aquarium. You might infer the cause to be the teacher tapping on the edge of the aquarium. In determining cause and effect, you have made a logical inference based on your observations.

Perhaps the fish swam to the surface because they reacted to the teacher's waving hand or for some other reason. When scientists are unsure of the cause of a certain event, they plan experiments to determine what causes the event. Although you have made a logical conclusion about the behavior of the fish, you would have to perform an experiment to be certain that it was the tapping that caused the effect you observed.

FIGURE 14

What cause-and-effect situations are occurring in this aquarium?

Practicing Scientific Processes

Scientists use an orderly approach to learn new information and to solve problems. The methods scientists may use include observing to form a hypothesis, designing an experiment to test a hypothesis, separating and controlling variables, and interpreting data.

Forming Operational Definitions

Operational definitions define an object by showing how it functions, works, or behaves. Such definitions are written in terms of how an object works or how it can be used; that is, what its job or purpose is.

Some operational definitions explain how an object, such as the car in **Figure 15,** can be used.

- A car is a vehicle that can move things from one place to another.

Or such a definition may explain how an object works.

- A car is a vehicle that can move from place to place.

FIGURE 16

What hypothesis could be made about these plants?

Forming a Hypothesis

Hypotheses A hypothesis is a prediction, based on observation, that can be tested. Hypotheses are often stated as if-and-then statements. A hypothesis needs to be testable. For example, a scientist has observed in **Figure 16** that plants that are fertilized grow taller than plants that are not. A scientist may form a hypothesis that says: If plants are fertilized, then they will grow taller. This hypothesis can be tested by an experiment.

FIGURE 15

What observations can be made about this car?

Designing an Experiment to Test a Hypothesis

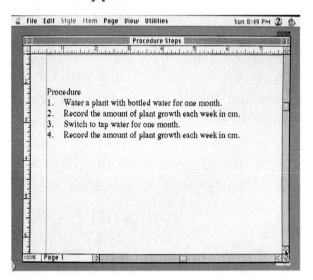

FIGURE 17

Possible procedural steps.

Once you have stated a hypothesis, you probably want to find out whether or not it explains an event or observation. In order to test a hypothesis, you must perform an experiment. When conducting an experiment, it is best to begin by writing out a procedure. A procedure is the plan that you follow in your experiment. A procedure tells you what materials to use and how to use them. After following the procedure, data are obtained. From this data, you can then draw a conclusion and make a statement about your results.

If the conclusion you draw from the data supports your hypothesis, then you can say that your hypothesis is reliable. Reliable means that you can trust your conclusion. If it did not support your hypothesis, then you would have to make new observations and state a new hypothesis—just make sure that it is one that you can test.

Planning a Procedure Suppose you hypothesize that a houseplant will grow better if watered with bottled water than with tap water. Let's figure out how to conduct an experiment to test the hypothesis: If a plant is watered with bottled water, then it will grow better and look healthier. An example procedure is seen in **Figure 17.** The data generated from this procedure are shown in **Figure 18.** The data show that the plant grew the same amount each month regardless of the type of water used. This conclusion does not support the original hypothesis made.

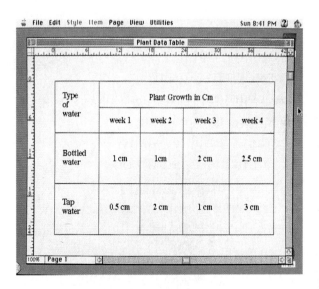

FIGURE 18

Data generated from procedural steps.

Separating and Controlling Variables

In any experiment, it is important to keep everything the same except for the item you are testing. The one factor that you change is called the independent variable. The independent variable in the experiment on the previous page was the type of water. The factor that changes as a result of the independent variable is called the dependent variable. The dependent variable in the experiment was the plant height. Always make sure that there is only one independent variable. If you have more than one, you will not know what caused the changes you observe in the independent variable. Many experiments have a control. A control is a treatment or an experiment that you can compare with the results of your test groups.

In the experiment with the plants, you made everything the same except the type of water being used. The soil, the amount of water given, the amount of light, and the temperature of the room should remain the same throughout the entire experiment. By doing so, you made sure that at the end of the experiment, any differences were the result of the type of water being used—bottled or tap. The type of water was the independent factor, and the height of the plant was the dependent factor.

Interpreting Data

The word *interpret* means "to explain the meaning of something." Look at the problem being explored in the plant experiment and find out what the data show. When you looked at the data you collected, you were checking to see if the variable had an effect. You were looking for an explanation. If there are differences in the data, the variable being tested may have had an effect. If there is no difference between the control and the test groups, the variable being tested apparently has had no effect.

Look back at **Figure 18** on page 553, which shows the results of this experiment. In this example, the use of tap water to water the plant was the control, while watering the plant with bottled water was the test. Data showed no difference in the amount of plant growth over a one-month period.

What are data? In the experiment described on these pages, measurements were taken so that at the end of the experiment, you had concrete numbers to interpret. Not every experiment that you do will give you data in the form of numbers. Sometimes data will be in the form of a description. At the end of a chemistry experiment, you might have noted that one solution turned yellow when treated with a particular chemical, and another remained clear when treated with the same chemical. Data, therefore, are stated in different forms for different types of scientific experiments.

Are all experiments alike? Keep in mind as you perform experiments in science that not every experiment makes use of all of the parts that have been described on these pages. For some situations, it may be difficult to design an experiment that will always have a control. Other experiments are complex enough that it may be hard to have only one dependent variable. Scientists use many variations in their methods of performing experiments. The skills in this handbook are here for you to use and practice. In real situations, their uses will vary.

Representing and Applying Data

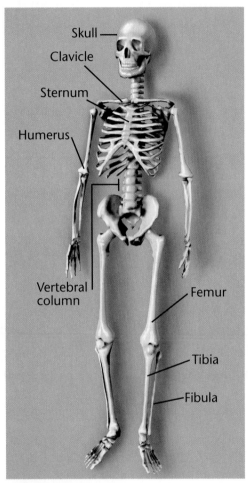

Skull
Clavicle
Sternum
Humerus
Vertebral column
Femur
Tibia
Fibula

FIGURE 19

Interpreting Scientific Illustrations

As you read a science textbook, you will see many drawings, diagrams, and photographs. Illustrations help you to understand what you read. Some illustrations are included to help you understand an idea that you can't see easily by yourself. For instance, we can't see atoms, but we can look at a diagram of an atom that helps us to understand

some things about atoms. Seeing something often helps you remember more easily. Illustrations also provide examples that clarify difficult concepts or give additional information about the topic you are studying. Maps, for example, help you to locate places that may be described in the text.

Most illustrations have captions. A caption identifies or explains the illustration. Some captions are short; others are longer and more descriptive. Diagrams often have labels that identify parts of the organism or the order of steps in a process, such as the labels in **Figure 19.**

Learning with Illustrations An illustration of an organism shows that organism from a particular side. In order to understand the illustration, you may need to identify the front (anterior) end, the tail (posterior) end, the underside (ventral), and the back (dorsal) side, as shown in **Figure 20.**

FIGURE 20

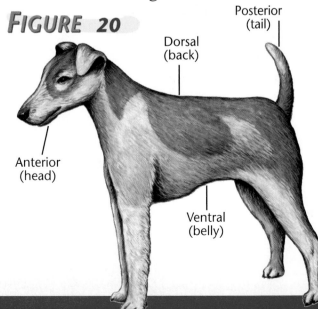

Posterior (tail)
Dorsal (back)
Anterior (head)
Ventral (belly)

Making Models

Have you ever worked on a model car, plane, or rocket? Models look, and sometimes work, much like the real thing, but they are often smaller or larger. In science, models are used to help simplify processes or structures that otherwise would be difficult to see and understand.

To make a model, you first have to get a basic idea about the structure or process involved. For example, make a model to show the differences in size of arteries, veins, and capillaries. First, read about these structures. All three are hollow tubes. Arteries are round and thick. Veins are flat and have thinner walls than arteries. Capillaries are small.

Now, decide what you can use for your model. Common materials are often best and cheapest to work with when making models. The different kinds and sizes of pasta shown in **Figure 21** might work for these models. Different sizes of rubber tubing might do just as well. Cut and glue the different noodles or tubing onto thick paper so the openings can be seen. Then label each. Now you have a simple, easy-to-understand model showing the differences in size of arteries, veins, and capillaries.

What other scientific ideas might a model help you to understand? A model of a molecule can be made from gumdrops (using different colors for the different elements present) and toothpicks (to show different chemical bonds). A working model of a volcano can be made from clay, a small amount of baking soda, vinegar, and a bottle cap. Other models can be devised on a computer.

FIGURE 21

Different types of pasta may be used to model blood vessels.

Skill Handbook

Measuring in SI

The International System (SI) of Measurement is accepted as the standard for measurement throughout most of the world. Four of the base units in SI are the meter, liter, kilogram, and second.

The size of the unit can be determined from the prefix used with the base unit name. Look at **Figure 22** for some common metric prefixes and their meanings. The prefix *kilo-* attached to the unit *gram* is kilogram, or 1000 grams. The prefix *deci-* attached to the unit *meter* is decimeter, or one-tenth (0.1) of a meter.

The metric system is convenient because its unit sizes vary by multiples of 10. When changing from smaller units to larger units, divide by 10. When changing from larger units to smaller units, multiply by 10. For example, to convert millimeters to centimeters, divide the millimeters by 10. To convert 30 millimeters to centimeters, divide 30 by 10 (30 millimeters equal 3 centimeters).

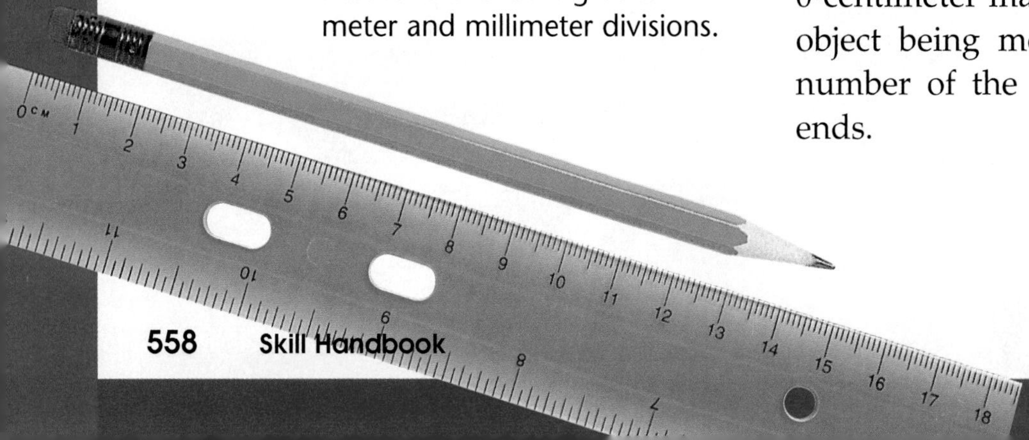

Metric Prefixes

Prefix	Symbol	Meaning	
kilo-	k	1000	thousand
hecto-	h	200	hundred
deka-	da	10	ten
deci-	d	0.1	tenth
centi-	c	0.01	hundredth
milli-	m	0.001	thousandth

FIGURE 22

Common metric prefixes

The meter is the SI unit used to measure length. A baseball bat is about one meter long. When measuring smaller lengths, the meter is divided into smaller units called centimeters and millimeters. A centimeter is one-hundredth (0.01) of a meter. A millimeter is one-thousandth of a meter (0.001).

Most metric rulers have lines indicating centimeters and millimeters, as shown in **Figure 23.** The centimeter lines are the longer, numbered lines; the shorter lines are millimeter lines. When using a metric ruler, line up the 0-centimeter mark with the end of the object being measured and read the number of the unit where the object ends.

FIGURE 23

Metric ruler showing centimeter and millimeter divisions.

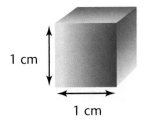

FIGURE 24

A square centimeter

Surface Area Units of length are also used to measure surface area. The standard unit of area is the square meter (m^2). A square that's one meter long on each side has a surface area of one square meter. A square centimeter, (cm^2), shown in **Figure 24,** is one centimeter long on each side. The surface area of an object is the total of the areas of its surfaces.

Volume The volume of a rectangular solid is also calculated using units of length. The cubic meter (m^3) is the standard SI unit of volume. A cubic meter is a cube one meter on each side. You can determine the volume of rectangular solids by multiplying length times width times height.

Liquid Volume During science activities, you will measure liquids using beakers and graduated cylinders marked in milliliters, as illustrated in **Figure 25.** A graduated cylinder is a cylindrical container marked with lines from bottom to top. Liquid volume is measured using a unit called a liter. A liter has the volume of 1000 cubic centimeters. Because the prefix *milli-* means thousandth (0.001), a milli-

liter equals one cubic centimeter. One milliliter of liquid would completely fill a cube measuring one centimeter on each side.

Mass Scientists use balances to find the mass of objects in grams. You will use a triple beam balance similar to the one shown in **Figure 26** on the next page. Notice that on one side of the balance is a pan and on the other side is a set of beams. Each beam has an object of a known mass, called a rider, that slides along the beam.

Before you find the mass of an object, set the balance to zero by sliding all the riders back to the zero point. Check the pointer on the right to make sure it swings an equal distance above and below the zero point on the scale. If the swing is unequal, find and turn the adjusting screw until you have an equal swing.

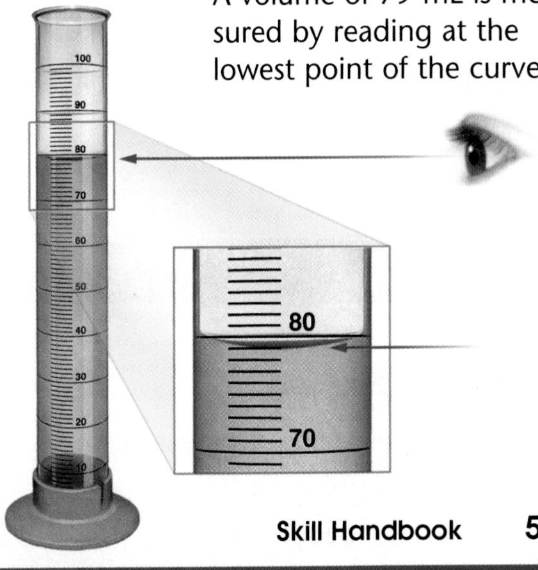

FIGURE 25

A volume of 79 mL is measured by reading at the lowest point of the curve.

Place an object on the pan. Slide the rider with the largest mass along its beam until the pointer drops below zero. Then move it back one notch. Repeat the process on each beam until the pointer swings an equal distance above and below the zero point. Add the masses on each beam to find the mass of the object.

You should never place a hot object or pour chemicals directly onto the pan. Instead, find the mass of a clean beaker or a glass jar. Place the dry or liquid chemicals in the container. Then find the combined mass of the container and the chemicals. Calculate the mass of the chemicals by subtracting the mass of the empty container from the combined mass.

Predicting

When you apply a hypothesis, or general explanation, to a specific situation, you predict something about that situation. First, you must identify which hypothesis fits the situation you

FIGURE 27

The daily high temperature is predicted every day.

are considering. People use prediction to make decisions every day. Based on previous observations and experiences, you may form a hypothesis that if it is wintertime, then temperatures will be low. From weather data in your area, temperatures are lowest in February. You may then use this hypothesis to predict specific temperatures and weather for the month of February. Someone could use these predictions to plan to set aside more money for heating bills during that month.

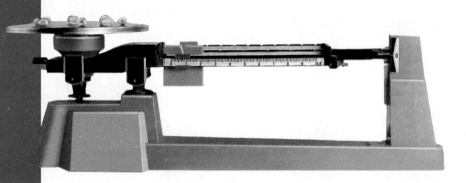

FIGURE 26

A beam balance is used to measure mass.

Using Numbers

When working with large populations of organisms, scientists usually cannot observe or study every organism in the population. Instead, they use a sample or a portion of the population. To sample is to take a small number of organisms of a population for research. Information discovered with the small sample may then be applied to the whole population. For example, scientists may take a small number of mice from a field to study the effects of day length on reproductive rate. This information could be applied to the population as a whole.

Estimating Scientific work also involves estimating. To estimate is to make a judgment about the size of something or the number of something without actually measuring or counting every member of a population. Here is a familiar example. Have you ever tried to guess how many kernels of popcorn were in a sealed jar?

If you did, you were estimating. What if you knew the jar of popcorn held one liter (1000 mL)? If you knew that 60 popcorn kernels would fit in a 100-milliliter jar, how many kernels would you estimate to be in the one-liter jar? If you said about 600 kernels, your estimate would be close to the actual number of popcorn kernels.

Scientists use a similar process to estimate populations of organisms from bacteria to buffalo. Scientists count the actual number of organisms in a small sample and then estimate the number of organisms in a larger area. For example, if a scientist wanted to count the number of black-eyed Susans, the field could be marked off in a large grid of 1-meter squares. To determine the total population of the field, the number of organisms in one square-meter sample can be multiplied by the total number of square centimeters in the field.

Using a Computerized Card Catalog

When you have a report or paper to research, you go to the library. To find the information, skill is needed in using a computerized card catalog. You use the computerized card catalog by typing in a subject, the title of a book, or an author's name. The computer will list on the screen all the holdings the library has on the subject, title, or author requested.

A library's holdings include books, magazines, databases, videos, and audio materials. When you have chosen something from this list, the computer will show whether an item is available and where in the library to find it.

Example You have a report due on dinosaurs, and you need to find three books on the subject. In the library, follow the instructions on the computer screen to select the "Subject" heading. You could start by typing in the word *dinosaurs.* This will give you a list of books on that subject. Now you need to narrow your search to the kind of dinosaur you are interested in, for example, *Tyrannosaurus rex.* You can type in *Tyrannosaurus rex* or just look through the list to find titles that you think would have information you need. Once you have selected a short list of books, click on each selection to find out if the library has the books. Then, check on where they are located in the library.

Using a CD-ROM

What's your favorite music? You probably listen to your favorite music on compact discs (CDs). But, there is another use for compact discs, called CD-ROM. CD-ROM means Compact-Disc–Read Only Memory. CD-ROMs hold information. Whole encyclopedias and dictionaries can be stored on CD-ROM discs. This kind of CD-ROM and others are used to research information for reports and papers. The information is accessed by putting the disc in your computer's CD-ROM drive and following the computer's installation instructions. The CD-ROM will have words, pictures, photographs, and maybe even sound and videos on a wide range of topics.

Example Load the CD-ROM into the computer. Find the topic you are interested in by clicking on the Search button. If there is no Search button, try the Help button. Most CD-ROMs are easy to use, but refer to the Help instructions if you have problems. Use the arrow keys to move down through the list of titles on your topic. When you double-click on a title, the article will appear on the screen. You can print the article by clicking on the Print button. Each CD-ROM is different. Click the Help menu to see how to find what you want.

Developing Multimedia Presentations

It's your turn—you have to present your science report to the entire class. How do you do it? You can use many different sources of information to get the class excited about your presentation. Posters, videos, photographs, sound, computers, and the Internet can help show our ideas. First, decide the most important points you want your presentation to make. Then sketch out what materials and types of media would be best to illustrate those points. Maybe you could start with an outline on an overhead projector, then show a video, followed by something from the Internet or a slide show accompanied by music or recorded voices. Make sure you don't make the presentation too complicated, or you will confuse yourself and the class. Practice your presentation a few times for your parents or brothers and sisters before you present it to the class.

Example Your assignment is to give a presentation on bird-watching. You could have a poster that shows what features you use to identify birds, with a sketch of your favorite bird. A tape of the calls of your favorite bird or a video of birds in your area would work well with the poster. If possible, include an Internet site with illustrations of birds that the class can look at.

Using E-Mail

It's science fair time and you want to ask a scientist a question about your project, but he or she lives far away. You could write a letter or make a phone call. But you can also use the computer to communicate. You can do this using electronic mail (E-mail). You will need a computer that is connected to an E-mail network. The computer is usually hooked up to the network by a device called a *modem.* A modem works through the telephone lines. Finally, you need an address for the person you want to talk with. The E-mail address works just like a street address to send mail to that person.

Example There are just a few steps needed to send a message to a friend on an E-mail network. First, select Message from the E-mail software menu. Then, enter the E-mail address of your friend. Next, type your message. Make sure you check it for spelling and other errors. Finally, click the Send button to mail your message and off it goes! You will get a reply back in your electronic mailbox. To read your reply, just click on the message and the reply will appear on the screen.

Using an Electronic Spreadsheet

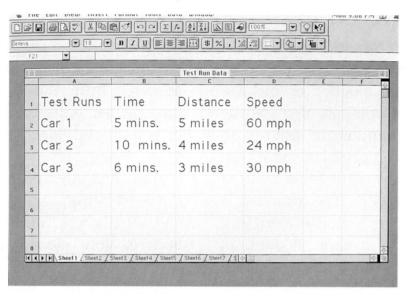

Your science fair experiment has produced lots of numbers. How do you keep track of all the data, and how can you easily work out all the calculations needed? You can use a computer program called a *spreadsheet* to keep track of data that involve numbers. A spreadsheet is an electronic worksheet. Type in your data in rows and columns, just as in a data table on a sheet of paper. A spreadsheet uses some simple math to do calculations on the data. For example, you could add, subtract, divide, or multiply any of the values in the spreadsheet by another number. Or you can set up a series of math steps you want to apply to the data. If you want to add 12 to all the numbers and then multiply all the numbers by 10, the computer does all the calculations for you in the spreadsheet.

Example Let's say that to complete your project, you need to calculate the speed of the model cars in your experiment. Enter the distance traveled by each car in the rows of the spreadsheet. Then enter the time you recorded for each car to travel the measured distance in the column across from each car. To make the formula, just type in the equation you want the computer to calculate, in this case, *speed 5 distance 3 time.* You must make sure the computer knows what data are in the rows and what data are in the columns so the calculation will be correct. Once all the distance and time data and the formula have been entered into the spreadsheet program, the computer will calculate the speed for all the trials you ran. You can even make graphs of the results.

English Glossary

This glossary defines each key term that appears in **bold type** in the text. It also shows the page number where you can find the word used.

Pronunciation Key

a...b**a**ck (bak)	oh...g**o** (goh)	sh...**sh**elf (shelf)
ay...d**ay** (day)	aw...s**o**ft (sawft)	ch...na**t**ure (nay chur)
ah...f**a**ther (fahth ur)	or...**or**bit (or but)	g...**g**ift (gihft)
ow...fl**ow**er (flow ur)	oy...c**oi**n (coyn)	j...**g**em (jem)
ar...c**ar** (car)	oo...f**oo**t (foot)	ing...s**ing** (sing)
e...l**e**ss (les)	ew...f**oo**d (fewd)	zh...vi**si**on (vihzh un)
ee...l**ea**f (leef)	yoo...p**ure** (pyoor)	k...**c**ake (kayk)
ih...tr**i**p (trihp)	yew...f**ew** (fyew)	s...**s**eed, **c**ent (seed, sent)
i (i + con + e)...**i**dea	uh...comm**a** (cahm uh)	z...**z**one, rai**s**e (zohn, rayz)
(i dee uh), l**i**fe (life)	u (+ con)...flow**er** (flow ur)	

abiotic (AY bi AH tihk) **factors:** all the non-living parts of an ecosystem, including air, soil, water, and temperature. (Chap. 6, p. 153)

acceleration (ak sel uh RAY shun): a change in speed or direction; depends on the mass of an object and the force pushing or pulling the object. (Chap. 10, p. 272)

acid rain: damaging acidic rain or snow formed when gases released by burning oil and coal mix with water in the air; can kill fish when it falls into rivers and lakes and can kill plants and trees when it falls to the ground. (Chap. 7, p. 193)

adaptation: any characteristic of an organism—a body shape, body process, or behavior—that helps the organism to survive in its environment and carry out life processes. (Chap. 5, p. 120)

air mass: a large body of air with the same properties as the area of Earth's surface over which it develops and moves. (Chap. 18, p. 505)

allele (uh LEEL): a different form that a gene may have for a specific trait. (Chap. 4, p. 106)

alternating current (AC): a kind of electricity that regularly reverses its direction, usually is generated by a power plant, and is used to run household appliances. (Chap. 13, p. 358)

anatomy: the physical makeup of a living organism. (Chap. 5, p. 132)

ancestry: a line of descent of a living organism. (Chap. 5, p. 132)

Archimedes (ar kih MEE deez): a Greek mathematician, inventor, and philosopher whose ideas included a compound machine used to pump water for irrigation more than 2000 years ago. (Chap. 11, p. 310) artesian well: a well drilled down into the groundwater layer of Earth's surface. (Chap. 17, p. 474)

asexual (ay SEK shuhl) reproduction: a type of reproduction in which an organism produces a new organism from one parent, and all the DNA of the new organism comes from that one parent. (Chap. 4, p. 92)

asteroids: pieces of rock found mostly in the asteroid belt that lies between Mars and Jupiter and that are made up of minerals similar to those that formed the planets. (Chap. 14, p. 388)

astronomical (as truh NAHM uh kul) unit (AU): a unit of measure equal to 150 million km, which is the average distance from Earth to the sun, and that is used to measure the distances between the planets. (Chap. 14, p. 385)

atmosphere (AT muh sfeer): the layer of air that surrounds Earth like a blanket, is made up of a mixture of gases plus small amounts of tiny solids and liquids, and keeps temperatures on Earth just right to support life. (Chap. 18, p. 496)

atom: an extremely small particle that is the basic unit of matter and is composed mainly of protons, neutrons, and electrons. (Chap. 8, p. 212)

average speed: describes the movement for an entire trip by dividing measured distance by measured time. (Chap. 10, p. 270)

axis: the imaginary line that Earth spins around and whose tilt is responsible for Earth's seasons. (Chap. 14, p. 379)

balance: a small, precise tool that is used to measure the mass of an object. (Chap. 10, p. 267)

battery: a device that converts internally stored chemical energy into electric current. (Chap. 13, p. 354)

biochemical rock: a type of sedimentary rock, such as coal or limestone, formed from dead organic matter that piled up and was compressed into rock over millions of years. (Chap. 15, p. 422)

biodiversity (bi oh duh VUR suh tee): the measure of the number of different species in a specific area. (Chap. 5, p. 119)

biosphere (BI oh sfeer): the part of Earth in which organisms can live; includes the topmost layer of Earth's crust, the surrounding atmosphere, and all the rivers, lakes, and oceans. (Chap. 6, p. 151)

biotic (bi AH tihk) factors: all the organisms that make up the living parts of an ecosystem, including trees, birds, bacteria, and frogs. (Chap. 6, p. 152)

captive breeding: the breeding of endangered species in captivity in order to build up the number of individuals in a population. (Chap. 5, p. 134)

cell: the smallest unit of life in a living thing; cells take in materials and release energy and waste products. (Chap. 2, p. 36)

cell membrane: the flexible structure that holds a plant or animal cell together, gives it shape, provides a boundary between the cell and its environment, and helps control what enters and exits the cell. (Chap. 3, p. 68)

cellular respiration: the process in which food and oxygen combine in the mitochondria to make carbon dioxide and water and release energy to do all of the cell's work. (Chap. 3, p. 70)

chemical change: the change of materials into other, new materials with different properties. (Chap. 9, p. 255)

chemical formula: a formula composed of symbols and numbers that tell what elements make up a compound and what their ratios are. (Chap. 8, p. 227)

chemical property: a characteristic of a substance that permits its change to a new substance—for example, the ability to burn and the ability to react with oxygen. (Chap. 9, p. 243)

chloroplasts (KLOR uh plasts): plant cell organelles that give plants their green color and that trap energy from sunlight and turn it into food during the process of photosynthesis. (Chap. 3, p. 72)

cichlid (SI klud): a type of African fish whose numerous species show a variety of differences yet still share many similarities. (Chap. 5, p. 132)

cinder cone volcano: a cone-shaped volcano whose steep sides are made up of blocks of solid lava that formed by explosive eruptions. (Chap. 16, p. 443)

cinders: small chunks of solid lava produced by a volcano. (Chap. 16, p. 443)

circuit (SUR kut): a complete, unbroken path that electrons can travel to do work. (Chap. 13, p. 351)

classification (klas if uh KAY shun): the grouping of objects based on common traits; for living organisms, can be based on common body parts, similarities in the materials that make up the bodies, or chemicals found inside the organisms' bodies. (Chap. 2, p. 45)

clay: a type of soil formed from the smallest grains of broken sedimentary rocks; holds moisture and is often found at the mouths of rivers. (Chap. 17, p. 472)

cleavage (KLEE vuj): property of some minerals to split into pieces with smooth, regular edges and surfaces. (Chap. 15, p. 409)

climate (KLIME ut): the pattern of weather that occurs in a particular area over many years based on such things as average temperature, precipitation, and humidity. (Chap. 18, p. 514)

cloning: the process of creating an organism that has exactly the same DNA as another organism; a clone receives all of its DNA from one parent and is genetically identical to its parent. (Chap. 4, p. 100)

comet: an object in the solar system made up of ice, dust, and frozen gases that develops a tail of light when it swings close to the sun and is pushed on by the solar wind. (Chap. 14, p. 386)

common ancestor (AN ses tur): the shared ancestor of new, different species that arose from one population. (Chap. 5, p. 129)

community (kuh MYEW nuh tee): all the populations living in an area who depend on each other for food, shelter, and other assorted needs. (Chap. 6, p. 158)

composite volcano: a volcano made of layers of thick, sticky lava and ash that is formed from alternating quiet and explosive eruptions; forms tall, pointed mountains. (Chap. 16, p. 443)

compound (KAHM pownd): a form of matter made by combining two or more different kinds of elements. (Chap. 8, p. 225)

compound machine: a machine made from the combination of two or more simple machines—for example, hand can openers and bicycles. (Chap. 11, p. 307)

condensation: the process in which water vapor changes back into its liquid form as the water vapor cools. (Chap. 17, p. 469)

conduction (kun DUK shun): the transfer of kinetic energy from faster-moving molecules to slower-moving molecules that do not travel from one place to another, but move in place. (Chap. 12, p. 334) conglomerates (kon GLAW muh rutz): rocks composed of large pebbles mixed and cemented together with other sediments. (Chap. 15, p. 421)

constellation (kahn stuh LAY shun): a group of stars, such as the Big Dipper, that forms a pattern in the sky and may be named after a real or imaginary animal, person, or object. (Chap. 14, p. 396)

consumer: an organism that eats other organisms. (Chap. 6, p. 168)

continental drift: a theory proposed by Wegener to explain the movement of the continents over millions of years from Pangaea to their present-day positions. (Chap. 16, p. 452)

convection (kun VEK shun): the transfer of thermal energy by the movement of molecules from one place to another in a liquid or gas; the type of transfer of thermal energy that produces weather. (Chap. 12, p. 336)

convection current: the cycle in which hot magma is forced upward toward Earth's surface, becomes cooler and denser as it rises, then starts to sink back down into the mantle, where it is heated again; as convection currents move around in the plastic-like layer just under the crust, they cause movement of the plates. (Chap. 16, p. 458)

crust: Earth's outer layer, which is firm and seems solid but is broken up into giant plates that are able to move around. (Chap. 16, p. 438)

crystals: solid materials, such as quartz and halite, whose atoms are arranged in repeating patterns and that have smooth surfaces, sharp edges, and points. (Chap. 15, p. 408)

curator: an individual who is responsible for the care of something—for example, a person in charge of a zoo. (Chap. 6, p. 165)

current: a mass of water moving in one direction in the ocean, either at the surface or flowing along the bottom or between the bottom and the surface; can carry seeds and plants from one continent to another and can affect the climate. (Chap. 17, p. 483)

decomposers: organisms, such as bacteria or fungi, that feed on dead organisms and the waste material of other organisms. (Chap. 6, p. 168)

density: a physical property of matter that can be found by dividing the mass of a material by its volume. (Chap. 9, p. 239)

deposition (dep uh ZIH shun): process in which rivers and streams change Earth by dropping sediments. (Chap. 17, p. 475)

desert: a hot, dry environment with low biodiversity. (Chap. 5, p. 119)

development: all the changes that take place during the life of an organism. (Chap. 2, p. 38)

direct current (DC): battery-powered electricity that travels in only one direction. (Chap. 13, p. 358)

dissolved: describes materials that have passed into solution. (Chap. 17, p. 479)

diversity: the variety of species shown by Earth's living organisms. (Chap. 5, p. 118)

DNA: a chemical that chromosomes are made of and that is found in the nucleus of every cell of all living things—the "genetic blueprint" that provides instructions for how organisms look and function. (Chap. 4, p. 103)

earthquake: the shaking of Earth produced when sudden movement occurs along a fault and quickly releases energy. (Chap. 16, p. 438)

eclipse (ee KLIHPS): an event that happens when the moon passes between the sun and Earth (a solar eclipse) or Earth passes between the sun and the moon (a lunar eclipse), producing a shadow. (Chap. 14, p. 382)

ecologist: a scientist who studies the interactions of organisms and their environment. (Chap. 6, p. 151)

ecology (ee KAH luh jee): the study of all the interactions between living and non-living parts of an ecosystem. (Chap. 6, p. 151)

ecosystem (EE koh sihs tum): a working unit made up of organisms interacting with each other and with nonliving factors. (Chap. 6, p. 148)

electrical energy: the energy of charges in motion. (Chap. 13, p. 346)

electricity: electric current or power; all forms depend on electrons moving from one place to another. (Chap. 13, p. 346)

electromagnet: a temporary magnet made from a current-carrying wire wrapped around an iron core, whose strength can be increased or decreased by changing the strength of the power source. (Chap. 13, p. 362)

electrons (ee LEK trahn): negatively charged particles that are the lightest of the three main particles found in an atom and that form a cloud around the atom's nucleus. (Chap. 8, p. 219)

element: natural or synthetic matter that is made up of only one type of atom; there are 112 known elements—for example, gold, aluminum, and oxygen. (Chap. 8, p. 223)

embryo: a fertilized egg that has begun dividing into more cells. (Chap. 4, p. 100)

endangered species: a species with so few living members that it is in danger of becoming extinct. (Chap. 6, p. 166)

energy (EN ur jee): the ability to bring about change in shape, temperature, speed, position, or direction of something; occurs in several different forms, such as heat energy and chemical energy, and often changes from one form to another. (Chap. 12, p. 320)

environment (en VI ur munt): everything in an organism's surroundings including other organisms, weather, water, sound, light, temperature, soil, and rocks. (Chap. 2, p. 35)

epicenter: the point at which the waves of an earthquake hit the surface of Earth, directly over the focus. (Chap. 16, p. 439)

erosion (ee ROH zhun): the wearing away of Earth's surface when soil or rock is loosened—for example, by wind and water—and then moved from one place to another. (Chap. 17, p. 476)

evaporation (ee vap uh RAY shun): the process that gradually changes water from a liquid state to water vapor. (Chap. 17, p. 468)

extinction: the dying out of a whole species. (Chap. 5, p. 141)

extrusive (EKS trew sihv): a type of igneous rock with an even, smooth texture and few or no visible crystals that forms when magma or melted rock from inside Earth cools at Earth's surface. (Chap. 15, p. 416)

fault: a feature of Earth's crust where plates rub against each other and build up energy that is released in waves during an earthquake. (Chap. 16, p. 439)

fault-block mountain: a type of mountain that is formed when whole sections of Earth's crust are faulted and broken and then some sections are uplifted. (Chap. 16, p. 461)

fertilization: the process in which sperm and egg unite, resulting in a new individual with a full set of chromosomes. (Chap. 4, p. 97)

fissures (FIHSH urs): large cracks in Earth's crust that can open up and allow lava to ooze out onto the ground and form extrusive igneous rocks. (Chap. 15, p. 416)

floodplain: a fertile, mostly flat area formed from sediments deposited by a river and that normally becomes flooded during a time of high water. (Chap. 17, p. 479)

focus: the exact spot inside Earth where an earthquake starts. (Chap. 16, p. 439)

folded mountain: a type of mountain formed when Earth's plates collided. (Chap. 16, p. 461)

foliated (FOH lee ay tud): a type of metamorphic rock, such as gneiss and slate, with bands of minerals that have been heated and squeezed into parallel layers. (Chap. 15, p. 428)

force: a push or a pull—for example, gravity and friction; can be balanced or unbalanced and can be found mathematically by multiplying mass times acceleration. (Chap. 10, p. 275)

forecast: a prediction of weather conditions based on observation and technology such as weather balloons, satellites, computers, and data from weather stations located around the world. (Chap. 18, p. 512)

fossil fuel: a fuel (coal, oil, natural gas) formed from the remains of ancient plants and animals that can be burned to produce energy. (Chap. 15, p. 424)

fossil record: all of the fossils that scientists have recovered from the ground; provides scientists with strong evidence that life on Earth has changed over time. (Chap. 5, p. 136)

fossils: the traces or remains of ancient plants and animals, which provide a history for life on Earth and help scientists form a picture of the past. (Chap. 5, p. 130)

fracture (FRAK chur): property of some minerals to break into pieces with irregular, jagged, or rough edges. (Chap. 15, p. 409)

freshwater: water in lakes, rivers, frozen in glaciers and polar ice caps, or underground and that is necessary for our survival. (Chap. 17, p. 467)

friction (FRIHK shun): the push or pull that opposes motion when two touching surfaces are sliding on each other. (Chap. 10, p. 273)

front: the place at which air masses meet and produce different kinds of weather conditions. (Chap. 18, p. 506)

fulcrum: the pivot point around which a lever turns. (Chap. 11, p. 299)

galaxy (GAL uk see): a group of stars, gas, and dust held together by gravity—for example, the Milky Way Galaxy, which contains 200 billion stars. (Chap. 14, p. 399)

gem: a rare, precious mineral that is clear, with no blemishes or cracks, and that can be cut and polished, making it ideal for jewelry. (Chap. 15, p. 413)

gene (JEEN): a small section of a chromosome that determines a trait; genes control all the traits of all organisms and provide all the information for the growth and life of a species. (Chap. 4, p. 103)

generator (JEN uh ray tur): a machine that changes mechanical energy into electric power. (Chap. 13, p. 366)

genetics (juh NET ihks): the study of how traits—the physical characteristics of an organism—are passed from parent to offspring. (Chap. 4, p. 102)

geologic time scale: a "diary" for life on Earth that is divided into time periods and that helps scientists keep track of when important events happened in Earth history, such as the appearance or disappearance of a species. (Chap. 5, p. 138)

geometric shapes: the shapes formed from straight lines, curves, and angles. (Chap. 11, p. 297)

global warming: an increase in average temperatures on Earth. (Chap. 18, p. 517)

gravel: loose, rounded rock fragments. (Chap. 17, p. 472)

gravity (GRA vuh tee): a pull that every object exerts on every other object and that pulls all things toward Earth in the same way. (Chap. 10, p. 264)

groundwater: water that soaks into the ground and collects in the spaces between soil and rock particles. (Chap. 17, p. 471)

habitat (HAB uh tat): the place in which an organism lives out its life and where resources such as food, space, and shelter are shared among the different species. (Chap. 6, p. 164)

heat: thermal energy that is transferred from a warmer object to a cooler object. (Chap. 12, p. 329)

hemoglobin (HEE muh gloh bun): the blood protein that carries oxygen to the body cells. (Chap. 5, p. 137)

hypothesis (hi POTH uh sus): a statement about a problem that can be tested. (Chap. 1, p. 18)

igneous (IG nee us) **rock:** a type of rock produced when magma from inside Earth cools and hardens on Earth's surface (extrusive igneous rock) or under Earth's surface (intrusive igneous rock). (Chap. 15, p. 415)

inclined plane: a simple machine with a sloped surface that makes it easier to lift objects—for example, a moving van ramp. (Chap. 11, p. 296)

inertia (ih NUR shuh): resistance to a change in motion. (Chap. 10, p. 276)

inference (IHN fuh runtz): an explanation of why something happened. (Chap. 1, p. 10)

inner core: the solid, innermost layer of Earth. (Chap. 16, p. 449)

inorganic: describes a substance that is not formed by plants or animals. (Chap. 15, p. 406)

intrusive (IN trew sihv): a type of igneous rock with large, beautiful crystals that is produced when magma cools under the crust inside Earth. (Chap. 15, p. 417)

invertebrate: an animal without a backbone, such as a sea urchin or sponge. (Chap. 17, p. 490)

joule (J): unit of measure used for work; a joule is equal to one newton-meter. (Chap. 11, p. 293)

kinetic (kih NET ihk) **energy:** the energy that comes from the motion of an object and that depends on the object's mass and speed. (Chap. 12, p. 323)

kingdom: a large group of organisms that share certain features; Earth's species are classified into six kingdoms—Archaebacteria, Eubacteria, Protista, Fungi, Plant, and Animal. (Chap. 2, p. 47)

landfill: an area in which garbage or trash is deposited, covered with a thin layer of dirt, and then watered down to keep the deposited materials from blowing around; 80 percent of U.S. trash and garbage goes to landfills. (Chap. 7, p. 187)

lava: the name given to magma when it erupts through an opening or vent at Earth's surface. (Chap. 16, p. 442)

law of conservation of matter: states that mass is never created or destroyed when things react with one another. (Chap. 8, p. 214)

levees: the natural banks of a river. (Chap. 17, p. 479)

lever: a simple machine made from a bar or rod that is free to rotate around a fulcrum; examples include seesaws and fishing rods. (Chap. 11, p. 299)

life-cycle analysis: a tool to help figure out the environmental impact of a product through its entire life, beginning with getting the natural resources to make the product and ending with disposal of the product in a landfill or by burning. (Chap. 7, p. 200)

light-year: a unit equal to the distance that light travels in a year, which is used to measure the distances between galaxies. (Chap. 14, p. 399)

limiting factor: a factor that controls the size of a population—for example, the availability of food or amount of rainfall. (Chap. 6, p. 159)

luster: a property of minerals that describes how light is reflected from the surface of a mineral—can be either shiny like a metal (metallic) or pearly, dull, or earthy (nonmetallic). (Chap. 15, p. 410)

magma: melted rock in which similar atoms combine as the rock cools, forming minerals whose grains fit together like a puzzle. (Chap. 15, p. 407)

magnet: a device that produces a magnetic field; has two poles that repel or attract and are similar to positive and negative charges. (Chap. 13, p. 359)

magnetic field: the area around a magnet in which magnetic forces act. (Chap. 13, p. 359)

magnitude: the size of something—for example, the size of a wave produced by an earthquake. (Chap. 16, p. 441)

mantle: the thickest layer inside Earth, located beneath the crust and above the outer core. (Chap. 16, p. 449)

mass: the amount of matter in an object. (Chap. 10, p. 266)

mass extinction: a large-scale dying out of many species over a short period of time. (Chap. 5, p. 141)

matter: anything that has mass and takes up space—for example, air; its basic unit is the atom. (Chap. 8, p. 212)

meanders (mee AN durz): wide bends in a stream formed when a stream begins to erode its banks more than the channel bottom. (Chap. 17, p. 477)

meiosis (mi OH sus): the process of sex cell formation during which the chromosome number in each sex cell becomes half the number found in other body cells. (Chap. 4, p. 96)

mesas (MAY sus): flat-topped, tablelike, natural rock structures that rise high above the surrounding land and were formed by the erosion of rock layers. (Chap. 15, p. 424)

metamorphic (met uh MOR fihk) **rock:** a type of rock, either foliated or non-foliated, that is formed when older rocks are heated and squeezed. (Chap. 15, p. 427)

meteorologist: a scientist who studies weather patterns to produce a forecast, or prediction of daily weather. (Chap. 18, p. 512)

mid-ocean ridge: an area where molten rock or magma oozes out onto the ocean floor. (Chap. 16, p. 457)

mineral (MIH nuh rool): an inorganic solid found in nature, having a definite, orderly pattern of atoms that are always arranged in the same way and that have the same chemical makeup throughout. (Chap. 15, p. 406)

mitochondria (mi tuh KAHN dree uh): the sausage-shaped cell organelles in which energy is released from food, providing power for all of the cell's activities. (Chap. 3, p. 70)

mitosis (mi TOH sus): the process of nuclear division that results in two nuclei containing identical information. (Chap. 4, p. 91)

mixture: two or more substances put together that do not combine to form a compound; can be the same throughout or made of several different items that can be seen. (Chap. 8, p. 228)

model: a representation or working version created to understand something that is too big, too small, or too complicated to understand otherwise. (Chap. 8, p. 214)

Modified Mercalli (mur KAH lee) **scale:** a scale used to measure the damage caused by earthquakes; ranges from a low of 1, where movement is not felt by people, to a high of 12, where buildings are completely destroyed. (Chap. 16, p. 440)

multiple alleles: two or more alleles that control a specific trait in an organism. (Chap. 4, p. 110)

mutation (myew TAY shun): a change in a gene or chromosome resulting from an error in mitosis or meiosis or an environmental factor; adds variation to the genes of a species and can be beneficial, harmful, or neutral. (Chap. 4, p. 111)

natural resources: things found in nature that organisms use to meet their needs; examples include vegetables, trees, oil, and minerals. (Chap. 7, p. 177)

natural selection: process in which organisms that are best suited to their environment survive, reproduce, and pass on their traits to their offspring; a mechanism for species' change over time. (Chap. 5, p. 125)

negatively charged: describes materials whose atoms easily gain electrons. (Chap. 13, p. 348)

neutron (NEW trahn): a particle without a charge that is found in the nucleus of an atom. (Chap. 8, p. 218)

newton (N): unit of force. (Chap. 11, p. 293)

niche (NIHCH): the role, or job, of an organism in an ecosystem. (Chap. 6, p. 164)

non-foliated: a type of metamorphic rock, such as marble and soapstone, that lacks distinct bands of minerals. (Chap. 15, p. 428)

nonrenewable (NAHN ree new uh bul) **resources:** resources, such as coal and oil, that form slowly over long periods of time and cannot be replaced by natural processes within 100 years. (Chap. 7, p. 182)

nucleus (NEW klee us): a cell organelle that contains chromosomes and that controls most of the cell's activities. (Chap. 3, p. 69)

observation: careful watching that can include measurements and descriptions and sketches or drawings of exactly what is seen. (Chap. 1, p. 10)

ore: a mineral that contains something that can be useful and sold for a profit. (Chap. 15, p. 414)

organ (OR gun): in many-celled organisms, a structure made up of two or more different types of tissue that work together to keep an organism alive—for example, the stomach is an organ that has muscle tissue, nerve tissue, and blood tissue. (Chap. 3, p. 80)

organ system: a group of different organs that work together to do a specific job—for example, a digestive system can be made up of a mouth, stomach, intestines, and liver. (Chap. 3, p. 81)

organelle: structure within the cytoplasm in cells that breaks down food, moves waste, and stores materials. (Chap. 3, p. 68)

organism: a living thing that has all five traits of life—it responds, moves, shows organization, reproduces, and grows and develops. (Chap. 2, p. 34)

outer core: the liquid layer inside Earth, located beneath the mantle and above the inner core. (Chap. 16, p. 449)

Pangaea (pan JEE uh): the supercontinent made up of all the continents joined together about 250 million years ago, before they broke apart and slowly moved to their present positions. (Chap. 16, p. 452)

particle: a tiny fragment; the three main particles in an atom are protons, neutrons, and electrons. (Chap. 8, p. 212)

periodic (peer ee AW dihk) **table:** a chart that organizes all known elements by the number of protons in the nucleus of each element and gives information such as the mass of an atom, the pattern of electrons, and whether an element is a solid, liquid, or gas. (Chap. 8, p. 223)

phase: the change in appearance of the moon when seen from Earth, depending on the position of the moon, the sun, and Earth in space. (Chap. 14, p. 380)

photosynthesis (foh toh SIHN thuh sus): the food-making process of plants, green algae, and many types of bacteria; in green plants, this process occurs inside the chloroplasts, where the sun's energy and water and carbon dioxide are converted into food and oxygen. (Chap. 3, p. 72)

physical change: a change in the shape, size, form, or state of matter that can be observed without changing the identity of the matter. (Chap. 9, p. 252)

physical property: a characteristic of matter that can be observed by the senses—for example, shape, size, taste, texture, color, and form. (Chap. 9, p. 238)

plate: a large section of Earth's crust and upper mantle that moves around on the plasticlike layer within the mantle. (Chap. 16, p. 456)

plate tectonics (tek TAW nihks): the theory developed to explain how Earth's plates move, based on evidence of seafloor spreading, convection currents, and Wegener's ideas about continental drift. (Chap. 16, p. 459)

pollutants (puh LEW tuntz): any materials that harm living things by interfering with life processes. (Chap. 7, p. 187); some are released by natural events such as erupting volcanoes, and others are produced by human activities such as burning fossil fuels. (Chap. 18, p. 497)

population: a group of the same type of organisms living in the same place at the same time. (Chap. 6, p. 157)

population density (DEN suh tee): the size of a particular population compared to the size of the area it occupies. (Chap. 6, p. 158)

positively charged: describes materials whose atoms easily lose electrons and as a result are left with more protons than electrons. (Chap. 13, p. 348)

potential (poh TEN shul) **energy:** the energy that is stored and comes from position or condition; can be transformed into kinetic energy when something acts to release the stored energy. (Chap. 12, p. 323)

precipitation (pree sihp uh TAY shun): water droplets that fall from the clouds to Earth—for example, rain and snow. (Chap. 17, p. 469)

predation (pruh DAY shun): the act of one organism feeding on another organism. (Chap. 6, p. 162)

predator (PRED uh tur): an organism that captures and eats other organisms. (Chap. 6, p. 162)

producer: an organism, such as a plant, that is able to use energy from the sun to make its own food. (Chap. 6, p. 168)

properties: the physical and chemical characteristics of matter. (Chap. 9, p. 238)

protein: a type of chemical that is used in the building of bones, muscles, and skin, and helps living things grow, digest food, and fight disease. (Chap. 5, p. 131)

proton (PROH tahn): a positively charged particle located in the nucleus of an atom. (Chap. 8, p. 218)

pulley: a simple machine composed of a surface with a chain or rope going around it and that can be either fixed or movable. (Chap. 11, p. 305)

radiation (ray dee AY shun): the energy that travels by waves in all directions from its source. (Chap. 12, p. 334)

ratio (RAY shee oh): a simple fraction that compares the amounts of two items. (Chap. 8, p. 214)

recycling (ree SIKE ling): reusing materials after they have been changed into another form, which often saves energy, water, and other resources. (Chap. 7, p. 198)

regeneration (ree jen uh RAY shun): the process by which some organisms can reproduce asexually or regrow missing body parts that have been lost because of an injury. (Chap. 4, p. 93)

renewable (ree NEW uh bul) **resources:** resources, such as water and trees, that can be replaced by natural processes in 100 years or less. (Chap. 7, p. 181)

reproduction: process by which organisms make more of their own kind—for example, by giving birth to live offspring, laying eggs, or producing flowers. (Chap. 2, p. 37)

revolution (rev oh LEW shun): the movement of Earth in a regular, curved path around the sun, which takes one year to complete. (Chap. 14, p. 379)

Richter (RICK tur) scale: a scale used to measure the amount of energy released by an earthquake. (Chap. 16, p. 440)

rock: a material, such as granite, that is usually composed of two or more minerals. (Chap. 15, p. 406)

rock cycle: a diagram that shows how the formation of igneous, sedimentary, and metamorphic rocks can be interrelated and how rocks are constantly recycled from one kind of rock to another. (Chap. 15, p. 430)

rotation (roh TAY shun): the spinning of Earth on its imaginary axis, which occurs once every 24 hours and causes day and night as Earth rotates toward or away from the sun. (Chap. 14, p. 378)

runoff: water that flows over the land and eventually flows into lakes, rivers, or streams; affected by the intensity and amount of rain, amount of vegetation, slope of the land, and whether the ground is already wet. (Chap. 17, p. 476)

salinity (suh LIH nuh tee): the saltiness of ocean water. (Chap. 17, p. 482)

salts: useful compounds formed along with water when acids and bases react with each other. (Chap. 9, p. 248)

sand: a kind of soil made up of fine particles of rock that have been weathered by glaciers, wind, or water. (Chap. 17, p. 472)

satellite: a spacecraft that is launched into space to gather information and send it back to Earth. (Chap. 14, p. 394)

science: the process of trying to understand the world around you; a set of steps that can be followed to help find out about something or to solve a problem. (Chap. 1, p. 8)

scientific law: a theory about the natural world that has been tested many times and produces the same results. (Chap. 10, p. 275)

screw: an inclined plane wrapped around a rod; this simple machine makes it easier to hold objects together. (Chap. 11, p. 297)

seafloor spreading: the process that forms new ocean crust by magma oozing out onto the ocean floor through a mid-ocean ridge. (Chap. 16, p. 457)

sedimentary rock: a type of rock formed in layers from pieces of other rocks, plant and animal matter, and minerals that evaporate or settle out of solution, usually over a period of thousands to millions of years. (Chap. 15, p. 420)

sediments (SED uh muntz): materials composed of pieces of rock, dissolved minerals, and organic matter that is carried and deposited or dropped by wind, water, ice, or gravity and collects in layers. (Chap. 15, p. 420)

seismograph (SIZE muh graf): an instrument that measures the wave magnitudes of earthquakes and records both faster P-waves (primary waves) and slower S-waves (secondary waves). (Chap. 16, p. 440)

sex cells: specialized cells involved in reproduction; eggs are female sex cells and sperm are male sex cells. (Chap. 4, p. 96)

sexual (SEK shul) reproduction: a type of reproduction in which a new organism is produced from two parents whose DNA combines to produce the new individual with its own new DNA. (Chap. 4, p. 96)

shield volcano: a volcano with long, sloping rock layers formed from quiet eruptions of fluid lava. (Chap. 16, p. 443)

silica: a material (liquid glass) found in lava that helps determine how the lava flows—the more silica, the stickier and slower the lava. (Chap. 16, p. 444) silt: a type of loose sedimentary rock made up of tiny rock grains that are slightly larger than clay. (Chap. 17, p. 478)

simple machine: a device that does work with only one movement—changes the size or direction of a force; a pulley, an inclined plane, a wheel and axle, and the three classes of levers. (Chap. 11, p. 296)

soil: the topmost surface layer of Earth in which plants grow and that is made up of a combination of minerals, water, air, and the decaying parts of plants and animals. (Chap. 17, p. 472)

solar (SOH lur) energy (EN ur jee): the energy from the sun, which is a renewable resource that can be used to reduce the need for nonrenewable energy resources such as coal and petroleum. (Chap. 12, p. 338)

solar system: the nine planets (Earth, Venus, Mars, Mercury, Jupiter, Saturn, Uranus, Neptune, and Pluto) and numerous other objects, such as asteroids, that circle the sun. (Chap. 14, p. 384)

solid waste: things that people throw away that are in solid or near-solid forms; examples include old newspapers, old plastic toys, and scrap metal. (Chap. 7, p. 195)

species: a group of organisms that can breed with one another and produce healthy offspring. (Chap. 2, p. 45)

speed: a measure of how far an object moves in a given amount of time. (Chap. 10, p. 269)

spring: a water source that forms where the top of the groundwater layer meets Earth's surface. (Chap. 17, p. 474)

star: dense, massive celestial body composed of gas; produces energy and has a life cycle. (Chap. 14, p. 398)

state of matter: a physical property that describes matter as a solid, a liquid, or a gas. (Chap. 9, p. 241)

static discharge: uncontrolled recombination of electric charges, producing, for example, lightning. (Chap. 13, p. 348)

static electricity: a form of potential electrical energy caused by the buildup or loss of electrons. (Chap. 13, p. 348)

stationary: describes a warm front or a cold front that has stopped moving. (Chap. 18, p. 506)

stratosphere (STRAT uh sfeer): the layer of Earth's atmosphere located above the troposphere; contains the ozone layer, which protects life on Earth from the sun's harmful ultraviolet rays. (Chap. 18, p. 498)

streak test: a test to help identify a mineral that involves scratching a mineral sample across a streak plate and producing a streak of color. (Chap. 15, p. 410)

submersibles (sub MUR suh bulz): underwater craft used to study the oceans. (Chap. 17, p. 485)

subscript: a little number written below and immediately after a letter in a chemical formula that tells how many atoms of an element are present. (Chap. 8, p. 227)

suspended: describes materials that are kept from falling or sinking. (Chap. 17, p. 479)

switch: a device that opens or closes a circuit to stop or start the movement of electrons. (Chap. 13, p. 352)

synthetic (sihn THET ihk): describes elements that are made in particle accelerators and are not found in nature—for example, technetium and neptunium. (Chap. 8, p. 223)

technology: the use of knowledge learned by science; examples include computers, video games, and robots. (Chap. 1, p. 9)

temperature: a measure of the kinetic energy of all the atoms of a material; commonly is measured by Celsius and Fahrenheit scales. (Chap. 12, p. 332)

terminals: the positive and negative connection points of a battery. (Chap. 13, p. 355)

texture: describes the size of the mineral crystal and how the crystals fit together. (Chap. 15, p. 407)

thermal (THUR mul) **energy:** the total amount of kinetic energy and potential energy in the atoms that make up a material; can be transferred from one place to another through conduction, convection, and radiation. (Chap. 12, p. 328)

threatened species: a species that is in need of protection but is not in immediate danger of extinction. (Chap. 6, p. 166)

tides: the changes in the level of ocean water during the course of a day—most places have two high tides and two low tides daily; one cause of tides is the pull of gravity between Earth, the moon, and the sun. (Chap. 17, p. 486)

tissue (TIH shew): in many-celled organisms, a group of similar cells that work together to do the same sort of work—for example, bone tissue is made up of just bone cells, and nerve tissue is made up of just nerve cells. (Chap. 3, p. 80)

toxic: describes something poisonous. (Chap. 17, p. 474)

trait: a specific feature of something, such as eye color, hair color, or height. (Chap. 2, p. 34)

transformation: a change of energy from one form to another with the total amount of energy staying the same—no energy is lost or gained. (Chap. 12, p. 321)

trilobites (TRI luh bites): the extinct relatives of today's lobsters, crabs, and insects. (Chap. 5, p. 130)

tropical rain forest: an environment with high biodiversity due to warm and steady temperatures and high rainfall, and which provides many resources for living organisms. (Chap. 5, p. 119)

troposphere (TROH puh sfeer): the lowest layer of Earth's atmosphere, which contains about 75 percent of the air we breathe and also is where weather, clouds, and air pollution are found. (Chap. 18, p. 498)

turbine (TUR bun): a large wheel that rotates and provides mechanical energy to a generator. (Chap. 13, p. 366)

upwarped mountains: mountains formed when Earth's crust was stretched and pushed up by forces inside Earth. (Chap. 16, p. 461)

vaccine: a preparation made from dead or weak viruses that is given by mouth or by a needle and which causes the body to make substances that resist particular viruses. (Chap. 2, p. 55)

variable: a factor that can change in an experiment. (Chap. 1, p. 19)

variations (vayr ee AY shuns): the different ways a certain inherited trait appears—for example, differences in height. (Chap. 4, p. 110)

virus (VI rus): a disease-causing particle that has some things in common with both living and nonliving things—it reproduces (in living cells) but it doesn't eat, grow, or respond to its environment. (Chap. 2, p. 54)

volcanic mountain: a mountain that forms when one plate is pushed beneath another plate, and as the plate melts, the magma rises and a volcano is created. (Chap. 16, p. 461)

volcano: a place where hot, liquid magma is forced up through the surface by pressure inside Earth and flows onto the ground; occurs most commonly along the boundaries of Earth's plates. (Chap. 16, p. 442)

voltage: a measure of the potential energy difference between the ends of a battery. (Chap. 13, p. 354)

volume: the amount of space taken up by matter. (Chap. 9, p. 239)

waning: the change in the moon's appearance when it seems to be getting smaller from night to night. (Chap. 14, p. 380)

water cycle: the constant cycling of water between Earth's surface and the atmosphere through evaporation, condensation, and precipitation, which is powered by energy from the sun and forms the basis for Earth's weather. (Chap. 17, p. 469)

wave: a movement of ocean water in which the water particles move in a circular path that makes the water rise and fall. (Chap. 17, p. 484)

waxing: the change in the appearance of the moon when it appears to be getting larger from night to night. (Chap. 14, p. 380)

weather: what is happening in the atmosphere right now—for example, rain, snow, clouds, and blue skies. (Chap. 18, p. 503)

weathering: the chemical and physical changes that alter the surface of Earth over time. (Chap. 9, p. 254)

wedge: an inclined plane with one or two sloping sides; examples of this simple machine include knives and chisels. (Chap. 11, p. 298)

weight: a measure of how much Earth's gravity pulls down on an object. (Chap. 10, p. 266)

wetland: a natural habitat whose soil contains much moisture and that may be completely or partially covered by water at different times of the year. (Chap. 7, p. 186)

wheel and axle: a simple machine that is made with two wheels of different sizes that are connected and turn together—for example, doorknobs and fishing reels. (Chap. 11, p. 304)

wind: the movement of air from warm, high-pressure areas into cooler, low-pressure areas due to differences in pressure. (Chap. 18, p. 504)

work: exerting a force on an object over a distance in the same direction as the object's motion. (Chap. 11, p. 292)

Glossary/Glosario

Este glosario define cada término clave que aparece en **negrillas** en el texto. También muestra el número de página en donde puedes encontrar la palabra usada.

abiotic factor / factor abiótico Cualquier cosa inanimada en un ecosistema. Los factores abióticos afectan el tipo y número de organismos que pueden sobrevivir en un ecosistema. (Cap. 6, pág. 153)

acceleration / aceleración Cambio en velocidad o dirección. (Cap. 10, pág. 272)

acid rain / lluvia ácida Ocurre cuando los gases que se liberan de la combustión del petróleo y del carbón se mezclan con agua en el aire, produciendo nieve o lluvia ácida. Este tipo de precipitación causa mucha contaminación y puede matar plantas y peces. (Cap. 7, pág. 193)

adaptation / adaptación Cualquier forma, proceso corporal o comportamiento que le permite a un organismo sobrevivir en su ambiente y llevar a cabo los procesos vitales. (Cap. 5, pág. 120)

air mass / masa de aire Gran cantidad de aire que tiene las mismas propiedades del área sobre la cual se forma y se mueve. (Cap. 18, pág. 505)

allele / alelo Diferente forma que presentan los genes que controlan un rasgo. (Cap. 4, pág. 106)

alternating current / corriente alterna Tipo de electricidad que usan los artefactos electrodomésticos, en el cual la electricidad viaja en una dirección por un rato y luego cambia de dirección. La dirección se alterna en un patrón regular. (Cap. 13, pág. 358)

Archimedes / Arquímedes Matemático, filósofo e inventor griego, que hace más de 2000 años propuso un sistema de bombeo de agua que usaba máquinas compuestas. (Cap. 11, pág. 310)

artesian well / pozo artesiano Tipo de pozo en que el agua fluye naturalmente hacia la superficie sin la ayuda de una bomba. (Cap. 17, pág. 474)

asexual reproduction / reproducción asexual Método de reproducción en que se crea un nuevo organismo a partir de un solo progenitor. En la reproducción asexual todo el DNA del organismo proviene de un solo progenitor. (Cap. 4, pág. 92)

asteroid / asteroide Pedazo de roca. Está compuesto de minerales similares a los que formaron los planetas. (Cap. 14, pág. 388)

astronomical unit (AU) / unidad astronómica (UA) Unidad que se usa para medir distancias en el espacio. Equivale a 150 millones de kilómetros, lo cual es la distancia promedio de la Tierra al sol. (Cap. 14, pág. 385)

atmosphere / atmósfera Capa de aire que rodea a la Tierra como si fuera una manta. El gas más común en la atmósfera es el nitrógeno. (Cap. 18, pág. 496)

atom / átomo Pequeña partícula que compone casi todo tipo de materia. (Cap. 8, pág. 212)

average speed / rapidez promedio Describe el movimiento de un viaje completo. (Cap. 10, pág. 270)

axis / eje Línea imaginaria, ligeramente inclinada, sobre la cual gira nuestro planeta. (Cap. 14, pág. 379)

balance / báscula Aparato que sirve para medir la masa. (Cap. 10, pág. 267)

bank / ribera Zona a ambos lados de un río. (Cap. 17, pág. 479)

battery / batería Fuente de energía en circuitos, la cual funciona mediante la separación química de las cargas positivas y negativas. (Cap. 13, pág. 354)

biochemical rock / roca bioquímica Tipo de roca que se forma cuando los materiales vivos se mueren, se apilan y se comprimen formando rocas. La tiza y el carbón son dos ejemplos. (Cap. 15, pág. 422)

biodiversity / biodiversidad Medida del número de diferentes especies en un área. Un bosque pluvial tropical tiene una biodiversidad alta, pero un desierto tiene una biodiversidad baja. (Cap. 5, pág. 119)

biosphere / biosfera La parte de la Tierra donde los organismos pueden vivir. Incluye la parte superior de la corteza terrestre, todos los océanos, ríos y lagos y la atmósfera que los rodea. La biosfera está compuesta de la combinación de todos los ecosistemas de la Tierra. (Cap. 6, pág. 151)

biotic factor / factor biótico Organismo que conforma la parte viviente de un ecosistema. (Cap. 6, pág. 152)

captive breeding / reproducción en cautiverio Apareamiento y reproducción en cautiverio de especies en peligro de extinción. (Cap. 5, pág. 134)

cell / célula Unidad básica de la vida en un ser viviente. (Cap. 2, pág. 36)

cell membrane / membrana celular Estructura flexible que mantiene unida a la célula y le da forma. (Cap. 3, pág. 68)

cellular respiration / respiración celular Proceso en el cual el alimento y el oxígeno se mezclan para formar dióxido de carbono y agua y así liberar energía. Este proceso ocurre dentro de la mitocondria. (Cap. 3, pág. 70)

chemical change / cambio químico Cambio que se produce cuando un material se convierte en otro con propiedades diferentes. Por ejemplo, el cambio que sufre una moneda de centavo brillante al volverse opaca y deslustrada. (Cap. 9, pág. 255)

chemical formula / fórmula química Se construye con símbolos o números que indican qué elementos hay en un compuesto y en qué razón se encuentran dichos elementos. (Cap. 8, pág. 227)

chemical property / propiedad química Característica de una sustancia que le permite convertirse en una nueva sustancia. La capacidad que tiene la madera de quemarse es un ejemplo de propiedad química. (Cap. 9, pág. 243)

chloroplast / cloroplasto Organelo verde que atrapa energía de los rayos solares y la convierte en alimento mediante el proceso de fotosíntesis. (Cap. 3, pág. 72)

cinder cone volcano / volcán de cono de carbonilla Tipo de volcán formado de erupciones explosivas de lava, las cuales pueden oírse a muchos kilómetros de distancia. (Cap. 16, pág. 443) cinders / carbonilla Pequeños trozos de lava sólida. (Cap. 16, pág. 443)

circuit / circuito Trayecto completo y cerrado que siguen los electrones. Por lo general, este trayecto es a través de alambres metálicos. (Cap. 13, pág. 351)

classification / clasificación Agrupamiento de objetos o información basándose en rasgos comunes. (Cap. 2, pág. 45)

cleavage / crucero Característica de un mineral que hace que al romperse, los pedazos del mineral presenten superficies uniformes, suaves y con bordes regulares. (Cap. 15, pág. 409)

climate / clima Patrón del tiempo que ocurre en un área en particular, durante muchos años. (Cap. 18, pág. 514)

cloning / clonación Significa la creación de un organismo, el cual contiene exactamente el mismo DNA que otro organismo. (Cap. 4, pág. 100)

comet / cometa Masa enorme compuesta de hielo, polvo y gases congelados. (Cap. 14, pág. 386)

common ancestor / antepasado común Individuo del cual provienen otros individuos de varias especies diferentes. Los individuos de cada especie poseen sus propios rasgos específicos y no pueden producir progenie al aparearse con los miembros de las otras especies, pero todos los miembros de las especies comparten algunos rasgos que heredaron del antepasado común. (Cap. 5, pág. 129)

community / comunidad Todas las poblaciones que viven en un área. Los miembros de una comunidad dependen mutuamente de los alimentos, refugio y otras necesidades. (Cap. 6, pág. 158)

composite volcano / volcán compuesto Tipo de volcán alto y escarpado compuesto de capas de lava viscosa y cenizas.

Los volcanes compuestos se forman de erupciones sin ruidos y explosivas, alternativamente. (Cap. 16, pág. 443)

compound / compuesto Forma de materia que resulta de la combinación de dos o más elementos. (Cap. 8, pág. 225)

compound machine / máquina compuesta Dos o más máquinas simples que funcionan juntas. (Cap. 11, pág. 307)

condensation / condensación Proceso en el cual el agua pasa nuevamente al estado líquido a medida que se enfría el vapor de agua. (Cap. 17, pág. 469)

conduction / conducción Transmisión de energía cinética de una molécula a otra. (Cap. 12, pág. 334)

conglomerate / conglomerado Roca sedimentaria formada por la consolidación de guijarros y otros sedimentos. (Cap. 15, pág. 421)

constelation / constelación Grupo de estrellas que forma un patrón en el firmamento. (Cap. 14, pág. 396)

consumer / consumidor Organismo que se alimenta de otros organismos vivos. (Cap. 6, pág. 168)

continental drift / deriva continental Movimiento de los continentes. (Cap. 16, pág. 452)

convection / convección Proceso de transmisión de energía térmica dentro de un líquido o un gas mediante el movimiento de moléculas de un lugar a otro. (Cap. 12, pág. 336)

convection current / corriente de convección Ciclo en que el magma caliente se enfría y se vuelve más denso a medida que asciende hacia la superficie. Luego, este magma más frío y denso comienza a hundirse de nuevo en el manto, en donde es calentado nuevamente. Este ciclo se repite continuamente. (Cap. 16, pág. 458) crust / corteza Capa externa de la Tierra. La corteza es firme y parece sólida, pero está dividida en trozos más grandes. (Cap. 16, pág. 438)

crystal / cristal Cuerpo sólido que tiene una estructura atómica con un patrón repetitivo. (Cap. 15, pág. 408)

curator / conservador Persona encargada de cuidar algo. (Cap. 6, pág. 165)

current / corriente Masa de agua que se mueve en una dirección. (Cap. 17, pág. 483)

decomposer / descomponedor Organismo que se alimenta de organismos muertos y de los materiales de desecho de otros organismos. (Cap. 6, pág. 168)

density / densidad Relación de la masa de un objeto con la cantidad de espacio que ocupa. (Cap. 9, pág. 239)

deposition / depositación Proceso en que el agua en movimiento deposita sedimentos. Es una de las maneras en que los ríos y arroyos cambian el relieve de la Tierra. (Cap. 17, pág. 475)

development / desarrollo Todos los cambios durante la vida de un organismo. (Cap. 2, pág. 38)

direct current / corriente directa Tipo de electricidad que viaja en una sola dirección y que permite el funcionamiento de una batería. (Cap. 13, pág. 358)

diversity / diversidad La variedad de la vida sobre la Tierra. (Cap. 5, pág. 118)

DNA / DNA Sustancia química que se encuentra en el núcleo de casi todas las células. Es el material del cual están hechos los cromosomas. Toda la información en el DNA de los cromosomas se llama información genética. (Cap. 4, pág. 103)

earthquake / terremoto Estremecimiento de la Tierra causado por un desatamiento rápido de energía. La energía viaja en forma de ondas a través de la Tierra y la cantidad de daño depende del tamaño de las ondas. (Cap. 16, pág. 438)

eclipse / eclipse Acontecimiento en que un cuerpo luminoso proyecta una sombra sobre otro cuerpo luminoso. Un eclipse puede ser lunar o solar. (Cap. 14, pág. 382)

ecologist / ecólogo Persona que se dedica al estudio de la ecología. (Cap. 6, pág. 151)

ecology / ecología Estudio de la relación entre los organismos vivientes y las partes inanimadas de un ecosistema. (Cap. 6, pág. 151)

ecosystem / ecosistema Está compuesto de organismos que interaccionan entre sí y con los factores inanimados del ambiente para formar una unidad funcional. (Cap. 6, pág. 148)

electrical energy / energía eléctrica Energía de las cargas eléctricas en movimiento. (Cap. 13, pág. 346)

electricity / electricidad Tipo de energía que depende del movimiento de electrones de un lugar a otro y que se usa para hacer funcionar artefactos, como por ejemplo, un reproductor de discos compactos. (Cap. 13, pág. 346)

electromagnet / electroimán Tipo de imán que se puede activar y desactivar. Se hace al enrollar un alambre que conduce corriente eléctrica alrededor de un núcleo de hierro. (Cap. 13, pág. 362)

electron / electrón Partícula del átomo la cual posee carga eléctrica negativa. (Cap. 8, pág. 219)

element / elemento Materia formada por un solo tipo de átomo. (Cap. 8, pág. 223)

embryo / embrión Huevo fecundado que ha comenzado a dividirse en más células. (Cap. 4, pág. 100)

endangered species / especie en peligro de extinción Se considera que una especie está en peligro de extinción, cuando la cantidad de miembros vivos es tan mínima que la especie completa puede desaparecer. (Cap. 6, pág. 166)

energy / energía Capacidad para causar cambio. La energía puede cambiar la temperatura, la forma, la velocidad, la posición o la dirección de un objeto. (Cap. 12, pág. 320)

environment / ambiente Todo lo que rodea a un organismo. Incluye a otros organismos, el agua, el tiempo, la temperatura, el tipo de suelo, las rocas, los sonidos y la luz; es decir, cualquier cosa con la que el organismo entre en contacto. (Cap. 2, pág. 35)

epicenter / epicentro Punto donde las ondas chocan contra la superficie terrestre, directamente sobre el foco de un terremoto. A menudo el epicentro es el lugar donde ocurre el peor daño, en un terremoto. (Cap. 16, pág. 439)

erosion / erosión Ocurre cuando se afloja el suelo o las rocas y son transportados de un lugar a otro. (Cap. 17, pág. 476)

evaporation / evaporación Proceso que convierte gradualmente el agua del estado líquido al estado gaseoso. (Cap. 17, pág. 468)

extintion / extinción La desaparición de una especie completa. (Cap. 5, pág. 141)

extrusive igneus rock / roca ígnea extrusiva Roca ígnea que se forma cuando la lava se enfría sobre la superficie terrestre. Las rocas ígneas intrusivas poseen cristales pequeños. (Cap. 15, pág. 416)

fault / falla Especie de grieta en la corteza, entre dos placas, donde ha habido movimiento. (Cap. 16, pág. 439)

fault-block mountain / montaña de bloque de falla Se forma cuando secciones completas de roca se convierten en fallas y se rompen; y algunas de estas secciones se elevan para formar una montaña. (Cap. 16, pág. 461)

fertilization / fecundación Proceso en el cual un óvulo y un espermatozoide se unen para formar un nuevo individuo, con un juego completo de 46 cromosomas. (Cap. 4, pág. 97)

fissure / grieta Hendidura de gran tamaño en la corteza terrestre. (Cap. 15, pág. 416)

floodplain / llanura aluvial Área que por lo general se inunda durante períodos de mucha precipitación. La llanura aluvial está formada por los sedimentos depositados por el río. (Cap. 17, pág. 479)

focus / foco Punto exacto donde comienza un terremoto. (Cap. 16, pág. 439)

folded mountain / montaña plegada Tipo de montaña que se forma del choque de las placas. (Cap. 16, pág. 461)

foliated rock / roca foliada Tipo de roca que presenta bandas de minerales que han sido calentados y compactados formando capas paralelas. (Cap. 15, pág. 428)

force / fuerza Cualquier empuje o atracción. Por ejemplo, la gravedad y la fricción son fuerzas. (Cap. 10, pág. 275)

fossil / fósil Resto o huella de una vida antigua. Provee a los científicos pruebas directas de que las especies cambian con el paso del tiempo. (Cap. 5, pág. 130)

fossil fuel / combustible fósil Restos de plantas y animales antiguos que podemos quemar para producir energía. (Cap. 15, pág. 424)

fossil record / récord fósil Todos los fósiles que los científicos han recobrado del suelo. Los fósiles de casi cada grupo principal de plantas o animales forman parte del récord fósil. (Cap. 5, pág. 136)

fracture / fractura Característica de un mineral que hace que al romperse, los pedazos del mineral presenten superficies desiguales, con bordes toscos e irregulares. (Cap. 15, pág. 409)

freshwater / agua fresca Agua que utilizamos para beber, cocinar y otras actividades. Es el agua que necesitamos para sobrevivir. (Cap. 17, pág. 467)

friction / fricción Es el resultado de movimientos opuestos entre dos superficies en contacto. (Cap. 10, pág. 273)

front / frente Lugar donde las masas de aire se juntan. Las interacciones entre las masas de aire ocasionan las condiciones del tiempo en el lugar del frente. (Cap. 18, pág. 506)

fulcrum / fulcro Punto de apoyo de una palanca. (Cap. 11, pág. 299)

genetics / genética Ciencia que estudia la manera en que los padres transfieren sus rasgos a la progenie. (Cap. 4, pág. 102)

geologic time scale / escala del tiempo geológico Especie de diario de la vida sobre la Tierra. Ayuda a los científicos a mantener un registro de cuándo una especie apareció y desapareció de la faz de la Tierra. Esta escala está dividida en cuatro largos lapsos de tiempo llamados eras y cada era está subdividida en períodos. (Cap. 5, pág. 138)

global warming / calentamiento global Aumento en las temperaturas medias en la Tierra. (Cap. 18, pág. 517)

gravity / gravedad Fuerza de atracción mutua que existe entre todos los objetos. (Cap. 10, pág. 264)

ground water / agua subterránea Agua que absorbe la superficie terrestre y que se va acumulando en pequeños espacios entre las piedras y el suelo. (Cap. 17, pág. 471)

galaxy / galaxia Conjunto de estrellas, gases y polvo que se mantienen unidos debido a la gravedad. (Cap. 14, pág. 399)

gem / gema Mineral raro que se puede cortar y pulir dándole una apariencia bella. Son ideales para la joyería. (Cap. 15, pág. 413)

gene / gene Pequeña sección de un cromosoma, que determina un rasgo. (Cap. 4, pág. 103)

generator / generador Máquina que convierte la energía mecánica en energía eléctrica. (Cap. 13, pág. 366)

habitat / hábitat Lugar donde vive un organismo. Diferentes especies comparten un hábitat. Los organismos comparten los recursos del hábitat, tales como alimentos, espacio y refugio. (Cap. 6, pág. 164)

heat / calor Energía térmica que se mueve desde un objeto caliente a uno más frío. (Cap. 12, pág. 329)

hypothesis / hipótesis Enunciado que se puede comprobar acerca de un problema. (Cap. 1, pág. 18)

igneus rock / roca ígnea Tipo de roca que se forma cuando la roca derretida o magma se enfría dentro de la Tierra. (Cap. 15, pág. 415)

inclined plane / plano inclinado Superficie inclinada que se usa para facilitar el traslado de la carga cuesta arriba. (Cap. 11, pág. 296)

inertia / inercia Resistencia al cambio de movimiento. Es decir, la tendencia de los objetos en reposo de permanecer en reposo y de los objetos en movimiento de continuar moviéndose en línea recta hasta que una fuerza actúe sobre ellos. (Cap. 10, pág. 276)

inference / inferencia Explicación, con base en la observación, de por qué sucedió algo. (Cap. 1, pág. 10)

inner core / núcleo interno La capa sólida más interna de la Tierra. (Cap. 16, pág. 449)

inorganic substance / sustancia inorgánica Sustancia que no está compuesta ni de material vegetal ni animal. (Cap. 15, pág. 406)

intrusive igneus rock / roca ígnea intrusiva Roca ígnea que se forma cuando el magma se enfría dentro de la corteza, en lugar de enfriarse sobre la superficie terrestre. Las rocas ígneas intrusivas poseen cristales grandes. (Cap. 15, pág. 417)

joule (J) / julio (J) Equivale a un newton-metro (N·m). (Cap. 11, pág. 293)

kinetic energy / energía cinética Energía que posee un objeto en movimiento. Dicha energía depende de la masa y de la velocidad del objeto. (Cap. 12, pág. 323)

kingdom / reino Grupo de gran tamaño formado por organismos que comparten ciertos rasgos comunes. (Cap. 2, pág. 47)

landfill / vertedero controlado Área donde se deposita la basura. (Cap. 7, pág. 187)

lava / lava Magma que ha llegado a la superficie terrestre. (Cap. 16, pág. 442)

law of conservation of matter / ley de conservación de la materia Ley que dice que la materia no puede ser ni creada ni destruida cuando una sustancia reacciona con otra. (Cap. 8, pág. 214)

lever / palanca Barra que gira libremente alrededor de un punto de apoyo llamado fulcro. (Cap. 11, pág. 299)

life-cycle analysis / análisis del ciclo de vida Una forma de averiguar el impacto de un producto, durante su existencia, en el ambiente. (Cap. 7, pág. 200)

light-year / año-luz Distancia que la luz viaja en un año. (Cap. 14, pág. 399)

limiting factor / factor limitante Factor que limita el crecimiento de una población; como por ejemplo, la cantidad de lluvia o alimento. (Cap. 6, pág. 159)

luster / lustre Manera en que la luz se refleja desde la superficie de un mineral. (Cap. 15, pág. 410)

M

magma / magma Roca caliente derretida. (Cap. 15, pág. 407)

magnet / imán Tipo de material que posee cierto número de electrones arreglados en ciertas maneras. La mayoría de los electrones apuntan en la misma dirección. (Cap. 13, pág. 359)

magnetic field / campo magnético Área alrededor de un imán donde actúan las fuerzas magnéticas. (Cap. 13, pág. 359)

magnitude / magnitud La fuerza destructiva de un terremoto. (Cap. 16, pág. 441)

mantle / manto La capa más gruesa de la Tierra. Posee características parecidas al plástico y puede moverse o fluir como si fuera brea caliente. (Cap. 16, pág. 449)

mass / masa Cantidad de materia en un objeto. La masa no cambia de un lugar a otro. (Cap. 10, pág. 266)

mass extinction / extinción masiva La desaparición en gran escala de muchas especies en un corto período de tiempo. (Cap. 5, pág. 141)

matter / materia Término que se usa para describir cualquier cosa que tenga masa y que ocupe espacio. Todo aquello que puedas tocar, saborear u oler es materia. (Cap. 8, pág. 212)

meander / meandro Serpenteo ancho que forman las corrientes de agua cuando no pueden adquirir más profundidad. La corriente de agua erosiona la parte de afuera del meandro y deposita los sedimentos en la parte de adentro. (Cap. 17, pág. 477)

meiosis / meiosis Proceso de la formación de células sexuales. Durante la meiosis, el número de cromosomas en cada célula es disminuido a la mitad. (Cap. 4, pág. 96)

mesa / meseta Estructura plana y rocosa con forma de mesa, más elevada que el resto del terreno que la rodea. (Cap. 15, pág. 424)

metamorphic non-foliated rock / roca metamórfica no foliada Tipo de roca que no presenta capas o bandas distintivas. (Cap. 15, pág. 428)

metamorphic rock / roca metamórfica Roca que se forma cuando las rocas más antiguas se calientan o se compactan. La palabra metamórfica significa "cambiar de forma". (Cap. 15, pág. 427)

meteorologist / meteorólogo Científico que estudia los patrones del tiempo para pronosticar el estado del tiempo diariamente. (Cap. 18, pág. 512)

mid-ocean ridges / dorsal medioceánica Áreas del océano por donde salen rocas derretidas o magma. Estas áreas tienen unas brechas o valles en el centro, en donde se forma el nuevo suelo oceánico a medida que se enfría la roca derretida. (Cap. 16, pág. 457)

mineral / mineral Sustancia química como el sodio y el cloro, que se encuentran en el aire, en el suelo y en el agua. (Cap. 2, pág. 44)

mineral / mineral Material sólido inorgánico que se encuentra en la naturaleza. Cada mineral posee rasgos o características que puedes usar para identificarlo. (Cap. 15, pág. 406)

mitochondrion / mitocondria Organelo que provee la energía necesaria para las funciones celulares. (Cap. 3, pág. 70)

mitosis / mitosis Proceso de división del núcleo que resulta en la formación de dos núcleos, los cuales poseen exactamente la misma información genética. (Cap. 4, pág. 91)

mixture / mezcla Resulta de juntar dos o más sustancias que no se combinan para formar un compuesto. (Cap. 8, pág. 228)

model / modelo Versión pequeña de algo más grande. (Cap. 8, pág. 214)

Modified Mercali scale / escala modificada Mercali Escala con que se mide el daño causado por un terremoto. Un 1 en esta escala indica que el terremoto no ha causado casi ningún daño. Un 12 indica que el terremoto ha causado la destrucción total de los edificios. (Cap. 16, pág. 440)

moon phases / fases lunares Cambios en la apariencia de la luna. Dependen de la posición de la luna en relación con el sol y con la Tierra. (Cap. 14, pág. 380)

mutation / mutación Cambio en un gene o en un cromosoma debido a un error en la meiosis o en la mitosis, o debido a un factor ambiental. Muchas mutaciones ocurren aleatoriamente, pero algunas otras son causadas por influencias externas, tales como los rayos X o las sustancias químicas peligrosas en el ambiente. (Cap. 4, pág. 111)

natural resource / recurso natural Todo lo que se encuentre en la naturaleza y que usen los seres vivos. (Cap. 7, pág. 177)

natural selection / selección natural Proceso en que los organismos que poseen rasgos que los hacen más aptos para un ambiente sobreviven, se reproducen y pasan esos rasgos a su progenie. (Cap. 5, pág. 125)

negative charge / carga eléctrica negativa Ocurre cuando un material gana electrones. (Cap. 13, pág. 348)

neutron / neutrón Partícula del átomo, la cual no posee carga eléctrica. (Cap. 8, pág. 218)

newton (N) / newton (N) Unidad de medida para la fuerza. (Cap. 11, pág. 293)

niche / nicho Papel que juega un organismo en su ecosistema. (Cap. 6, pág. 164)

nonrenewable resource / recurso no renovable Recurso que no se puede reemplazar mediante procesos naturales en un período de 100 años o menos. (Cap. 7, pág. 182)

nucleus / núcleo Organelo que controla la mayor parte de las funciones celulares. (Cap. 3, pág. 69)

nucleus / núcleo Centro del átomo. Constituye casi toda la masa del átomo. (Cap. 8, pág. 218)

observation / observación Acción de mirar algo cuidadosamente para poder anotar exactamente lo que ocurre. (Cap. 1, pág. 10)

ore / mena Cualquier mineral que contiene algo que pueda ser útil y que se pueda vender para obtener una ganancia. Por ejemplo, el hierro que se usa para hacer acero proviene de la mena del mineral hematita. (Cap. 15, pág. 414)

organ / órgano Estructura formada por dos o más tipos de tejidos que trabajan conjuntamente. (Cap. 3, pág. 80)

organ system / sistema de órganos Grupo de órganos que trabajan conjuntamente para realizar cierta función. (Cap. 3, pág. 81)

organelle / organelo Parte de la célula que realiza funciones especializadas. Todos los organelos de la célula se mueven dentro del citoplasma. (Cap. 3, pág. 68)

organism / organismo Ser viviente que presenta todos los rasgos de la vida, tales como ser capaz de responder a estímulos, de moverse, de producir progenie, de crecer y de desarrollarse, además de mostrar organización,. (Cap. 2, pág. 34)

outer core / núcleo externo Capa líquida que rodea el núcleo interno de la Tierra. (Cap. 16, pág. 449)

P

pangaea / pangaea Idea propuesta por el científico alemán Alfred Wegener, quien propuso en 1912 que todos los continentes estuvieron una vez unidos formando un supercontinente, que al separarse formó los continentes actuales. (Cap. 16, pág. 452)

particle / partícula Parte diminuta que forma la materia. (Cap. 8, pág. 212)

periodic table / tabla periódica Tabla que organiza los elementos de acuerdo con el número de protones en el núcleo de cada elemento. (Cap. 8, pág. 223)

photosynthesis / fotosíntesis Proceso mediante el cual las plantas, las algas verdes y otros tipos de bacterias atrapan la energía solar y la convierten en alimento. (Cap. 3, pág. 72)

physical change / cambio físico Cualquier cambio de un material en tamaño, forma o estado físico, en que la identidad de la materia no cambia. (Cap. 9, pág. 252)

physical property / propiedad física Término que usan los científicos para describir todas las características de la materia que se puedan detectar con los sentidos. (Cap. 9, pág. 238)

plate / placa Cada una de las 12 secciones enormes que forman la parte superior del manto y la corteza terrestre. Las placas pueden moverse porque descansan sobre una capa del manto que parece plástico. (Cap. 16, pág. 456)

plate tectonics / tectónica de las placas Teoría que resultó de la combinación de ideas sobre la deriva continental y el desplazamiento del fondo oceánico y la cual explica el movimiento de las placas. (Cap. 16, pág. 459)

pollutant / contaminante Todo material que pueda causar daño a los seres vivos, al interferir con los procesos vitales. (Cap. 7, pág. 187) (Cap. 18, pág. 497)

population / población Grupo del mismo tipo de organismos que vive en el mismo lugar, al mismo tiempo. (Cap. 6, pág. 157)

population density / densidad demográfica Comparación del tamaño de una población con relación al tamaño del área en que vive. (Cap. 6, pág. 158)

positively charged / cargado positivamente Ocurre cuando un material pierde electrones y se queda con más protones que electrones. (Cap. 13, pág. 348)

potential energy / energía potencial Energía almacenada que posee un objeto en reposo. Esta energía depende de la posición o condición del objeto. (Cap. 12, pág. 323)

precipitation / precipitación Caída a la Tierra de las gotas grandes de agua suspendidas en las nubes. La lluvia, la nieve, el granizo y el aguanieve son todos ejemplos de precipitación. (Cap. 17, pág. 469)

predation / predación Acto mediante el cual un organismo se alimenta de otro organismo. (Cap. 6, pág. 162)

predator / predador Animal que captura y come otros animales. (Cap. 6, pág. 162)

producer / productor Organismo que produce su propio alimento, como por ejemplo una planta. (Cap. 6, pág. 168)

property / propiedad Característica que describe alguna cosa. (Cap. 9, pág. 238)

protein / proteína Sustancia química que lleva a cabo una variedad de funciones en los seres vivos. Algunas proteínas se usan en la construcción de material viviente, como los músculos, los huesos y la piel. Otras proteínas ayudan a los seres vivos a crecer, a digerir los alimentos y a combatir enfermedades. (Cap. 5, pág. 131)

proton / protón Partícula del átomo, la cual posee carga eléctrica positiva. (Cap. 8, pág. 218)

pulley / polea Superficie con una cuerda o cadena a su alrededor. (Cap. 11, pág. 305)

radiation / radiación Energía que viaja, desde su fuente, en todas direcciones por medio de ondas. (Cap. 12, pág. 334)

ratio / razón Fracción simple que compara las cantidades de dos artículos. (Cap. 8, pág. 214)

recycling / reciclaje Significa volver a usar un material después de que dicho material ha sido convertido en un material diferente. (Cap. 7, pág. 198)

regeneration / regeneración Proceso mediante el cual algunos organismos son capaces de reemplazar las partes corporales perdidas debido a lesiones. (Cap. 4, pág. 93)

renewable resource / recurso natural renovable Recurso que se puede reemplazar mediante procesos naturales en un período de 100 años o menos. (Cap. 7, pág. 181)

reproduction / reproducción La capacidad de un organismo de producir progenie. (Cap. 2, pág. 37)

revolution / traslación Movimiento de la Tierra alrededor del sol. La Tierra demora un año en trasladarse, una vez, alrededor del sol. (Cap. 14, pág. 379)

Richter scale / escala Richter Escala con que se mide la energía liberada durante un terremoto. Esta escala mide el tamaño de las ondas producidas por un terremoto. (Cap. 16, pág. 440)

rock / roca Sustancia hecha de dos o más minerales. (Cap. 15, pág. 406)

rock cycle / ciclo de las rocas Proceso de cambio en el cual un tipo de roca se convierte en otro tipo de roca. (Cap. 15, pág. 430)

rotation / rotación Movimiento de la Tierra sobre su eje. Cada rotación dura 24 horas. (Cap. 14, pág. 378)

runoff / agua de escorrentía Agua que fluye sobre la superficie terrestre y, que a la larga, desemboca en los ríos, arroyos o lagos. El agua de escorrentía es una de las causas de la erosión. (Cap. 17, pág. 476)

salinity / salinidad Medida de la cantidad de sal en el agua. (Cap. 17, pág. 482)

salt / sal Compuesto que resulta de la reacción entre un ácido y una base. (Cap. 9, pág. 248)

sand / arena tipo de suelo compuesto de partículas finas de rocas que han sido meteorizadas por los glaciares, el viento o el agua. (Cap. 17, pág. 472)

satellite / satélite Nave espacial lanzada al espacio que recoge información y la transmite a la Tierra. (Cap. 14, pág. 394)

science / ciencia Es el proceso mediante el cual tratamos de comprender el mundo que nos rodea. Esto quiere decir que la ciencia es un conjunto de pasos que podemos seguir para averiguar más acerca de algo o para resolver un problema. (Cap. 1, pág. 8)

scientific law / ley científica Ley que resulta después de que una teoría ha sido probada muchas veces y ha dado los mismos resultados. Es una descripción exacta de algo importante en la naturaleza. (Cap. 10, pág. 275)

screw / tornillo Plano inclinado enrollado alrededor de una barra. (Cap. 11, pág. 297)

seafloor spreading / desplazamiento del fondo oceánico Proceso mediante el cual se forma el nuevo suelo oceánico. A medida que el magma llega a la superficie, este empuja y separa los dos lados de la dorsal. (Cap. 16, pág. 457)

sediment / sedimento Pedazos de rocas y otros materiales arrastrados por los ríos, los océanos, las olas, los glaciares y el viento. (Cap. 15, pág. 420)

sedimentary rock / roca sedimentaria Tipo de roca que se forma cuando los pedazos de rocas, plantas y animales o minerales disueltos se acumulan formando capas rocosas. (Cap. 15, pág. 420)

seismograph / sismógrafo Aparato que registra la magnitud de las ondas de un terremoto. (Cap. 16, pág. 440)

sex cell / célula sexual Célula especializada que está involucrada en la reproducción. Las células sexuales femeninas se llaman óvulos y las masculinas se llaman espermatozoides. (Cap. 4, pág. 96)

sexual reproduction / reproducción sexual Método de reproducción en que dos progenitores producen un nuevo organismo. En este proceso, el DNA de ambos padres se combina para formar un individuo con su propio DNA. (Cap. 4, pág. 96)

shale / esquisto arcilloso Roca sedimentaria formada por partículas pequeñísimas de arcilla. (Cap. 15, pág. 421)

shield volcano / volcán de escudo Volcán formado de largas capas inclinadas de rocas que se acumulan debido a las erupciones sin ruidos de lava fluida. (Cap. 16, pág. 443)

silt / roca de cieno y de arcilla Roca sedimentaria formada por partículas más grandes que la arcilla, pero más pequeñas que la arena. (Cap. 17, pág. 478)

simple machine / máquina simple Dispositivo que realiza trabajo mediante un solo movimiento. Es la forma más básica de una herramienta útil. (Cap. 11, pág. 296)

solar energy / energía solar Energía que proviene del sol. Es un ejemplo de un recurso renovable. (Cap. 12, pág. 338)

solar system / sistema solar Está formado por nueve planetas y numerosos objetos que giran en órbitas alrededor del sol. (Cap. 14, pág. 384)

solid waste / desecho sólido Cualquier objeto en forma sólida o casi sólida que una persona bote. (Cap. 7, pág. 195)

species / especie Grupo de organismos que pueden aparearse entre sí y producir progenie saludable. (Cap. 2, pág. 45)

speed / rapidez Una medida del grado de velocidad con que se mueve un objeto, en un tiempo dado. (Cap. 10, pág. 269)

spring / manantial Flujo de agua fresca que brota de la tierra cuando la capa superior de las aguas subterráneas llega a la superficie. (Cap. 17, pág. 474)

star / estrella Inmensas nubes compuestas de polvo y gases que se calientan y producen energía. Son cuerpos celestes con luz propia. (Cap. 14, pág. 398)

state of matter / estado de la materia Indica si una muestra de materia es sólida, líquida o gaseosa. Por ejemplo, un cubo de hielo es agua en estado sólido. (Cap. 9, pág. 241)

static discharge / descarga estática Resulta cuando muchas cargas eléctricas se mueven al mismo tiempo. Los rayos son un ejemplo. (Cap. 13, pág. 348)

static electricity / electricidad estática Forma de energía eléctrica potencial que se presenta cuando se acumulan o se pierden electrones. (Cap. 13, pág. 348)

stationary / estacionario Tipo de frente que se ha detenido sobre una región. (Cap. 18, pág. 506)

stratosphere / estratosfera Capa de la atmósfera que contiene el gas ozono. Esta capa importante de gas protege la vida en la Tierra de los dañinos rayos ultravioletas del sol. (Cap. 18, pág. 498)

streak test / prueba de la veta Prueba que ayuda a identificar un mineral aunque el mineral parezca otro mineral diferente. La veta no necesariamente es del mismo color del mineral. (Cap. 15, pág. 410)

submersibles / sumergibles Naves submarinas experimentales, tales como ALVIN que se utilizan para el estudio del fondo oceánico. También pueden estudiar la vida en el océano y los efectos humanos sobre el ambiente marino. (Cap. 17, pág. 485)

subscript / subíndice Número pequeño que se escribe un poco más abajo en una fórmula y el cual indica el número de átomos del elemento. (Cap. 8, pág. 227)

switch / interruptor Dispositivo que abre o cierra un circuito. (Cap. 13, pág. 352)

synthetic / sintético Tipo de elemento que no se encuentra en la naturaleza, pero que se fabrica en máquinas llamadas aceleradores de partículas. (Cap. 8, pág. 223)

technology / tecnología Es el uso de los conocimientos aprendidos a través de la ciencia. (Ch. 1, pág. 9)

temperature / temperatura Promedio de la energía cinética de todos los átomos de un material. (Cap. 12, pág. 332)

terminal / terminal Cada uno de dos puntos de conexión de una batería. (Cap. 13, pág. 355)

texture / textura Se refiere al tamaño del cristal y a la manera cómo encajan los granos del cristal o mineral. (Cap. 15, pág. 407)

thermal energy / energía térmica Cantidad total de energía, tanto cinética como potencial, de los átomos que forman un material. (Cap. 12, pág. 328)

threatened species / especie amenazada de extinción Especie bajo protección pero que no está en peligro inmediato de extinción. (Cap. 6, pág. 166)

tide / marea Cambio en el nivel del agua del océano durante el curso de un día. Normalmente, el nivel del agua del océano alcanza un nivel alto y un nivel bajo dos veces al día. (Cap. 17, pág. 486)

tissue / tejido Grupo de células semejantes que realizan el mismo tipo de trabajo. (Cap. 3, pág. 80)

trait / rasgo Característica específica de algo. Por ejemplo, algunos rasgos humanos incluyen ojos marrones, cabello rojizo y la capacidad de caminar. (Cap. 2, pág. 34)

transformation / transformación Cambio de la energía de una forma a otra. (Cap. 12, pág. 321)

trilobite / trilobita Parientes extintos de animales actuales, tales como los cangrejos, las langostas y los insectos. (Cap. 5, pág. 130)

troposphere / troposfera La capa más baja de la atmósfera terrestre. Es donde ocurren las condiciones del tiempo, las nubes y la contaminación del aire. Esta capa contiene aproximadamente un 75% del aire que respiramos. (Cap. 18, pág. 498)

turbine / turbina Rueda giratoria gigante que recibe energía de diferentes fuentes y que provee energía mecánica a un generador. (Cap. 13, pág. 366) upwarped mountain / montaña plegada anticlinal Tipo de montaña que se forma cuando el magma ascendente hace que la corteza terrestre se estire. (Cap. 16, pág. 461)

vaccine / vacuna Medicamento hecho de virus muertos o debilitados. Se administra por vía oral o por inyección. Las vacunas hacen que el cuerpo produzca sustancias que resisten ciertos virus. (Cap. 2, pág. 55)

variable / variable Factor que se puede cambiar en un experimento. (Cap. 1, pág. 19)

variation / variación Las diferentes formas en que puede presentarse cierto rasgo. (Cap. 4, pág. 110)

variation / variación Diferencias en los rasgos de los miembros de una especie y que los distingue a unos de los otros. (Cap. 5, pág. 124)

virus / virus Partícula que comparte características tanto con los seres vivos como con los seres inanimados. Se parece a los seres vivos porque es capaz de reproducirse. Se parece a los seres inanimados porque no crece, no come y no responde a su ambiente. (Cap. 2, pág. 54)

volcanic mountain / montaña volcánica Se forma en áreas donde una placa es empujada debajo de otra placa. A medida que una placa es empujada dentro del manto, la placa comienza a derretirse, empujando magma a través de la corteza y formando volcanes. (Cap. 16, pág. 461)

volcano / volcán Lugar en la Tierra donde el magma caliente y líquido, el cual es roca que se ha derretido debajo de la corteza, sube y sale a la superficie terrestre. (Cap. 16, pág. 442)

voltage / voltaje Mide la diferencia de energía potencial que hay entre los extremos de una batería. (Cap. 13, pág. 354)

volume / volumen Espacio que ocupa un objeto. (Cap. 9, pág. 239)

waning moon / luna menguante Cuando la luna parece que disminuye de tamaño noche tras noche. (Cap. 14, pág. 380)

water cycle / ciclo del agua Movimiento constante del agua desde la superficie terrestre hasta la atmósfera y nuevamente a la superficie terrestre. (Cap. 17, pág. 469)

wave / ola Movimiento del agua oceánica que hace que el agua suba y caiga. Dicho movimiento transmite energía. (Cap. 17, pág. 484)

waxing moon / luna creciente Cuando la luna parece que aumenta de tamaño noche tras noche. (Cap. 14, pág. 380)

weather / tiempo Lo que sucede en la atmósfera en todo momento. Comprende los cambios en la atmósfera causados por la presión atmosférica, los vientos, la temperatura y el agua. (Cap. 18, pág. 503)

weather forecast / pronóstico del tiempo Estudio que hace un meteorólogo sobre las condiciones del tiempo. (Cap. 18, pág. 512)

weathering / meteorización Cambio físico responsable de la mayor parte de la forma de la superficie terrestre. (Cap. 9, pág. 254)

wedge / cuña Máquina simple compuesta de uno o dos lados inclinados. (Cap. 11, pág. 298)

weight / peso Una medida de la fuerza que ejerce la gravedad sobre un objeto. El peso cambia de un lugar a otro, de acuerdo con la fuerza de gravedad. (Cap. 10, pág. 266)

weight / peso Una medida de la fuerza de gravedad. (Cap. 11, pág. 294)

wetland / ciénaga Área parcialmente cubierta con agua. (Cap. 7, pág. 186)

wheel and axle / rueda y eje Máquina simple que consiste en dos ruedas de diferentes tamaños, conectadas por un eje, y las cuales giran al mismo tiempo. (Cap. 11, pág. 304)

wind / viento Movimiento del aire que resulta de las diferencias en presión atmosférica. (Cap. 18, pág. 504)

work / trabajo Es el producto de una fuerza ejercida sobre un objeto, a través de una distancia, cuando la fuerza y el movimiento son en la misma dirección. (Cap. 11, pág. 292)

Index

The index for *Glencoe Science* will help you locate major topics in the book quickly and easily. Each entry in the Index is followed by the numbers of the pages on which the entry is discussed. A page number given in **boldface type** indicates the page on which that entry is defined. A page number given in *italic type* indicates a page on which the entry is used in an illustration or photograph. The abbreviation *act.* indicates a page on which the entry is used in an activity.

Mushrooms, 50, *50*, 51

Mutations, 110-**111,** *111*

Photo Credits

vi Dr. Gopal Murti/SPL/Photo Researchers; vii (tl)Sinclair Stammers/SPL/Photo Researchers, (tr)Betty Barford/Photo Researchers, (bl)Phil Degginger/Color-Pic; (br)Philip & Karen Smith/Tony Stone Images; viii (t, br)Dr. Mitsu Ohtsuki/ SPL/Photo Researchers, (bl)Henry Groskinsky/Peter Arnold; ix (t)Tim Davis/ Tony Stone Images, (b)Phil Jude/SPL/Photo Researchers; x E.R. Degginger/ Bruce Coleman, (tr)NASA/JPL/TSADO/Tom Stack & Associates, (c)C.Dani-I. Jeske/Earth Scenes, (bl)Mark Jones/Minden Pictures, (br)Tui De Roy/Minden Pictures; xi (t)Tracy Aiguier/The Picture Cube, (c)I.M. House/Tony Stone Images, (b)Charles D. Winters/Photo Researchers; xii (t)Glencoe photo; (b) Cabisco/Visuals Unlimited; xiv (t)Matt Meadows, (b)Morrison Photography; xv (t)Matt Meadows, (b)Morrison Photography; xvi (t)Morrison Photography, (b)Francois Gohier/Photo Researchers; xvii Morrison Photography; xviii (t c)Matt Meadows, (b)Breck P. Kent/Earth Scenes; xix (l)Phil Degginger/Color-Pic, (c)Morrison Photography, (r)E.R. Degginger/Color-Pic; xx (t)Leonard Lessin/FBPA, (b)David Parker/SPL/Photo Researchers; xxii (t)Mike Rustad Photography, (c)Courtesy Dr. Isidro Bosch, (b)Steve Skjold; xxiii (t)Historical Picture Archive/Corbis, (c)Morrison Photography, (b) Courtesy Addison Bain; 2 Ralph Lee Hopkins/The Wildlife Collection; 4 (l)Karl Weatherly/Corbis, (r)David Schultz/Tony Stone Images; 5 Pete Saloutos/Tony Stone Images; 6 Lawrence Migdale/Stock Boston; 7 Stephen Frisch/Photo 20-20; 8 Morrison Photography; 9 (t) David Parker/PL/Photo Researchers, (b)Jeff Greenberg/PhotoEdit; 10 Morrison Photography; 12 Will & Deni McIntyre/Photo Researchers; 13 Courtesy Amanda Shaw & FocusOne; 15 Jack Demuth; 16 (t)Morrison Photography, (b)Tony Freeman/PhotoEdit; 18 Donald Johnston/Tony Stone Images; 19 20 22 23 24 25 Morrison Photography; 26 Jack Demuth; 27 (l)Morrison Photography, (r)Ralph Lee Hopkins/The Wildlife Collection; 30-31 Karen Tweedy-Holmes/ Corbis; 31 Morrison Photography; 32 Cabisco/Visuals Unlimited; 33 Morrison Photography; 34 E.R. Degginger/Color-Pic; 35 Shin Yoshino/Minden Pictures; 39 (t)Picture Perfect, (c,b)Zefa/The Stock Market; 41 Morrison Photography; 42 Jonathan Elderfield/Gamma Liaison; 43 (t)Morrison Photography, (bl, br) E.R Degginger/Color-Pic; 45 (t) Pat & Tom Leeson/Photo Researchers, (b)E.R. Degginger/Color-Pic; 46 (t)Dr.G.J. Chafaris/Color-Pic, (b) H.Reinhard/ Okapia/Photo Researchers; 47 (t)Flip Nicklin/Minden Pictures, (b)Dave B. Fleetham/Tom Stack & Associates; 48 Morrison Photography; 49 (t, c) David M. Phillips/Visuals Unlimited, (b)Dr.Gopal Murti/ SPL/ Photo Researchers; 50 (tl, bl)Andrew Syred/SPL/Photo Researchers, (tr, br)Paul Skelcher/Rainbow; 51 (t, c) Picture Perfect, (b)Aaron Haupt; 52 (t)Zefa/The Stock Market, (c) Glencoe photo, (b)K.G. Preston Mafham/Animals Animals; 53 Zefa-Ziesler/The Stock Market; 54 Chris Bjornberg/Photo Researchers; 55 Matt Meadows; 56 Doug Martin; 57 (tl) E.R. Degginger/Color-Pic, (tr) Cabisco/Visuals Unlimited, (bl)Picture Perfect, (br)Dr.Gopal Murti/SPL/Photo Researchers; 59 Dietrich Gehring/The Wildlife Collection; 60 Manfred Kage/Peter Arnold; 61 Matt Meadows; 62 (l)Brian Parker/Tom Stack & Associates, (tr)SusanVan Etten/PhotoEdit, (br)John Cancalosi/Peter Arnold; 63 (l)David M.Dennis/Tom Stack & Associates, (c)Eric Grave/Photo Researchers, (r)David Young-Wolff/ PhotoEdit; 64 (t)Morrison Photography, (c)Dr.Tony Brain/ SPL/Photo Researchers, (b)Leonard Lessin/FBPA; 68 Biophoto Associates/Photo Researchers; 69 Michael Abbey/Photo Researchers; 71 (t)Robert & Linda Mitchell, (b)John Walsh/SPL/Photo Researchers; 72 Doug Sokell/Tom Stack & Associates; 74 (l)M.Abbey/Photo Researchers, (r) Biodisc; 74-75 Wendy Shattil & Bob Rozinski/Tom Stack & Associates; 75 (t)Dr.Dennis Kunkel/Phototake, (b)Andrew Syred/Tony Stone Images; 76 (l)Larry Mulvehill/Photo Researchers; 76-77 Morrison Photography; 77 (t)Carolina Biological Supply Co./Oxford Scientific Films/Earth Scenes, (c)R.Kessel-G.Shih/Visuals Unlimited, (b)Bruce Iverson; 78 Morrison Photography; 81 H.H. Sharp/Photo Researchers; 82 Courtesy Genzyme Corporation; 84 Courtesy Eyes On You Magazine; 85 (t)John Walsh/SPL/Photo Researchers, (bl)Dr.Dennis Kunkel/Phototake, (br)Morrison Photography; 86 Mike Eichelberger/Visuals Unlimited; 88 J-C Carton/Bruce Coleman; 89 Morrison Photography; 90 N. Smythe/Photo Researchers; 92 Holt Studios International/Photo Researchers; 93 (t)Betty Barford/Photo Researchers, (bl)Biophoto Associates/Photo Researchers, (br)Jerome Wyckoff/Earth Scenes; 94 95 Matt Meadows; 96 (t)Prof. P. Motta/SPL/Photo Researchers, (b)Andrew Syred/SPL/Photo Researchers; 98 Heather Angel Photography; 100 Dwight R. Kuhn; 101 AP/Wide World Photos; 102 Walter Hodges/Tony Stone Images; 103 Biophoto Associates/Photo Researchers; 104 Matt Meadows; 106 (t)Donald Specker/Earth Scenes, (b)Jane Grushow/Grant Heilman Photography; 108 Mark E. Gibson; 110 (t)Matt Meadows, (b)Morrison Photography; 111 (l)Michael P. Gadomski/Earth Scenes, (r)Gregory K. Scott/Photo Researchers; 112 (l)Bob Daemmrich/Stock Boston, (r)Derrick Ditchburn/Visuals Unlimited; 113 (tl)Betty Barford/Photo Researchers, (tr)AP/ Wide World Photos, (bl)Walter Hodges/Tony Stone Images, (br)Gregory K. Scott/Photo Researchers; 114 E.R.Degginger/ Color-Pic; 115 Michelle Garrett/Corbis; 116 Gerry Ellis/ENP Images; 117 Morrison Photography; 119 (t)Lynn M.Stone/Earth Scenes, (c)Mike Bacon/Tom Stack & Associates, (b)Kennan Ward; 120 (t)E.R. Degginger/Color-Pic, (c) Gregory G. Dimijian/Photo Researchers, (bl)Rod Planck /Photo Researchers, (br)Leonard Lee Rue III/Photo Researchers; 122 E.R. Degginger/Color-Pic; 123 Maslowski/Photo Researchers; 126 127 Morrison Photography; 130 (l)Sinclair Stammers/SPL/Photo Researchers, (c)Francois Gohier/Photo Researchers, (r) Gary Retherford/Photo Researchers; 131 (t)George E. Jones III/Photo Researchers, (b)Michael Durham/ENP Images; 134 Tom Brakefield/Bruce Coleman; 135 (t)Wendy Shatti/Bob Rozinski, (b) Wendy Shattil/Bob Rozinski/Tom Stack & Associates; 136 Patrick Aventurier/Gamma Liaison; 137 Richard T. Nowitz/Photo Researchers; 139 Morrison Photography; 142 Tom McHugh/Photo Researchers; 143 (t)Lynn M. Stone/Earth Scenes, (b)Gerry Ellis/ENP Images; 146 David M. Dennis; 147 Aaron Haupt; 148 (t) David M. Dennis/Tom Stack & Associates, (b)Todd Gipstein/Photo Researchers; 148-149 Carr Clifton/Minden Pictures; 149 (l)Colin Milkins/Oxford Scientific Films/ Animals Animals, (tr)Harold R.Hungerford/Photo Researchers, (br)Ed Reschke/Peter Arnold; 150 Matt Meadows; 151 (t)Morrison Photography, (b)SSEC/University of Wisconsin, Madison; 154 (l) F. Stuart Westmorland/ Photo Researchers, (r)Jim Zipp/Photo Researchers; 155 Francois Gohier/Photo Researchers; 156 (t)David Woodfall/ENP Images, (b)Gerry Ellis/ENP Images; 157 Flip Nicklin/Minden Pictures; 158 Fred Bavendam/Minden Pictures; 159 Glencoe photo; 160 Matt Meadows; 161 Doug Martin; 163 Lynn M. Stone; 164 Matt Meadows; 165 Courtesy Sea World of Ohio; 166 Raymond Gehman/Corbis; 167 Rosemary Calvert/Tony Stone Images; 170 Glencoe photo; 171 (t)David M. Dennis, (b)Kit Latham/FPG International; 173 Morrison Photography, (t)Jim Zipp/Photo Researchers, (b)Mark Newman/Photo Researchers; 174 Nancy Linden; 175 Morrison Photography; 176 Doug Armand/Tony Stone Images; 177 (t)Morrison Photography, (bl)David R. Frazier Photolibrary, (bc)John Mead/ SPL/Photo Researchers, (br)Philip & Karen Smith/Tony Stone Images; 178 (t)Steven Weinberg/Tony Stone Images, (bl)World View/Tony Stone Images, (bc)E.R. Degginger/Color-Pic, (br)Camcar; 179 (l)Mark E. Gibson, (r)Jack Demuth; 180 Michael Dwyer/Stock Boston; 181 (t)E.R.Degginger/Color-Pic, (b)Nancy Ross-Flanigan; 182 (clockwise from top)Kristin Finnegan/Tony Stone Images, Simon Fraser/SPL/Photo Researchers, Phil Degginger/Color-Pic;

185 David R. Frazier Photolibrary; 186 Jeff Greenberg/David R. Frazier Photolibrary; 187 Lonnie Duka/Tony Stone Images; 188 189 Morrison Photography; 190 (l)Elaine Comer-Shay, (r)Morrison Photography; 193 Jack Demuth; 194 Michael Giannechini/Photo Researchers; 196 Jack Demuth; 197 (l)Morrison Photography, (r)Fred Bavendam/Peter Arnold; 198 (l)Morrison Photography; (c)David R. Frazier Photolibrary, (r)Will McIntyre/Photo Researchers; 199 (l)E.R. Degginger/Color-Pic, (cl)F.Pedrick/The Image Works, (tr)Hank Morgan/Photo Researchers, (cr)Morrison Photography; 200 Morrison Photography; 201 John Elk III/Stock Boston; 202 Mark E. Gibson/Visuals Unlimited; 203 (tl) David R. Frazier Photolibrary, (tr)John Elk III/Stock Boston, (bl)Jeff Greenberg/David R. Frazier/Photolibrary, (br)Jack Demuth; 206 Johnny Johnson; 207 Susan Marquart; 208-209 George Steinmetz; 209 Courtesy Prof. H. Kazerooni, University of California at Berkeley; 210 Kaz Mori/The Image Bank; 211 Morrison Photography; 212 Ancient Art & Architecture Collection; 213 David R. Frazier Photolibrary; 215 Morrison Photography; 216 David R. Frazier Photolibrary; 217 (t)Dr. Mitsuo Ohtsuki/SPL/Photo Researchers, (c) ESA/TSADO/Tom Stack & Associates, (b)Morrison Photography; 220 221 222 Morrison Photography; 223 Fermi National Accelerator Laboratory/Photo Researchers; 225 Sandy King/The Image Bank; 227 228 Morrison Photography; 229 Ronnie Kaufman/The Stock Market; 230 Morrison Photography; 231 Simon Fraser/SPL/Photo Researchers; 232 Todd Gipstein/Corbis; 233 (t)Ancient Art & Architecture Collection, (b)Kaz Mori/The Image Bank; 236 Norbert Rosing/Earth Scenes; 237 Breck P. Kent/Earth Scenes; 238 239 Morrison Photography; 240 (t)E.R. Degginger/Color-Pic, (b)Morrison Photography; 241 (t)E.R. Degginger/ Color-Pic, (b)Morrison Photography; 242 Morrison Photography; 243 (l)E.R. Degginger/Color-Pic, (r)Alan L. Detrick/Color-Pic; 244 (t)Morrison Photography, (b)Bob Daemmrich/Stock Boston; 245 (t)Morrison Photography, (b)Simon Fraser/SPL/Photo Researchers; 247 248 Morrison Photography; 249 Courtesy Addison Bain; 250 Mark Joseph/Tony Stone Images; 251 Phil Degginger/Color-Pic; 252 Morrison Photography; 253 (tl) Art Montes de Oca/ FPG International, (tr) Morrison Photography,(bl)E.R. Degginger/Color-Pic, (br) Amy C. Etra/PhotoEdit; 254 (t)C.C. Lockwood/Earth Scenes, (b)Brenda Tharp/Photo Researchers; 255 Gerry Ellis/ENP Images; 256 257 Morrison Photography; 258 Charles Benes/FPG International; 259 (t, c)Morrison Photography, (b)Gerry Ellis/ENP Images; 261 (t, cl, cr)Morrison Photography, (b)Michael Nelson/FPG International; 262 William Sallaz/Duomo; 263 Morrison Photography; 264 Henry Groskinsky/Peter Arnold; 265 Werner H. Muller/Peter Arnold; 266 Michael Newman/PhotoEdit; 267 (t)Morrison Photography, (b)Matt Meadows; 268 (t)NASA, (c)Morrison Photography, (b)BLT Productions; 269 (t)Matt Meadows, (b)Chuck Kuhn/The Image Bank; 270 David Young-Wolff/ PhotoEdit; 271 272 273 274 Morrison Photography; 275 Lawrence Migdale/ Stock Boston; 276 (t)NASA, (b)Scott Markewitz/FPG International; 279 Matt Meadows; 280 281 Morrison Photography; 282 (l)NASA, (r)David Young-Wolff/PhotoEdit; 283 David Young-Wolff/PhotoEdit; 285 Insurance Institute for Highway Safety; 286 Steve Skjold; 287 (t)Werner H. Muller/Peter Arnold, (c) Insurance Institute for Highway Safety, (b)Scott Markewitz/FPG International; 290 Erick Bakke/Allsport; 291 Morrison Photography; 292 (t)Peter Steiner/The Stock Market, (b)Amwell/Tony Stone Images; 293 Peter Arnold/Peter Arnold; 296 Bob Daemmrich/Stock Boston; 300 Wiley & Wales/Adventure Photo & Film; 301 Mark C. Burnett/Photo Researchers; 302 303 304 307 Morrison Photography; 308 (t)Morrison Photography, (b)Sarah Putnam/The Picture Cube; 309 Matt Meadows; 310 Stephen R. Brown/The Picture Cube; 311 Bill Gallery/ Stock Boston; 312 Scott Robinson/Tony Stone Images; 313 Clive Brunskill/ Allsport; 314 Rube Goldberg,™ and © of Rube Goldberg Inc. Distributed by United Media; 315 (t)Amwell/Tony Stone Images, (b)Mark C. Burnett/Photo Researchers; 316 Morrison Photography; 318 Mark Junak/Tony Stone Images; 319 Morrison Photography; 320 Jim Cummins/FPG International; 322 (t) Morrison Photography, (b)Tim Thompson/Corbis; 323 (t)Tim Davis/Tony Stone Images, (b)Chris Harvey/Tony Stone Images; 324 SW.Productions; 325 (l)SW Productions, (r)John Eastcott/Photo Researchers; 326 Morrison Photography; 327 L & M Photos/FPG International; 328 Steve Skjold/PhotoEdit; 329 330 331 332 335 Morrison Photography; 337 Tom & Pat Leeson/Photo Researchers; 339 Miro Vintoniv/Stock Boston; 340 (t)David R. Frazier Photolibrary, (b)Michael Giannechini/Photo Researchers; 341 (l)Steve Skjold/ PhotoEdit, (r)John Eastcott/Photo Researchers; 342 343 Morrison Photography; 344 David W. Hamilton/The Image Bank; 345 Matt Meadows; 346 E.R. Degginger/Color-Pic; 349 Matt Meadows; 350 Phil Jude/SPL/Photo Researchers; 351 Ben Van Hook/Duomo; 355 Morrison Photography; 356 357 Matt Meadows; 359 363 Morrison Photography; 364 (t)Bob Kramer/The Picture Cube, (b)Morrison Photography; 366 (t)E.R. Degginger/Color-Pic, (b)Russell D.Curtis/Photo Researchers; 367 (t)Mark C.Burnett/Photo Researchers, (b)John Mead/SPL/PhotoResearchers; 368 369 Morrison Photography; 372 Jack Demuth; 373 Morrison photography; 374-375 Adriel Heisey/Photographers/Aspen; 375 NASA/JPL; 376 NASA; 377 381 Morrison Photography; 382 NASA; 383 Jerry Lodriguss/Photo Researchers; 386 TSADO/NOAO/Tom Stack & Associates; 387 (t)Alan Stern(Southwest Research Institute), Marc Buie(Lowell Observatory), ESA and NASA; (b)NASA/JPL/Tom Stack & Associates; 388 (t) E.K.Karkoschka (LPL) and NASA, (c)NASA/JPL/TSADO/Tom Stack & Associates, (b) NASA/ Photo Researchers; 389 (t)USGS/NASA/TSADO/Tom Stack & Associates, (c,b)NASA; 390 Morrison Photography; 391 Pat Rawlings; 392 Mark S. Robinson, Northwestern University; 393 NASA; 394 NASA/Photo Researchers; 395 NASA; 398 (t)D.Figer (UCLA) and NASA, (b)NASA/SPL/Photo Researchers; 400 Mike O'Brine/Tom Stack & Associates; 401 (l) D. Figer (UCLA) and NASA, (r) Alan Stern (Southwest Research Institute), Marc Buie (Lowell Observatory), ESA and NASA; 403 Bill & Sally Fletcher/Tom Stack & Associates; 404 Galen Rowell/Mountain Light; 405 Matt Meadows; 406 through 408 Morrison Photography; 409 (c)Charles D. Winters/Photo Researchers, (tr)(bl)(br)Morrison Photography; 410 Morrison Photography; 412 (t)Morrison Photography, (bl) Runk/ Schoenberger/Grant Heilman Photography, (br)E.R. Degginger/Color-Pic; 413 (t)E.R. Degginger/Bruce Coleman, (b)David R.Frazier Photolibrary; 414 Harvey Lloyd/Peter Arnold; 415 Otto Hahn/Peter Arnold; 416 Morrison Photography; 417 (l)Morrison Photography, (r)Breck P. Kent/Earth Scenes; 418 Matt Meadows; 420 Morrison Photography; 420-421 Jim Corwin/Photo Researchers; 421 (t)Bob Evans/Peter Arnold, (b)Morrison Photography; 422 (l)Alan D. Carey/Photo Researchers, (c)Morrison Photography; 422-423 Fred Bavendam/Minden Pictures; 423 (tr)James L. Amos/Photo Researchers, (bl)Mark C. Burnett/Photo Researchers, (br)E.R. Degginger/Color-Pic; 424 Milton Rand/Tom Stack & Associates; 425 Steve Hamblin/Liaison International; 426 Derek Karp/Earth Scenes; 427 (t)Phil Degginger/Color-Pic, (b)Alice Q. Hargrave/Gamma Liaison; 428 (t)E.R. Degginger/Color-Pic, (cl)Breck P. Kent/ Earth Scenes, (cr)Jerome Wyckoff/Earth Scenes, (bl)Ancient Art & Architecture Collection, (br)E.R. Degginger/Color-Pic; 429 Matt Meadows; 430 (l)G.I. Bernard/Earth Scenes,(tr)E.R. Degginger/Color-Pic, (br)Jerome Wyckoff/Earth Scenes; 431 Magrath Photography/SPL/Photo Researchers; 432 Mike Rustad Photography; 433 (tl)Morrison Photography, (tr)Otto Hahn/Peter Arnold, (bl) E. R. Degginger/Color-Pic, (br)Matt Meadows; 435 (c,r)Morrison Photography, (l)Doug Martin; 436 Lysaght/Gamma Liaison; 437 Matt Meadows; 440 Jean-Marc Giboux/Gamma Liaison; 440-441 John T. Barr/Gamma Liaison; 442 C.Dani-I. Jeske/Earth Scenes; 442-443 David Muench Photography; 443 Mark Jones/Minden Pictures; 445 Tui de Roy/Minden Pictures; 446 Francois Gohier/ Photo Researchers; 447 David Weintraub/Photo Researchers; 448 Krafft/Photo Researchers; 450 Matt Meadows; 452 Martin Land/SPL/Photo Researchers; 454 (t)Carr Clifton/Minden Pictures, (b)David Parker/SPL/Photo Researchers; 459 Ann Duncan/Tom Stack & Associates; 460 (tl)Jeff Lepore/Photo Researchers, (tr)Michael Durham/ENP Images, (bl)E.R.

Degginger/Color-Pic, (br)John Lemker/Earth Scenes; **463** (t)Lysaght/Gamma Liaison, (c)Jean-Marc Giboux/ Gamma Liaison, (b)Jeff Lepore/Photo Researchers; **465** (t)Breck P. Kent/Earth Scenes, (b)Dr. Ian Robson/SPL/Photo Researchers; **466** David M. Barron/ Animals Animals; **467** Matt Meadows; **468** David R. Frazier Photolibrary; **470** (t)Frank Siteman/The Picture Cube, (b)Tracy Aiguier/The Picture Cube; **471** (l)Mary Kate Denny/PhotoEdit, (r)CLEO Photography/The Picture Cube; **472 473** Morrison Photography; **474** John Lemker/Earth Scenes; **475** Charlie Palek/Earth Scenes; **477** Gerry Ellis/ENP Images; **479** Phil Degginger/Color-Pic; **480** George H.H.Huey/Earth Scenes; **481** Tess Young/Tom Stack & Associates; **482** W. Gregory Brown/Animals Animals; **485** (t)Vince Cavataio/Allsport, (b)Amos Nachoum Photography; **488** John Pontier/Earth Scenes; **489** David Young-Wolff/PhotoEdit; **490** Courtesy Dr. Isidro Bosch; **491** Vince Cavataio/Allsport; **493** Kennan Ward; **494** David Lawrence/The Stock Market; **495** Morrison Photography; **498** NOAA/SPL/Photo Researchers; **500** Morrison Photography; **501** Matt Meadows; **502** I.M. House/Tony Stone Images; **503** (l)Manfred Mehlig/Tony Stone Images, (c)Lou Jones/The ImageBank, (r)Michael Marten/SPL/Photo Researchers; **507** Kent Wood/Photo Researchers; **508** Matt Meadows; **509** Charles D. Winters/Photo Researchers; **510** John Lund/Tony Stone Images; **511** The Purcell Team/Corbis; **512** Phil Degginger/Earth Scenes; **513** Stephen Frisch/Stock Boston; **516** (t)Rhonda Klevansky/Tony Stone Images, (b)Jack Demuth; **517** (t)Joseph Van Os/The Image Bank, (b)James H. Carmichael Jr./The Image Bank; **518** Historical Picture Archive/Corbis; **519** (t)I.M. House/Tony Stone Images, (b)The Purcell Team/Corbis; **530** (t)NIBSC/Science Photo Library/PR, (bl)Dr. Richard Kessel, (br)David John/Vu; **531** (t)Runk/Schoenberger, (bl)Andrew Syred/Science Photo Library/PR, (br)Rich Brommer; **532** (bl)Ralph Reinhold/ES, (t)G.R. Roberts, (br)Scott Johnson/AA; **533** Martin Harvey/DRK; **541** Jack Demuth; **542** Morrison Photography; **543** Jack Demuth; **544** Morrison Photography; **546** David Shopper/Stock Boston; **549** Jeff Hetler/Stock Boston; **550** Morrison Photography; **551** Glencoe photo; **552** (t)Morrison Photography, (b)Bob Abraham/The Stock Market; **553 554 555 556 557 558** Morrison Photography; **560** Charles Krebs/The Stock Market; **561** (t)Morrison Photography, (b)Michael P. Gadomski/Photo Researchers; **562** Jack Demuth; **563** Morrison Photography; **564 565** Jack Demuth; **566** Morrison Photography

Art Credits

3, John Edwards & Associates; **6,** Thomas J. Gagliano/Gagliano Graphics; **10,** John Edwards & Associates; **14,** Preface, Inc.: **17,** Preface, Inc.; **22,** Preface, Inc.; **27,** John Edwards & Associates; **35,** Laurie O'Keefe; **36,37,** Barbara Hoopes Ambler; **38,** John Edwards & Associates; **44,** John Edwards & Associates; **65,** John Edwards & Associates; **66,** Rolin Graphics, Inc.; **67,** Rolin Graphics, Inc.; **68,** Rolin Graphics, Inc.; **69,** Rolin Graphics, Inc.; **70,** Rolin Graphics, Inc.; **72,** Rolin Graphics, Inc.; **80,81,** Rolin Graphics, Inc.; **85,** Rolin Graphics, Inc. **86,** Rolin Graphics, Inc. **87,** Preface, Inc.; **91,** Thomas J. Gagliano/Gagliano Graphics; **97,** Barbara Hoopes Ambler; **99,** Pond & Giles; **103,** Thomas J. Gagliano/Gagliano Graphics; **105,** John Edwards & Associates(cr), Thomas J. Gagliano/Gagliano Graphics(bl); **106,** Thomas J. Gagliano/Gagliano Graphics; **107,** Pond & Giles; **109,** Pond & Giles; **118,** Laurie O'Keefe; **121,** Laurie O'Keefe; **124,125,** Laurie O'Keefe; **128,129,** Laurie O'Keefe; **131,** Preface, Inc.; **132,** Barbara Hoopes Ambler; **133,** Barbara Hoopes Ambler; **136,** Laurie O'Keefe; **138,** Laurie O'Keefe; **140,141,** Barbara Hoopes Ambler; **143,** Barbara Hoopes Ambler(l), Laurie O'Keefe(r); **152,153,** Laurie O'Keefe; **155,** John Edwards & Associates; **162,** Laurie O'Keefe; **168,169,** Barbara Hoopes Ambler; **171,** Laurie O'Keefe(tr), Barbara Hoopes Ambler(b); **173,** Preface, Inc.; **176,** Preface, Inc.; **178,179,** John Edwards & Associates; **182,** Preface, Inc. **184,** John Edwards & Associates; **190,** Preface, Inc.; **191** John Edwards & Associates; **192,** John Edwards & Associates; **195,** Thomas J. Gagliano/Gagliano Graphics; **213,** Preface, Inc.; **214,** Preface, Inc.; **218,** Barbara Hoopes Ambler; **219,** Thomas J. Gagliano/Gagliano Graphics; **233,** Thomas J. Gagliano/Gagliano Graphics; **234,** John Edwards & Associates; **235,** John Edwards & Associates; **246,** Preface, Inc.; **248,** Preface, Inc.; **270,** Precision Graphics; **277,** John Edwards & Associates; **278,** Precision Graphics; **284,** Preface, Inc.; **287,** Precision Graphics; **294,** John Edwards & Associates; **297,** Precision Graphics; **298,299,** John Edwards & Associates; **300,** Precision Graphics; **301,** Precision Graphics; **304,** Precision Graphics; **305,** Thomas J. Gagliano/Gagliano Graphics; **306,** Precision Graphics; **307,** Precision Graphics; **310,** Precision Graphics; **311,** Thomas J. Gagliano/Gagliano Graphics; **315,** Precision Graphics; **321,** Thomas J. Gagliano/Gagliano Graphics; **324,** Thomas J. Gagliano/Gagliano Graphics; **327,** Thomas J. Gagliano/Gagliano Graphics; **333,** Thomas J. Gagliano/Gagliano Graphics; **334,** Thomas J. Gagliano/Gagliano Graphics; **336,** Precision Graphics; **338,** Thomas J. Gagliano/Gagliano Graphics; **341,** John Edwards & Associates; **342,** Thomas J. Gagliano/Gagliano Graphics; **347,** Thomas J. Gagliano/Gagliano Graphics; **348,349,** John Edwards & Associates; **352,** Thomas J. Gagliano/Gagliano Graphics; **353,** John Edwards & Associates; **354,** John Edwards & Associates; **355,** Thomas J. Gagliano/Gagliano Graphics; **358,** Preface, Inc.; **360,** Thomas J. Gagliano/Gagliano Graphics; **362,** Thomas J. Gagliano/Gagliano Graphics; **363,** Thomas J. Gagliano/ Gagliano Graphics; **365,** Thomas J. Gagliano/Gagliano Graphics; **369,** John Edwards & Associates; **371,** Thomas J. Gagliano/Gagliano Graphics; **378,** John Edwards & Associates; **380,** John Edwards & Associates; **382,** John Edwards & Associates; **383,** John Edwards & Associates; **384,385,** John Edwards & Associates; **396,** John Edwards & Associates; **399,** John Edwards & Associates; **401,** John Edwards & Associates; **416,417,** Chris Forsey; **430,** Thomas J. Gagliano/ Gagliano Graphics; **438,** Chris Forsey; **439,** Precision Graphics; **440,** Precision Graphics; **441,** Preface, Inc. **449,** Chris Forsey; **452,** John Edwards & Associates; **453,** John Edwards & Associates; **456,457,** Chris Forsey; **458,** Chris Forsey; **460,** Chris Forsey; **463,** Chris Forsey; **469,** John Edwards & Associates; **471,** John Edwards & Associates; **476,** John Edwards & Associates; **477,** John Edwards & Associates; **481,** Preface, Inc.; **482, 483, 484, 486, 487, 491, 496, 497, 499, 504, 505, 506, 510, 514, 515, 519, 521, 522,** John Edwards & Associates; **529,** Thomas J. Gagliano/Gagliano Graphics; **544,** Preface, Inc.(tl), Thomas J. Gagliano/Gagliano Graphics(br); **545,** Preface, Inc.; **547,** Preface, Inc.; **548,** Preface, Inc,; **549,** Preface, Inc.; **556,** John Edwards & Associates; **559,** Preface, Inc.(tl), John Edwards & Associates(tr); **560,** John Edwards & Associates.

Permissions

Page 56 Reprinted with the permission of Simon & Schuster Books for Young Readers, an imprint of Simon & Schuster Children's Publishing Division from *HATCHET* by Gary Paulsen. Copyright ©1987 Gary Paulsen.

Page 123 Excerpt from *National Audubon Society Field Guide to North American Birds—Eastern Region* by John Bull and John Farrand, Jr. (1994), published by Alfred A Knopf, Inc. By permission of Chanticleer Press.

Page 368 Excerpt from *Promise Me the Moon* by Joyce Annette Barnes. Copyright ©1997 by Joyce Annette Barnes. Used by permission of Dial Books for Young Readers, a division of Penguin Books USA, Inc.

Page 393 Excerpt from *Barbary* reprinted by Permission of Francis Collin, literary agent, ©1986 by Vonda N. McIntyre.

Page 462 From *Pacific Crossing*, copyright ©1992 by Gary Soto, reprinted by permission of Harcourt Brace & Company.

PERIODIC TABLE OF THE ELEMENTS

Columns of elements are called groups. Elements in the same group have similar chemical properties.

Element	Hydrogen
Atomic number	1
Symbol	H
Atomic mass	1.008

State of matter

Gas

Liquid

Solid

Synthetic

The first three symbols tell you the state of matter of the element at room temperature. The fourth symbol identifies human-made, or synthetic, elements.

1

1	Hydrogen 1 H 1.008

2

	2
2	Lithium 3 Li 6.941	Beryllium 4 Be 9.012
3	Sodium 11 Na 22.990	Magnesium 12 Mg 24.305

3 **4** **5** **6** **7** **8** **9**

	3	4	5	6	7	8	9
4	Scandium 21 Sc 44.956	Titanium 22 Ti 47.867	Vanadium 23 V 50.942	Chromium 24 Cr 51.996	Manganese 25 Mn 54.938	Iron 26 Fe 55.845	Cobalt 27 Co 58.933
5	Yttrium 39 Y 88.906	Zirconium 40 Zr 91.224	Niobium 41 Nb 92.906	Molybdenum 42 Mo 95.94	Technetium 43 Tc (98)	Ruthenium 44 Ru 101.07	Rhodium 45 Rh 102.906
6	Lanthanum 57 La 138.906	Hafnium 72 Hf 178.49	Tantalum 73 Ta 180.948	Tungsten 74 W 183.84	Rhenium 75 Re 186.207	Osmium 76 Os 190.23	Iridium 77 Ir 192.217
7	Actinium 89 Ac (227)	Rutherfordium 104 Rf (261)	Dubnium 105 Db (262)	Seaborgium 106 Sg (266)	Bohrium 107 Bh (264)	Hassium 108 Hs (277)	Meitnerium 109 Mt (268)

(Potassium 19 K 39.098, Calcium 20 Ca 40.078 — period 4, groups 1–2)
(Rubidium 37 Rb 85.468, Strontium 38 Sr 87.62 — period 5, groups 1–2)
(Cesium 55 Cs 132.905, Barium 56 Ba 137.327 — period 6, groups 1–2)
(Francium 87 Fr (223), Radium 88 Ra (226) — period 7, groups 1–2)

The number in parentheses is the mass number of the longest lived isotope for that element.

Rows of elements are called periods. Atomic number increases across a period.

The arrow shows where these elements would fit into the periodic table. They are moved to the bottom of the page to save space.

Lanthanide series

Cerium 58 Ce 140.116	Praseodymium 59 Pr 140.908	Neodymium 60 Nd 144.24	Promethium 61 Pm (145)	Samarium 62 Sm 150.36

Actinide series

Thorium 90 Th 232.038	Protactinium 91 Pa 231.036	Uranium 92 U 238.029	Neptunium 93 Np (237)	Plutonium 94 Pu (244)